D1498081

Isotope Chronostratigraphy: Theory and Methods

Academic Press Geology Series

Mineral Deposits and Global Tectonic Settings—A. H. G. Mitchell and M. S. Garson—*1981*

Applied Environmental Geochemistry—I. Thornton (ed.)—*1983*

Geology and Radwaste—A. G. Milnes—*1985*

Mantle Metasomatism—M. A. Menzies and C. J. Hawkesworth (eds.)—*1987*

The Structure of the Planets—J. W. Elder—*1987*

Fracture Mechanics of Rock—B. K. Atkinson (ed.)—*1987*

Isotope Chronostratigraphy: Theory and Methods—Douglas F. Williams, Ian Lerche, and W. E. Full—*1988*

Isotope Chronostratigraphy: Theory and Methods

DOUGLAS F. WILLIAMS
Department of Geology
University of South Carolina
Columbia, South Carolina

IAN LERCHE
Department of Geology
University of South Carolina
Columbia, South Carolina

W. E. FULL
Department of Geology
Wichita State University
Wichita, Kansas

QE
652.5
. W55
1988

Indiana University
Library
Northwest

ACADEMIC PRESS, INC.
Harcourt Brace Jovanovich, Publishers
San Diego   New York   Berkeley   Boston
London   Sydney   Tokyo   Toronto

INDIANA
UNIVERSITY
NORTHWEST

LIBRARY

COPYRIGHT © 1988 BY ACADEMIC PRESS, INC.
ALL RIGHTS RESERVED.
NO PART OF THIS PUBLICATION MAY BE REPRODUCED OR
TRANSMITTED IN ANY FORM OR BY ANY MEANS, ELECTRONIC
OR MECHANICAL, INCLUDING PHOTOCOPY, RECORDING, OR
ANY INFORMATION STORAGE AND RETRIEVAL SYSTEM, WITHOUT
PERMISSION IN WRITING FROM THE PUBLISHER.

ACADEMIC PRESS, INC.
1250 Sixth Avenue
San Diego, California 92101

United Kingdom Edition published by
ACADEMIC PRESS INC. (LONDON) LTD.
24-28 Oval Road, London NW1 7DX

Library of Congress Cataloging-in-Publication Data

Williams, Douglas F.
    Isotope chronostratigraphy : theory and methods / Douglas F.
Williams, Ian Lerche, William Full.
        p.   cm. — (Academic Press geology series)
    Includes index.
    ISBN 0-12-754560-3   (alk. paper)
    1. Stratigraphic correlation.   2. Oxygen—Isotopes.   3. Carbon-
-Isotopes.   I. Lerche, I. (Ian)   II. Full, William.   III. Title.
IV. Series.
QE652.5.W55   1988
551.7'01—dc19   87-27129
                                                                       CIP

PRINTED IN THE UNITED STATES OF AMERICA
88 89 90 91   9 8 7 6 5 4 3 2 1

CONTENTS

viii   **Contents**

PREFACE

Isotope Chronostratigraphy has a long and involved history, the exact beginning of which is difficult to pinpoint. In some ways it is the logical outgrowth of a chapter on Pleistocene oxygen isotope stratigraphy written for the book *Principles of Pleistocene Stratigraphy Applied to the Gulf of Mexico*, edited by N. Healy-Williams (1984, IHRDC Press). Publication of that book, and subsequent pilot projects with several enterprising petroleum companies, enabled a small group of faculty, students, and technicians at the University of South Carolina to form the Isotope Stratigraphy Group for the purpose of refining the stratigraphy of Plio-Pleistocene exploration wells from offshore Gulf of Mexico. Because of the proprietary nature of that work, however, very little of the exciting results could be published in the open literature. Thus, the industry as a whole remained relatively unaware of this innovative technology.

More directly, this book is derived from a short course that the senior author presented to seven of the major petroleum companies in February and November 1985. As he does in this book, Williams attempted to show in his short course the potential of using stable oxygen and carbon isotope records for making detailed stratigraphic correlations of exploration sections of Tertiary and, quite possibly, Mesozoic age. The short course was designed to familiarize petroleum geologists and stratigraphers with this new form of chemical stratigraphy. Its intent was also to remove some of the mystery about stable isotopes as they had been used in the past and to establish some principles for properly interpreting stable isotope records in stratigraphic and chronostratigraphic frameworks. One of the outgrowths of these short course presentations, however, was the realization that a more quantitative approach was necessary if isotope chronostratigraphy was to be carried to its ultimate potential. Thus Lerche and Full entered the picture.

As mathematical geologists, Lerche and Full have a grasp of ways to manipulate geophysical, geological, and geochemical data. We have teamed together in an effort to document some of the available approaches for analyzing isotope data as time series information. The combination of the empirical models with the quantitative approaches described herein makes isotope chronostratigraphy a more powerful tool both in exploration sections, where pure chronostratigraphic information may be the desired end product, and in marine sections where the primary aim is to gain further understanding of the paleooceanographic history of ocean basins. We do not claim to have monopolized all the potential approaches either empirically or mathematically. Developments in the study of isotope distributions in sedimentary rocks, and in the refinement of mathematical treatments of data, are sufficiently rapid that even as quickly as the publishers bring this book to print, new editions will be needed to keep abreast of new advances.

This work would not have been possible without the aid and cooperation of a great number of our colleagues in industry, at the University of South Carolina, at Wichita State University, and at other institutions. In particular, we acknowledge the encouragement and insights we have received from Robert Ehrlich, Richard Fillon (Texaco), Barry Kohl (Chevron), Christopher Kendall, and Nancy Healy-Williams. Nick Pisias, Nick Shackleton, Ken Miller, James P. Kennett, and Maurice Renard are thanked for generously sharing not yet published data. We also thank the staff of the Isotope Stratigraphy Group (David Mucciarone, Dwight Trainor, Judy McClendon, Mary Evans, Steve Hardin, and Jeff Corbin) for their unselfish efforts in generating the data and thereby producing the exciting results we have obtained to date in our exploration work. Tom McKenna is thanked for the thankless job of getting the computer programs, written in Wichita by Full, to run on the USC mainframe. We thank John Haramut and David Mucciarone for drafting the figures. David Krantz helped significantly with editing portions of the book. Donna Black, Sheri Howell, and Joyce Goodwin did outstanding jobs of typing various parts of the book. We thank our families for the forbearance and tolerance that only families can give during the many weekends, nights, and holidays we worked on this book. Last but not least, Williams and Lerche thank James R. Durig, Dean of the College of Science and Mathematics, for his ardent support of our research programs.

Chapter 1 | Introduction

I. Rationale for a New Chemical Stratigraphy

Dwindling reserves and oil production quotas in traditional producing areas and generally shallow water depositional systems have combined to force the petroleum industry to undertake unprecedented efforts to streamline its petroleum exploration efforts. Many of these efforts have been directed toward (1) developing new techniques for more accurately locating hydrocarbon source rocks and their related oil and gas reservoirs and (2) exploring deeper water environments in both traditional and frontier exploration areas. Unconventional and innovative approaches are therefore needed; very often these approaches lead to unexpected breakthroughs in exploration efforts, particularly in deeper water environments where some interpretations of standard foraminiferal zonations (Fig. 1) have come in conflict with interpretations of seismic sequence boundary patterns and other geophysical interpretations. While many of these foraminiferal zonations have worked exceedingly well in zones 1–3 (Fig. 2) during decades of exploration in onshore and shallow-water drill sites, the stratigraphic behavior of many shallow-dwelling benthic species or assemblages becomes less reliable in the deeper zones 4–6. At the very least, their behavior becomes less understood in many exploration wells when combined with other data. A number of benthic foraminiferal zonations used by the petroleum industry, primarily in sections from shallow-water environments, commonly yield stratigraphic intervals with average durations in excess of 500,000 yr (Fig. 1). In addition, some of the benthic datums are strongly controlled by facies changes and thus tend to be regionally diachronous.

As it becomes necessary to explore the deep-water blocks of the Gulf of Mexico for major petroleum and gas reservoirs in the late Tertiary and

1

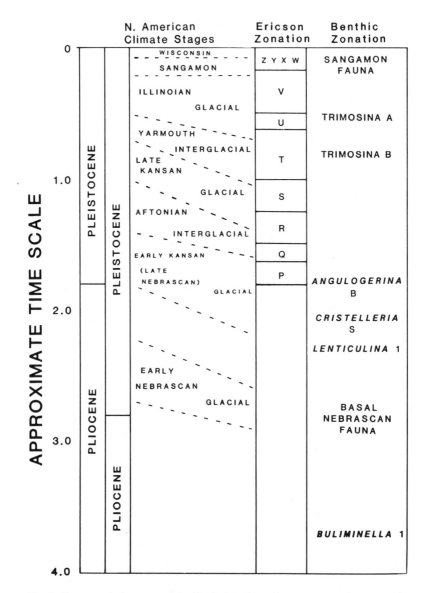

Fig. 1. Conceptual placement of the North American climate stages and a commonly used foraminiferal zonation for the Plio-Pleistocene of the Gulf of Mexico Basin based on selected index species. (From Stainforth *et al.*, 1975.)

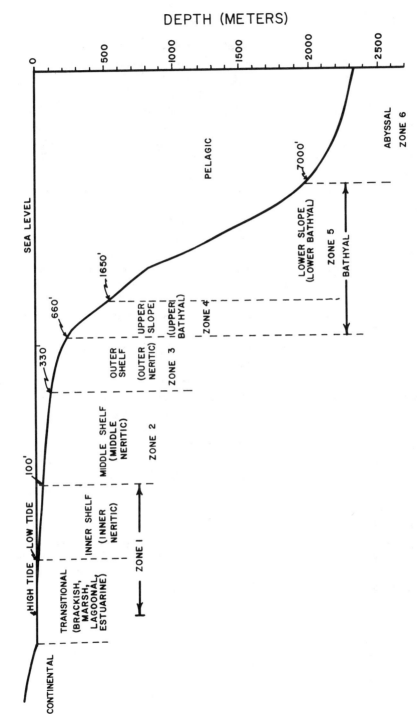

Fig. 2. Schematic representation of the bathymetry and informal depth classification of a passive continental margin.

Pleistocene sediments of the continental slope, time-stratigraphic schemes of a high resolution are needed for precise intercorrelation of exploration wells. Conventional stratigraphic techniques simply cannot provide the necessary resolution in many cases because (1) the Pleistocene is a relatively brief period of geologic time and lacks an adequate number of biostratigraphic markers; (2) the Gulf of Mexico has received enormous accumulations of terrigenous sediments during the Pleistocene because of glacioeustatic sea level fluctuations; and (3) shifting depocenters and regional unconformities, whether because of salt tectonics or sea level changes, make it difficult to accurately correlate wells in time and space dimensions.

While the petroleum industry is being forced to explore deeper water environments, part of the academic marine geological community is involved with stratigraphic and paleoceanographic studies of deep-water marine sediments. The Deep Sea Drilling Project (DSDP) has made available numerous marine Tertiary sequences from the continental rise and abyssal depths of the world's oceans. Representative sections for much of the Tertiary have enabled deep-water biostratigraphers, paleomagnetists, and geochemists working with siliceous and calcareous sediments to derive new techniques, such as isotope stratigraphy, paleomagnetic stratigraphy, tephrochronology, and quantitative biostratigraphy. The use of these new techniques to study DSDP sections has led to a fairly detailed picture of the oceanographic and biological history of the oceans during the Tertiary. However, the academic community has made only limited investigations with these techniques in continental margin sections from shelf and slope depths, and this is largely because of the lack of deep drilling capabilities. Core materials from the Gulf of Mexico margin that do exist, such as the Shell *Eureka* boreholes, are regarded quite wrongly by many in the academic community as not suitable for detailed study. Certainly paleomagnetic work is not possible, but we have found many of the *Eureka* cores to contain excellent material for geochemical and micropaleontological studies.

One of the primary incentives for this book and our research, therefore, is the idea that the petroleum geology and paleoceanography communities have a great deal to benefit from each other. In particular, it is our belief that the oxygen and carbon isotopic records for the Tertiary, as identified by the DSDP and in exploration boreholes, have the potential to become indispensable tools for the petroleum geologist and biostratigrapher in their efforts to identify and define new oil and gas reservoirs in deep-water and frontier environments. Also, stratigraphic sections from exploration wells provide unparalleled opportunities to study thick continental margin sections that otherwise would be unavailable to the academic community.

Toward this goal, the Isotope Stratigraphy Group at the University of South Carolina has been working to meet some of the petroleum industry's needs for precise stratigraphic correlations of exploration wells from deep-water blocks of the Gulf of Mexico by developing a new form of chemo-stratigraphy: isotope chronostratigraphy. Most of our work, which began in 1981, has been directed toward obtaining a high-resolution stratigraphy for the Pliocene–Pleistocene sections (Healy-Williams, 1984). Our basic approach is to integrate biostratigraphic and lithostratigraphic information (preferably quantitative data if available) with oxygen isotope stratigraphy. From this approach we establish the empirical models for a time-strati-graphic framework with a resolution that is far superior to conventional biostratigraphic techniques alone (i.e., "tops" of benthic and planktonic foraminifera and calcareous nannofossils).

As this text will show, isotope chronostratigraphy offers substantial improvements in the stratigraphic resolution of exploration wells when integrated with quantitative biostratigraphy and interpreted using some of the quantitative methods of data analysis discussed in later chapters. It should be stressed that these approaches are not restricted to the Gulf of Mexico or to the stratigraphic interpretation of isotope data. The approaches developed in this book should work equally well in all marine sedimentary basins of Tertiary age. It is also our opinion that the use of stable carbon and oxygen isotope data in stratigraphic sections offers unrealized potential in identifying chemically defined stratigraphic horizons and mapping diagenetic gradients in ancient sedimentary basins.

Therefore, our primary purposes in this book are to synthesize the oxygen and carbon isotopic records for the Tertiary and introduce the concept of isotope chronostratigraphy. We show how this new technology can be used in exploration wells in particular and in stratigraphic studies of sedimentary rocks in general. In addition, we describe some of the quantitative tech-niques, some standard and some not, for analyzing isotope data as time series information to improve correlations of complete and incomplete stratigraphic sections.

II. The Model of Isotope Chronostratigraphy

As shown schematically in Figs. 3 and 4, the oxygen and carbon isotopic records for Tertiary marine carbonates have distinct signals as functions of time. The fact that many of the particular features of the records are global events, driven by large-scale changes in the interconnected ocean–atmo-sphere–cryosphere system, offers the possibility of deriving a comprehensive age dating framework to:

1. Make detailed correlations and provide precise time lines between boreholes using isotope chronostratigraphy;

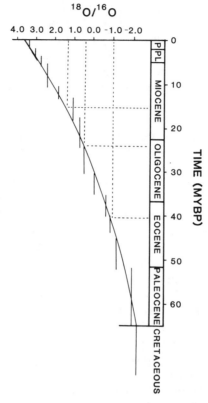

Fig. 3. Schematic representation of oxygen isotope chronostratigraphy utilizing calibrated steplike shifts and long-term trends in the oxygen isotope record for Tertiary marine sediments.

2. Develop a chronology for exploration wells linked to an absolute time scale through integration with available paleontological control;
3. Detect regional unconformities and local diastems that are unresolvable using biostratigraphy alone;
4. Test the time equivalency of commonly used biostratigraphic, lithostratigraphic, and seismic horizons;
5. Establish regional time lines linking shallow-water and deep-water environments.

For example, Fig. 3 illustrates a schematic conceptualization of the use of an oxygen isotope record from overlapping Tertiary sections for a given sedimentary basin as a function of time. The record exhibits a number of distinct features including steplike shifts in the average isotopic composition from one time period to another and a long-term trend toward more positive

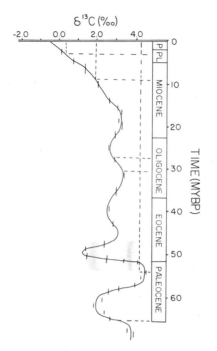

Fig. 4. A schematic representation of carbon isotope chronostratigraphy utilizing calibrated shifts and long-term trends in the carbon isotopic record for Tertiary marine sediments.

$\delta^{18}O$ values from the early to late Tertiary. This stepwise evolution of the oxygen isotopic record could be used to resolve and thus provide absolute age estimates by determining the particular oxygen isotopic value for a time period. For example, with a calibrated record, an oxygen isotopic value of ±0.4‰ would translate into an absolute age of approximately 24 MY ±1 MY (late Oligocene). Another example in Fig. 3 shows that an oxygen isotopic composition of approximately −1‰ would translate into a late Eocene age of 40 MY ± 1 MY. Once such a record has been calibrated for a particular sedimentary basin, the record can then be used to make age estimates in other sequences in which the biostratigraphic resolution is unavailable or unacceptable. Evidence not shown indicates that for particular time periods, such as the mid-Miocene, resolution greater than 100 KY can be attained.

A schematic representation of the carbon isotopic $\delta^{13}C$ record from Tertiary marine carbonates (Fig. 4) shows additional stratigraphic or age modeling possibilities not shown in the oxygen isotope record of the Tertiary (Fig. 3). From the middle Miocene to the Pleistocene, for example, the carbon isotope record of marine carbonates becomes steadily more negative, while the oxygen isotope record becomes more positive. Other significant changes occur in a series of gradual but large carbon isotopic shifts. Ideally, the Neogene trend toward more negative carbon isotope values could be used to make age estimates in a calibrated basin. For example, values between 0–+1‰ would correspond to the Plio-Pleistocene interval (Fig. 4). Values between 2–2.5‰ would correspond to the middle Miocene. As will be discussed later, good evidence exists for global carbon cycles from the Eocene-Oligocene through the Miocene. In the late Paleocene, the most positive $\delta^{13}C$ values of the entire Tertiary are observed (>3.5‰) (Fig. 4). The premise of isotope chronostratigraphy is that characteristics such as these can provide independent verification of an age model between the oxygen and carbon isotope chronostratigraphies.

Traditionally, biostratigraphic control within petroleum exploration relied almost exclusively upon information from benthic foraminifera (Fig. 1). Although such qualitative paleontological data are available for thousands of wells, especially in the Gulf of Mexico, much of this information is useful only in a loosely constrained time framework. Many of the zones are regional and some index faunas have different synonyms, which are loosely defined or vary widely among paleontological groups in the petroleum industry. Many of the datums are widely spaced in time and in many cases diachronous.

These problems often limit the usefulness of various datums for establishing regional time lines, especially as exploration efforts shift from shallow-water to deep-water environments. In shelf to upper slope environments (Fig. 2), many microfossil groups lack sufficient diversity or phylogenetic change to establish a high-resolution stratigraphic framework. A particular problem with benthic foraminiferal zonations is the fact that many key index species of shallow-water assemblages do not occur in deep-water environments.

One of the purposes of this book is to show how the adoption of isotope chronostratigraphy can help overcome many problems presently faced by petroleum geologists. First, the Tertiary oxygen and carbon isotopic records from foraminifera provide *global datums*. The isotope records are characterized by well-calibrated offsets or shifts and other isotopic changes that can be utilized to establish stratigraphic correlations of unprecedented resolution. It is our belief that oxygen and carbon isotopic records, integrated with seismic and biostratigraphic information, provide the missing link to tie shallow-water stratigraphic zonations, established after decades of petroleum exploration, to the zonations now being developed for deep-water environments.

Second, isotope chronostratigraphy provides a new chemical stratigraphy for comparing the microfossil zonations from one sedimentary basin to another, for example, from a traditional exploration area such as the Gulf of Mexico to frontier areas such as those in Indonesia or offshore China. At the outset, however, we emphasize that the approaches we are using do not necessarily replace any of the conventional biostratigraphic borehole evaluation techniques. We have adopted an approach that supplements those stratigraphies, achieving an increased level of stratigraphic resolution necessary for more accurate exploration of petroleum reservoirs. As discussed, the industrywide need for this increased stratigraphic resolution is particularly necessary in the deep-water tracts of the Gulf of Mexico.

Third, the chronostratigraphic framework derived from isotope chronostratigraphy eliminates much of the confusion that exists with the use of the conceptual glacial–interglacial stages of North America (Fig. 1) and the arguments that abound in the literature over placement of the Pliocene-Pleistocene boundary. In its most ideal sense, the framework of isotope chronostratigraphy is an absolute time frame calibrated with magnetostratigraphy, radiometric absolute age dating, and globally defined biostratigraphic events. In basins or time periods where this calibration is presently not available, such as in the Mesozoic–Paleozoic and terrigenous clastic depositional systems, we believe that useful stratigraphic information is still possible in some cases.

Fourth, sufficient data are now available to demonstrate that isotope chronostratigraphy has tremendous potential in older parts of the Tertiary and possibly in parts of Mesozoic sedimentary basins as well. Due to the foresight of several companies, our first attempts at utilizing this new strategy in various projects have been extremely successful. In many wells, we have found significant hiatuses, established precise time lines, and determined important accumulation-rate variations that would have remained transparent to conventional stratigraphic techniques.

III. The Format of This Synthesis

With these thoughts in mind, we begin introducing the concept of isotope chronostratigraphy by defining some principles used in the interpretation of isotope records and by describing the generation of stable isotopic data. We also establish criteria for characterizing the nature and magnitude of the oxygen and carbon isotopic signals for stratigraphic sections. We then describe the relationship of the oxygen and carbon isotopic records of Tertiary marine carbonate to an absolute time frame and discuss some of the driving mechanisms for controlling those signals. Most previous work using stable isotopes of Tertiary marine carbonate has been paleoceanographic in emphasis. This work will be reviewed but not emphasized because our primary approach is the chronostratigraphic nature of the signals and their use in stratigraphic correlations of exploration wells. We will show examples of our efforts to test the applicability of isotope chronostratigraphy in exploration wells. Unfortunately, the proprietary nature of much of this information makes it necessary to withhold many of the specifics about well locations, water depths, etc. We can state with confidence, however, that four years of effort in working with various depositional environments and periods of the Tertiary have, in most cases, placed our approaches beyond the feasibility phase to a more intense application phase.

In addition, we will integrate some of the historical aspects in the development of the Tertiary isotope record and discuss some of the new technological advances that make it possible for this chemical stratigraphy technique to become more routine in exploration. Much work still remains to be done on the possible effects of diagenesis, either through deep burial depths, pressurized fluids, or the regional thermal history of the region, but this work is in progress. In the northern Gulf of Mexico, we have successfully used our approach in some wells to subbottom target depths (TDs) greater than 15,000 ft ($>$4500 m) with little apparent overprint on the primary oceanic signals.

Most important, we describe the theoretical basis for quantitatively treating isotope data as part of a digital signal that can be used as time series information for locating common (also missing) isotopic events, correlating isotope records of varying quality and completeness, properly determining the spectral character of isotope records, and correcting for changes in sediment accumulation and compaction rates through time. We give the mathematical basis for the techniques used to achieve these goals and we have spent considerable time coding the mathematics for computer manipulation of the isotope data. Details on the computer programs are available from the authors.

Chapter 2 | Principles of Interpretation

I. An Empirical Approach for Establishing Interwell Correlations and Zonations

This chapter establishes (standardizes) common terminology for describing the features of isotope records. It also familiarizes the reader with some of the empirical ways in which stable oxygen and carbon isotopic data can be viewed in a stratigraphic sense. One can then submit the data for isotope chronostratigraphy into the most appropriate method for quantitatively processing the information (described in Chapter 11). Standardization of the terminology and criteria for describing isotope stratigraphic data will help the reader to understand the types of oxygen and carbon isotopic changes that are recorded in marine carbonates from various time periods in the Tertiary.

To assist in the description of these records, consider Fig. 5, which illustrates how most of the data are presented in this book. First, note that the scales for the oxygen ($\delta^{18}O$) and carbon ($\delta^{13}C$) records are reversed with negative values to the right of zero for oxygen and to the left for carbon. This convention (discussed in greater detail in Williams, 1984) started early in the development of oxygen isotope records of marine carbonates as paleotemperature records of the Cretaceous and Pleistocene oceans were determined by Harold Urey, Cesare Emiliani, and their co-workers. To a first approximation, carbonate formed in warm water has a more negative oxygen isotopic value than carbonate from colder waters (on the order of 0.22‰ per °C), thereby explaining the direction of the $\delta^{18}O$ scale. Temperature has only a small role in the carbon isotope distribution in marine carbonate (on the order of 0.035‰ per °C; Emrich *et al.*, 1970).

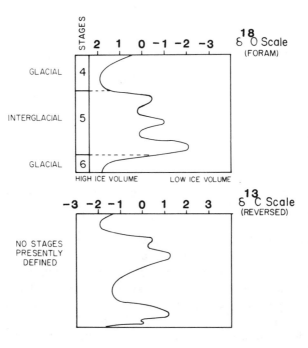

Fig. 5. Representation of the descriptive terminology commonly used to characterize oxygen and carbon isotope records of the Tertiary, depending on the carbonate fraction used in generating the isotopic signals.

We now believe that most of the $\delta^{18}O$ and $\delta^{13}C$ signals are functions of global changes in ocean chemistry caused by different but related processes, thereby forming the basis of isotope chronostratigraphy. It is perhaps appropriate that the isotope scales remain reversed for now to keep their differences clear. Isotopic values described as less than 0‰ indicate a relative decrease of the heavy isotope in the material being analyzed; the terms light, negative, and depleted are used interchangeably in this case. The terms heavy, positive, and enriched indicate a relative increase in the heavy isotope (either ^{18}O or ^{13}C). The concept of isotope stages, which are so important in the Plio-Pleistocene, is also shown in Fig. 5 and will be discussed in greater detail later.

CYCLE	REPEATING, QUASIPERIODIC PART OF AVERAGE SIGNAL; UNIFORM TO VARIABLE DURATION
EVENT	UNIQUE EVENT, I.E., DISTINCTLY ABOVE/BELOW AVERAGE SIGNAL;
SHIFT	VERY ABRUPT CHANGE (200 KY) REVERSIBLE OR NONREVERSIBLE
LONG—TERM TRENDS	STEADY CHANGE IN DELTA VALUES WITH TIME IN NEGATIVE OR POSITIVE DIRECTION
ABSOLUTE VALUES	ESTABLISH BOUNDARY CONDITIONS GLACIAL—INTERGLACIAL—MELTWATER INDENTIFICATION OF EVENTS

Fig. 6. Interpretative definitions of some features or characteristics of oxygen and carbon isotope signals.

From this point, we will define some criteria, given in Fig. 6, for interpreting isotope records and, we will illustrate these interpretative definitions in Figs. 7–11:

1. The overall patterns exhibited in the isotope records, i.e., the occurrence of long-term trends, events, and cycles.
2. The absolute values of the $\delta^{18}O$ and $\delta^{13}C$ records at particular depths or time periods.
3. The occurrence of particular shifts (usually nonreversible) in the isotopic records.

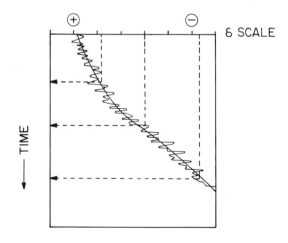

Fig. 7. A schematic example of long-term trends in an isotope record with time (or depth) increasing downward.

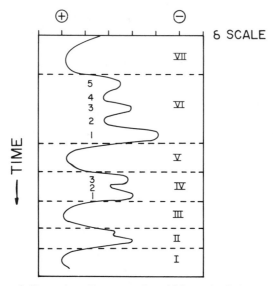

Fig. 8. A schematic illustration of isotope cycles, which may be distinguished by their shape, amplitude, or duration into isotope stages (Emiliani, 1966). Presently, only oxygen isotope stages are designated for the Plio-Pleistocene.

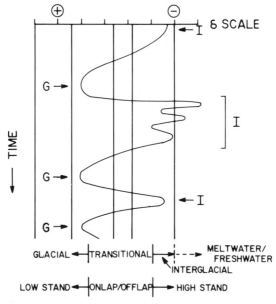

Fig. 9. An illustration of how the absolute δ values in a calibrated isotope record can be used to establish boundary conditions for a particular time period, basin, or carbonate fraction. In this example, the record would best represent an oxygen isotope record for planktonic foraminifera from the Pleistocene of the Gulf of Mexico.

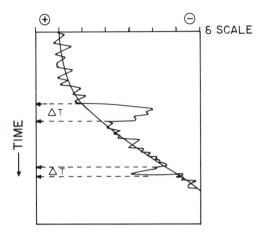

Fig. 10. A schematic representation of how isotope events can be empirically defined. Events may be variable in duration and occur as negative or positive excursions from the overall isotope record.

4. The magnitude (per mil change) of the shifts, long-term trends, cycles, and events as a function of time.
5. The direction of the shifts (i.e., in a positive or negative direction).

The exact order in which these criteria are used in applying isotope chronostratigraphy is dictated first by the form of the carbonate analyzed (i.e., benthic foraminifera, planktonic foraminifera, calcareous nannofossils,

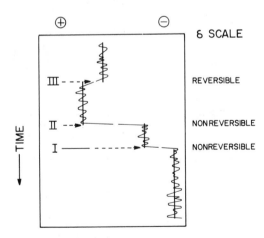

Fig. 11. An illustration of how isotope shifts are commonly observed in isotope records of the Tertiary. Shifts may be in the negative or positive direction and may be repeating (reversible) or nonreversible in a section with time.

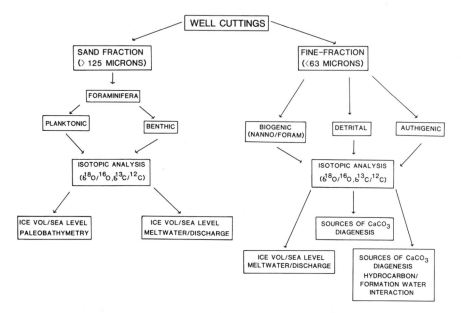

Fig. 12. Flow chart illustrating how different fractions of the sediment or rock record can be analyzed to determine isotope signals with different types of information: stratigraphic, diagenetic, or chronostratigraphic.

bulk carbonate, or a particular size fraction from the sediment) (Fig. 12). The next logical step is to calibrate the isotope features to a time scale, if this has not been done already. Some general idea as to the time frame from which the sample came (i.e., Plio-Pleistocene, Miocene, Paleogene) is required. This information enables the stratigrapher (1) to choose the appropriate DSDP type section for comparison with the exploration section (if foraminifera are providing the isotope signal) and (2) to select the appropriate biostratigraphic or lithostratigraphic information to be integrated with the isotope records, particularly if carbonate data from a detrital fraction or whole rock are being used to construct the stratigraphic framework. Of course, the exact process to be followed depends on whether a new type section is being constructed for a particular sedimentary basin and a particular carbonate phase or whether a type section is currently available for that area.

Having assembled this information, the initial approach should be to examine the overall pattern or character of the isotope record. Are there long-term trends (i.e., occurring over 1–10 MY periods) during which systematic changes occur in the isotopic record (Fig. 7)? Is there evidence for any particular cyclic changes in the isotope signals (Fig. 8)? Do these cycles appear to be quasiperiodic, i.e., do the maximum and minimum values in the cycles achieve the same levels in the record? At what frequency do the cycles occur? Do they occur at regular or irregular intervals of depth or time? Of course, one of the purposes of this book is to show that much of this empirical approach, while valuable in itself, can be greatly supplemented by utilizing quantitative methods for handling the isotope data as time-series information.

Evidence for the types of isotope features shown schematically in Figs. 7–11 can be seen in real data. For example, the presence of $\delta^{13}C$ cycles can be seen very nicely in the Oligocene Atlantic and Pacific records shown in Fig. 59. An example of long-term shifts in the isotope record is the monotonic ^{18}O enrichment that occurs in the Neogene portions of the oxygen isotopic records shown in Figs. 17, 18, and 27. An example of cycles that are not strictly periodic or repeating can be seen in the $\delta^{13}C$ cycles of the early Paleogene shown in Fig. 64.

The next criteria that should be examined are the absolute values of both the oxygen and carbon isotopic records (Fig. 9). For example, what is the range of the values? Do any unusual values, either positive or negative, occur at any particular depth or time intervals in the stratigraphic record? Are these absolute values related to any particular part of the lithological or biostratigraphic zonation? Reference to the Tertiary $\delta^{13}C$ record in Fig. 24 and the schematics in Figs. 3 and 4 show that in portions of any particular record, absolute values are found to occur uniquely during particular time intervals. For example, the extremely positive $\delta^{13}C$ values approaching +3.5‰ in the late Paleocene are unique to the Tertiary isotopic record of Fig. 24. Similarly, $\delta^{13}C$ values less than +1‰ are restricted to the latest Pliocene–Pleistocene record.

The absolute values of a particular record also come into play when one is looking for regional isotopic effects caused by such phenomena as the input of a distinct detrital carbonate fraction or the input of large amounts of fresh meltwater into a marine basin. Isotope signals from either of these phenomena can provide valuable stratigraphic, and possibly chronstratigraphic, information. A good example of the meltwater effect is found in Gulf of Mexico exploration wells. Large $\delta^{18}O$ anomalies from these major events are prominent features in the Plio-Pleistocene $\delta^{18}O$ record of planktonic foraminifera from the Gulf of Mexico. The Pleistocene $\delta^{18}O$ record of the Gulf of Mexico is sufficiently well known to state that $\delta^{18}O$ values in planktonic foraminifera (i.e., *Globigerinoides ruber*), which are more negative than −1.55‰, are very likely associated with meltwater events in the Gulf of Mexico (see the Orca Basin record shown in Fig. 36). Values that fall in the range between −1.0–−0.5‰ are related to transitions between interglacial and glacial sea levels in the Pleistocene. The $\delta^{18}O$ values that are more positive than 0‰ are associated strictly with glacial intervals. In this manner the absolute values in a calibrated record can be used for interpretations in terms of glacio-eustatic changes, meltwater events, etc. These values apply only to the Pleistocene of the Gulf of Mexico.

It is also possible that particular isotopic values for a specific carbonate phase at certain depth intervals can be used as "fingerprints" with which to geochemically trace the occurrence of a horizon from section to section. This particular horizon (or event) (Fig. 10) may be caused by the input and dispersal of a different type of carbonate into a sedimentary basin, or it may be related to a regional diagenetic trend or similar phenomena.

The third type of criteria refers to the occurrence of shifts, or particularly large changes in the isotopic record, which usually occur in one direction (either positively or negatively) (Fig. 12). Examples of positive shifts in the $\delta^{18}O$ record can be seen for the mid-Miocene interval and across the Eocene-Oligocene boundary in Figs. 17-19 and 23. Such shifts usually relate to global changes in some physiochemical property of the oceans, which in turn can be used to provide globally synchronous datums for isotope chronostratigraphy. An example of a carbon isotope shift is seen in the late Miocene at approximately 6.1-5.8 MYBP in numerous DSDP sites (Figs. 56 and 57). Another example of the type of shifts useful for constructing isotope chronostratigraphic frameworks is reflected in the stepwise changes or shifts in the Pliocene-Pleistocene $\delta^{18}O$ record (Fig. 41, from Thunell and Williams, 1983).

Also included in the third criterion is attention to the magnitude of the oxygen and carbon shifts. For example, if a well-documented shift greater than 1‰ is represented in a section under study by a shift of only 0.5‰ either the particular change has been incorrectly correlated or part of the section is missing. Shackleton and Hall (1984a) utilized this strategy to show that the various $\delta^{13}C$ records for the Paleocene-Eocene sections from the south Atlantic (Fig. 63) were less complete than the more heavily sampled site 527 (Fig. 64) record. Similarly, the $\delta^{18}O$ change from the middle Miocene to the Quaternary is on the order of 2.5‰ (Figs. 17-19). Analyses of samples over this time interval in an unknown section might reveal that the isotopic change with time is only on the order of 1‰. With a small number of carefully chosen samples, the stratigrapher would know that large parts of the Neogene record in the section were missing. Most biostratigraphic zonations are also capable of such broad determinations, but as will be discussed in the chapter on methodology (Chapter 3), isotopic data can be produced very quickly for either rough age estimates or complete chrono-stratigraphic analysis to make high-resolution correlations.

Last, but not least, the direction of isotopic changes exhibited in particular shifts is also important, i.e., whether or not the shifts of a particular magnitude, and from a particular time, are in the positive or negative direction (Fig. 11). Examination of the Neogene $\delta^{18}O$ change from the mid-Miocene to Pleistocene shows that the 2.5‰ change occurs from negative to positive, representing the climatic deterioration during that time interval (Figs. 17–19, 23, and 27). The $\delta^{13}C$ change during that time is in the negative direction, opposite to the $\delta^{18}O$ change.

In practice, all of these criteria are utilized to evaluate the character and stratigraphic usefulness of the stable isotopic records. Our experience has been that the best and most sensible approach to isotope chronostratigraphy is an integrated approach: other available stratigraphic data are consulted but not necessarily used to dictate the interpretation of the isotopic records. Isotope chronostratigraphy is used in this manner to supplement conventional biostratigraphic evaluation techniques to achieve the best level of stratigraphic resolution possible. Linking these approaches to the information retrieveable using the semblance, filter, deconvolution, and frequency domain methods (autocorrelation, cross-correlation, spectral analysis, etc.) makes isotope chronostratigraphy a powerful technique indeed.

II. Integration of Isotope Chronostratigraphy and Biostratigraphy

Our experience has shown that integration of biostratigraphic information with the isotope records is by far the best approach to enhancing the chronostratigraphic resolution of exploration wells. A few examples from the published literature will illustrate how isotope records have been used to refine particular biostratigraphic zonations and the timing of either the last or first appearance of datums (LADs and FADs) in various microfossil groups. These examples are similar to the approaches that can be used to calibrate the zonation for a particular company for the basin or time period of highest exploration priority.

Figure 13 shows how the first appearance datum in the nannofossil genus *Amaurolithous* appears to be globally synchronous due to its position relative to the late Miocene $\delta^{13}C$ shift (Haq *et al.*, 1980). This same type of refinement is shown in Fig. 14 (modified from Thierstein *et al.*, 1977), which shows how oxygen isotope stratigraphy in the Pleistocene is used to determine the timing of two important nannofossil datum levels in Pleistocene sediments (the FAD of *Emiliania huxleyi* and the LAD of *Pseudoemiliania lacunosa*). *Pseudoemiliania lacunosa* became extinct in the middle of oxygen isotope stage 12, approximately 458 KYBP, and *E. huxleyi* reached its first consistent appearance during isotope stage 8, approximately 268 KYBP (Thierstein *et al.*, 1977).

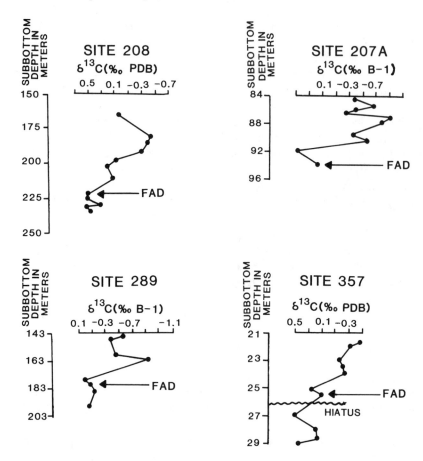

Fig. 13. The use of carbon isotope shifts at approximately 6.2 MY in the late Miocene to determine the time of the first appearance datum of the calcareous nannofossil *Amaurolithus spp.* in DSDP sections from the Pacific (sites 207A, 208, and 289) and Atlantic (site 357) (modified from Haq *et al.*, 1980).

Our studies of exploration wells from the Gulf of Mexico using isotope chronostratigraphy show that many stratigraphically useful datums from open ocean records occur nearly synchronously in the Gulf of Mexico. A similar approach has been used for planktonic foraminifera across the Eocene–Oligocene boundary by Keigwin and Corliss (1986) and for diatoms by Burcke *et al.* (1982).

Another example of how the isotope record can be used to refine the stratigraphy within a microfossil assemblage zone is illustrated in Fig. 15.

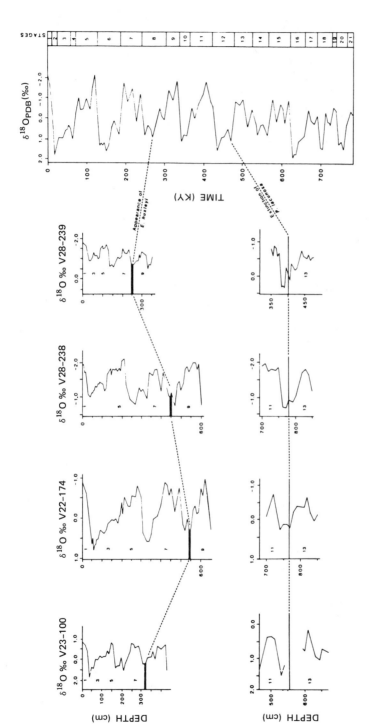

Fig. 14. The use of oxygen isotope stratigraphy to test the synchronous nature of important biostratigraphic datums in calcareous nannofossils. The first appearance level of *Emiliania huxleyi* consistently appears in the middle part of isotope stage 8 (approximately 268 KYBP) in four cores from the Atlantic (V23-100, V22-174) and Pacific (V28-238, V28-239). The last appearance datum of *Pseudoemiliania lacunosa* occurs near the boundary between isotope stages 12 and 13, approximately 475 KYBP (adapted from Thierstein *et al.*, 1977) (Williams, 1984). The occurrence of the nannofossil datums is shown with respect to the stacked composite isotope record from Imbrie *et al.* (1984).

The oxygen isotope records in Caribbean DSDP site 502B and Gulf of Mexico borehole E67-135 are used to subdivide the deep-water planktonic foraminiferal zonation based on the Ericson *Globorotalia menardii* complex and other elements of the fauna from the Yarmouthian and Kansan intervals of the Gulf of Mexico Pleistocene. Note the number of isotopic events defineable within the biostratigraphic zonal boundaries in both the Caribbean and Gulf of Mexico (Fig. 15). A quantitative study of the S zone by Neff (1983) was able to subdivide the foraminiferal S zone into subzones S1, S2 and S3, whereas the oxygen isotope record is able to subdivide zone S into 12–13 isotopic substages. These isotope events can then be calibrated to formal isotope stages as shown in Fig. 32.

   In summary, the integration of oxygen and carbon isotope chronostratigraphy with biostratigraphy leads to more accurate biostratigraphic and lithostratigraphic correlations. Isotope chronostratigraphy may also be used to calibrate biostratigraphic zonations based on shallow-water assemblages to zonations based either on deep-water benthic assemblages or calcareous nannofossils and planktonic foraminifera that do not extend into the shallow water. With this approach it is possible to compare and integrate these zonations so that the best stratigraphic correlations and

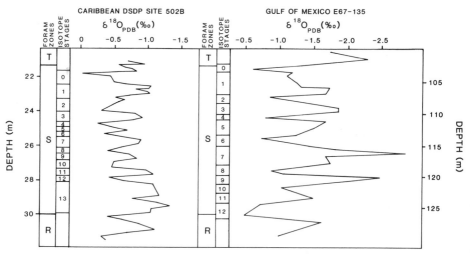

Fig. 15. A comparison of the oxygen isotope record of Caribbean core site 502B (Prell, 1982) with that of Gulf of Mexico borehole E67-135 within the Ericson S biostratigraphic zone for each section (Williams, 1984).

regional time lines can be drawn between wells from varying water depths.

III. Effects of Diagenesis on Isotope Records

In most cases and in most depositional systems, the $^{18}O/^{16}O$ and $^{13}C/^{12}C$ ratios in marine fossil carbonates (i.e., foraminifera, calcareous nanno-fossils) are fairly stable and resistant to diagenetic change at low temperatures and pressures. The $^{18}O/^{16}O$ and $^{13}C/^{12}C$ ratio in fossil foraminifera and nannofossils are not significantly affected by isotopic exchange with bottom or pore waters at shallow burial depths. Although diagenetic alteration from recrystallization and severe dissolution may affect the oxygen isotopic data (Baker *et al.*, 1982; Elderfield and Gieskes, 1982; Killingley, 1983); diagenesis is usually negligible in sediments of Plio-Pleistocene age. Very often microscopic or scanning electron microscopic examinations of the carbonate phase can be used to anticipate an isotopic effect. In general, the $\delta^{13}C$ values of marine carbonates are thought to be less affected by diagenesis than the oxygen isotopic values because of the differences in the respective reservoirs for potential isotopic changes (Anderson and Arthur, 1983).

In general, the calcareous ooze–chalk–limestone transformation related to depth of burial and time generally increases the chances of diagnenetic effects (Scholle, 1977; Schlanger and Douglas, 1974). In most cases, lithification in marine carbonate sediments leads to small isotopic effects (Hudson, 1977). Killingley (1983) recently proposed that diagenesis in a closed system could lead benthic foraminiferal $\delta^{18}O$ values (which are generally more positive) to become negative and more like the $\delta^{18}O$ values of planktonic foraminifera. Recystallization of marine carbonates in the presence of pore waters, whose $\delta^{18}O$ values have been altered because of high heat flow or basalt–clay mineral alterations, may be significant in some cases.

In short, the exact magnitude of $\delta^{18}O$ isotopic changes because of recrystallization is not known in all cases. One must always be aware of potential diagenetic effects in the stratigraphic or chronostratigraphic use of isotope data. We have found, however, that unless diagenesis is severe, the $\delta^{18}O$ pattern or signal is offset and not obliterated in many sections.

In general, $\delta^{13}C$ values of marine carbonates are less susceptible to the usual forms of diagenesis through compaction, etc., except in the presence of significant amounts of organic matter. The temperature coefficient for carbon isotope fractionation during carbonate precipitation ($\sim 0.035‰$ per

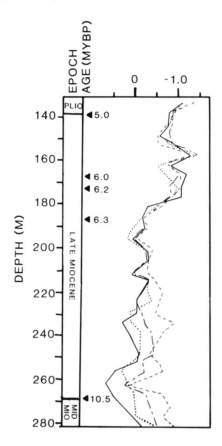

Fig. 16. A comparison of the Miocene carbon isotope records based on different planktonic and benthic species from Indian Ocean DSDP site 238 (modified from Vincent *et al.*, 1980). *O, umbonatus,* ——; *P. Wuellerstorfi,* — - —; *C. subglobosa,* · · · ·; *G: sacculifer,* - - -.

°C) is small relative to that for oxygen isotopes (~0.22‰ per °C). Therefore, a large carbon isotope effect is likely when recrystallization occurs in the presence of pore water bicarbonate in which the $\delta^{13}C$ has been significantly changed (i.e., the addition of ^{12}C from organic matter oxidation and microbial activity).

More work is clearly needed in this area, particularly to test the effect of burial depths exceeding 30,000 ft. Tests that we have done so far indicate that diagenesis modifies but does not totally eliminate the primary stratigraphic signal, particularly with respect to the $\delta^{13}C$ of carbonate phases.

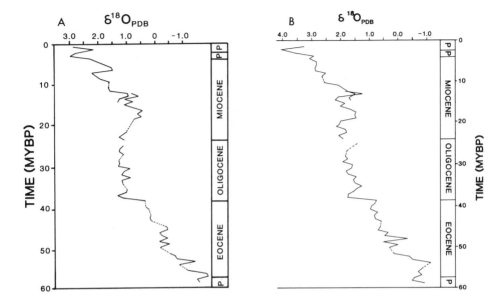

Fig. 17. A comparison of the oxygen isotopic record for (A) planktonic and (B) benthic foraminifera from the Tertiary sections of south Pacific DSDP sites 277, 279A, and 284 (modified from Shackleton and Kennett, 1975a).

IV. Species Effects

It has been well documented that some foraminifera and coccolith species do not secrete their carbonate materials in strict isotopic equilibrium with the ambient temperature, salinity, and $\delta^{13}C$ of the inorganic carbon reservoir of ocean waters. Carbon isotope effects have been documented in planktonic foraminifera (Williams *et al.*, 1977; Berger *et al.*, 1978), benthic foraminifera (Duplessy *et al.*, 1970; Bélanger *et al.*, 1981; Graham *et al.*, 1981), and coccoliths (Goodney *et al.*, 1980). Morphometric studies of foraminifera at the University of South Carolina indicate the presence of isotopic differences between different morphotypes of the same species (Healy-Williams *et al.*, 1985). Although these species and morphotype effects are important in detailed investigations of the paleoceanographic or physiochemical changes occurring in oceanic waters, for most stratigraphic purposes, it has been shown that $\delta^{18}O$ or $\delta^{13}C$ signals of different species generally parallel one another (Fig. 16, for example). In fact, the close correspondence between the Tertiary isotopic records shown in Figs. 17 and 18, based on analyses

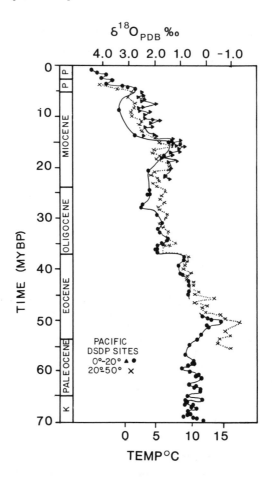

Fig. 18. Oxygen isotopic records of benthic foraminifera of Tertiary age from several Pacific DSDP sites differentiated according to their latitudinal location. The pre-mid-Miocene temperature scale is based on the assumption that the $\delta^{18}O$ of the pre-middle-Miocene bottom water was $-1.10‰$ relative to SMOW (modified from Savin, 1977, and Douglas and Savin, 1976).

of mixed foraminiferal assemblages and calcareous nannofossils (Fig. 19), indicates that species and morphotype effects are part of the signal but do not play the major role in the application of isotope chronostratigraphy to exploration wells. We are continuing to evaluate these potential effects.

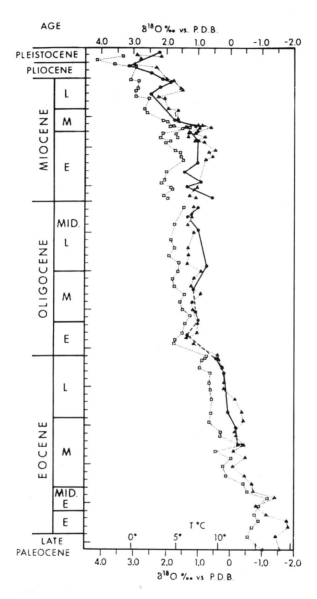

Fig. 19. A comparison of the oxygen isotopic composition of the calcareous nannofossils in the <44 μm fraction compared to the [18]O of benthic foraminifera (squares) and planktonic foraminifera (triangles) (from Margolis *et al.*, 1975).

Chapter 3 | Methodology

I. Generation of the Stable Isotope Data

The analytical techniques for determining the $^{18}O/^{16}O$ or $^{13}C/^{12}C$ ratio in calcium carbonate ($CaCO_3$) from marine sediments are now uniformly established and can be employed cheaply and quickly. Details of the techniques can be found in Williams (1984, and references cited), but some basics are repeated here for completeness.

Carbonate samples (usually 0.1–1 mg by weight) may be in the form of whole rock or bulk carbonate, foraminifera, calcareous nannofossils, or carbonate grains of detrital origin, depending on the depositional basin, the time period, and whether stratigraphic or chronostratigraphic information is desired. The carbonate material is first roasted at 380°C *in vacuo* to remove any organic contaminants. Samples with drilling mud or large quantities of hydrocarbons, as in an oil shale, may require special pretreatment. Although sidewall cores or core samples have the least problems in terms of subbottom depth placement, we have had excellent results from ditch cuttings when proper logging has been done. Knowledge of the casing points and micropaleontological information on displaced faunas is of great help in making interpretations of the isotope records at the first level.

The carbonate fraction is then reacted with purified phosphoric acid (H_3PO_4) in an evacuated reaction vessel held at a constant temperature of 50, 60, or 70°C, depending on the reaction time required to achieve total digestion of all $CaCO_3$:

$$2H_3PO_4 + 3CaCO_3 \rightleftharpoons Ca_3(PO_4)_2 + 3H_2O \uparrow + 3CO_2 \uparrow.$$

The resultant carbon dioxide gas is purified of water vapor by fractional freezing. Liquid nitrogen is used to transfer the purified CO_2 into an isotope ratio mass spectrometer. For specific information on the various procedures available, the Stable Isotope Laboratory at the University of South Carolina has prepared a detailed procedures manual.

In the mass spectrometer, the CO_2 molecules are ionized and separated into ion beams of three different masses ($44 = {}^{12}C\ {}^{16}O\ {}^{16}O$; $45 = {}^{13}C\ {}^{16}O\ {}^{16}O$; $46 = {}^{12}C\ {}^{16}C\ {}^{18}O$). Calculating ratios of the intensities of the 46/44 ion beams in the mass spectrometer yields the ${}^{18}O/{}^{16}O$ ratio of the CO_2 gas. The mass 45/44 ratio yields the ${}^{13}C/{}^{12}C$ ratio. The ${}^{18}O/{}^{16}O$ and ${}^{13}C/{}^{12}C$ ratios of the sample CO_2 are then compared repeatedly to the ratios in a laboratory reference CO_2 gas under identical operating conditions. By convention, the ratio is then expressed as a delta (δ) value as the parts per thousand (per mil, ‰) enrichment or depletion in the minor isotope, ${}^{18}O$ or ${}^{13}C$, respectively:

$$\delta^{18}O = \frac{{}^{18}O/{}^{16}O \text{ sample} - {}^{18}O/{}^{16}O \text{ reference}}{{}^{18}O/{}^{16}O \text{ reference}} \times 1000$$

$$\delta^{13}C = \frac{{}^{13}C/{}^{12}C \text{ sample} - {}^{13}C/{}^{12}C \text{ reference}}{{}^{13}C/{}^{12}C \text{ reference}} \times 1000$$

All δ values are described relative to the heavier and less abundant isotope in terms of *positive, enriched,* or *heavy,* i.e., containing more ${}^{18}O$ or ${}^{13}C$ than in the reference CO_2 (Fig. 5).

Today's isotope ratio mass spectrometers can determine $\delta^{18}O$ and $\delta^{13}C$ values with a precision of ±0.02‰ or to much better than 2% of the total isotope change observed in biogenic carbonate from Tertiary sediments.

To guarantee that δ values can be compared between laboratories, every major isotope laboratory corrects the measured δ values for machine effects (Craig, 1957; Deines, 1970) and calibrates its reference CO_2 gas to the universal PDB (Pee Dee Belemnite) standard. All values for foraminifera are reported relative to PDB, which was powdered belemnite rostra from the Pee Dee Formation of South Carolina (Urey *et al.*, 1951; Epstein *et al.*, 1953). To ensure the integrity of the isotopic data, a laboratory reference carbonate is analyzed daily prior to the analysis of each set of unknown carbonates.

II. Preparation of Well Samples for Isotopic Analyses

A. Foraminifera or Calcareous Nannofossil Analyses

Figures 19 and 20 show the $\delta^{18}O$ and $\delta^{13}C$ signals recorded in carbonate from calcareous nannofossils of undifferentiated species composition. The

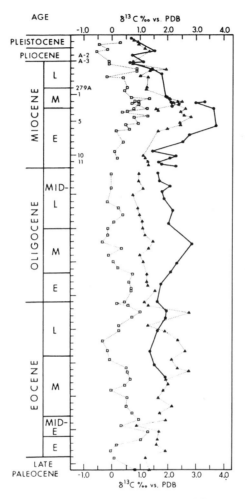

Fig. 20. A comparison of the Tertiary carbon isotope record from calcareous nannofossils from the <44 μm fraction with the records for benthic foraminifera (squares) and planktonic foraminifera (triangles) (from Margolis *et al.*, 1975).

nannofraction signals are very comparable to those based on mixed benthic foraminiferal assemblages (Fig. 18) and planktonic foraminifera (Figs. 17 and 21). Besides faithfully recording the overall character of the Tertiary isotope record, calcareous nannofossils have been shown to accurately reflect the details of the late Pleistocene $\delta^{18}O$ signal as recorded by data from a single species of planktonic foraminifera (Anderson and Steinmetz, 1983) (Fig. 22). These data illustrate that the primary factors controlling the oxygen and carbon isotopic signal of biogenic carbonate in marine

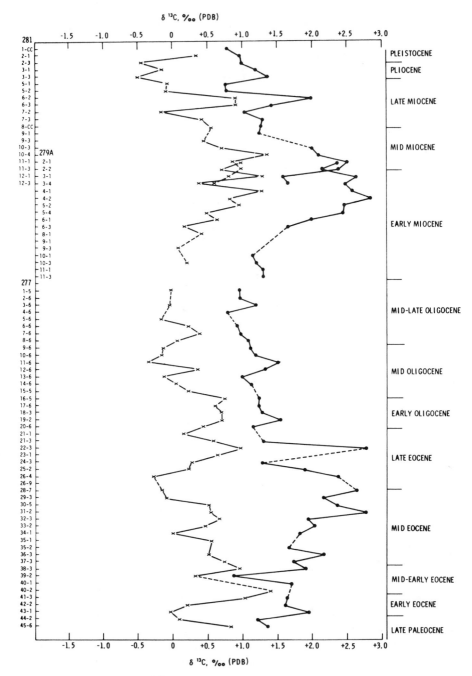

Fig. 21. Carbon isotope data from planktonic and benthic foraminifera at southwest Pacific DSDP sites 277, 279, and 281 (modified from Shackleton and Kennett, 1975a). The samples are arbitrarily plotted at constant divisions.

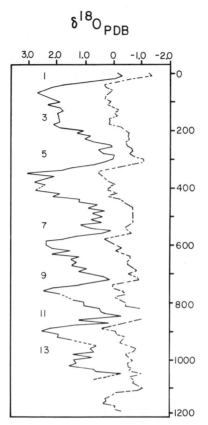

Fig. 22. Comparison of the oxygen isotope record of the planktonic foraminiferal species *Globigerinoides sacculifer* with that derived from the calcareous nannofossil fraction of the same late Pleistocene core from the Caribbean (modified from Anderson and Steinmetz, 1983).

sediments are not limited to any particular species or group (i.e., benthic, planktonic foraminifera, or calcareous nannofossils).

This point is important because in some instances it may be more desirable to analyze the calcareous nannofossil fraction than the foraminiferal fraction for isotope chronostratigraphy (Fig. 12). Less time is often involved in separating the nannofraction from a sediment sample than in separating a particular species or mixed assemblage of bethnic and planktonic foraminifera. In addition, it is sometimes not possible to get sufficient specimens of foraminifera for an isotopic analysis although present mass spectrometers can analyze CO_2 from as little carbonate as 0.1–0.3 mg (approximately equivalent to 1–40 specimens of foraminifera). Calcareous nannofossil material may sometimes be more resistant to dissolution than planktonic foraminifera. This is not a major consideration, however, since both the calcareous nannofossil and foraminifera comprise the primary calcareous components of most continental margin sediments.

One advantage of working with foraminifera is that they are easy to identify taxonomically, and gross signs of recrystallization, calcite over-crusts, or the effects of dissolution can be determined using a microscope. An addition advantage occurs if the samples are heavily saturated with hydrocarbons. In this case, foraminifera may be easier to clean of the contaminating hydrocarbon than the carbonate from a particular size fraction.

Although the nannofossil ^{18}O curve exhibits an offset with respect to the foraminiferal *Globigerinoides sacculifer* curve (Fig. 22), the displacements are systematic. The patterns in the two records are visually very similar, particularly in the timing of the isotopic changes, which is our primary interest for chronostratigraphy. The differences in amplitude could also weigh in favor of analyses of the calcareous nannofossil fraction. An advantage of using the foraminiferal fractions is that it is not necessary to perform any type of X-ray diffraction analyses on the sediment, as would be the case in some instances to determine if the coccolith fraction contains any carbonate other than the desired carbonate phase.

The procedures are fairly well established for separating foraminifera from unlithified sediment. The samples are dried, disaggregated of floccu-lated clay in a hot soap solution or weak hydrogen peroxide solution, and washed through a 63-μm sieve. It is our standard procedure to save all the material that passes through the 63-μm sieve for possible isotopic analyses at a later time. If foraminifera are to be analyzed, the >63-μm residue is desired; if detrital carbonate or calcareous nannofossil carbonate is to be analyzed, then the <63-μm fraction can be processed for analysis. Both residues are dried at temperatures usually <50°C in order to eliminate any potential risks of exchange with atmospheric oxygen at high temperatures.

Whenever possible, monospecific samples of benthic and planktonic foraminifera are picked from a split of the entire fraction >150 μm or from incremental size divisions in order to select specimens of a uniform size (Curry and Matthews, 1981). In some laboratories, the foraminiferal speci-mens are placed in distilled water or methanol and briefly cleaned ultrasoni-cally of any adhering calcareous and clay debris. The supernatant is removed with a syringe, and the samples are dried again at <50°C. Once dried, the specimens are ready for the isotopic analyses as described in Section I of this chapter. In many cases, it is advantageous to crush benthic specimens in order to ensure that they react quickly in the phosphoric acid.

The procedures for separating the nannofossil fraction from foraminiferal carbonate usually involve wet sieving or a timed centrifugation technique to achieve a relatively pure but polyspecific nannofossil assemblage consisting of coccoliths, coccospheres, and discoasters within a particular size range [<44 μm as in Margolis *et al.*, 1975 (Figs. 19 and 20) or the 3–25-μm fraction as in Anderson and Steinmetz, 1981 (Fig. 22)]. Other details of the procedures can be found in the literature.

B. Analyses of Unspecific Carbonate Phases from Whole Rock (Bulk) Carbonate

In some cases, it is desirable that samples be processed quickly for whole rock analyses of the total carbonate present in the sample. There are some disadvantages to this approach, however. Without doing a carbonate determination of the sample material, it is difficult to know if sufficient carbonate is present in the sample to obtain the CO_2 necessary for an isotopic analysis. Without knowledge of the particular carbonate phases present in the sample, uncertainty occurs as to the appropriate reaction procedure (i.e., the temperature and duration of the reaction). For example, if dolomite is present, it is necessary to utilize a longer digestion time than if only calcite and aragonite are present. In addition, it is often difficult to determine which component is carrying the isotope signal in rocks of mixed mineralogies. Knowledge of the mineralogies present in the whole rock analysis is necessary to understand the factors controlling the isotopic signal.

However, as shown by the results on bulk carbonate in Figs. 23 and 24, if whole rock analysis will serve the purpose, then such an approach usually saves time in processing and handling the samples. The data are therefore more quickly and cheaply obtained from an exploration standpoint (i.e., often the "bottom line"). At the outset one must choose (1) the type of information to be sought, (2) the level of effort (or budget) to achieve the desired results, and (3) the type of chronostratigraphic record to be sought in the initial phase of the isotopic work. For example, it might make sense in an initial study of a Plio-Pleistocene section from the Gulf of Mexico to analyze both the foraminiferal and detrital carbonate fractions, first to document which type of signal is being recorded in the detrital carbonate and then to use the foraminiferal record to calibrate the detrital record. After this calibration, extension of the results into other wells could proceed using the detrital fraction. Of course, one would not attempt to correlate detrital isotope records from exploration wells across great distances, but for correlations within blocks or fields, this approach may yield perfectly acceptable results.

C. Effects of Sample Handling on δ Values

Basically, very few things can be done to significantly affect the δ values from a particular carbonate fraction. Complications can arise if (1) the

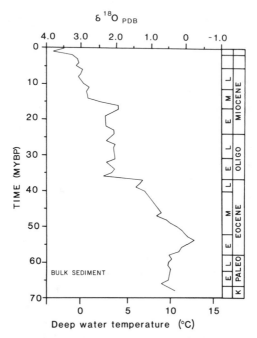

Fig. 23. The Tertiary $\delta^{18}O$ record based on isotopic analyses of the bulk carbonate in South Atlantic DSDP sites 525 and 528 (Shackleton, 1987). Reproduced with permission from Oxford University Press.

sample is heated in an oven to temperatures exceeding 300°C; (2) the sample is treated with an extremely hot, or strong, corrosive solvent that may dissolve or contaminate the $CaCO_3$; (3) the presence of unusual drilling muds cause problems in the washing procedure and in the roasting; (4) particularly lithified samples have to be carefully crushed to disaggregate the foraminiferal fraction; or (5) the presence of large amounts of hydrocarbons make it necessary to pretreat the sample with a solvent to remove the petroleum fraction (i.e., triple mix, carbon tetrachloride, or some other chemical that does not significantly alter the isotopic composition of carbonate fractions present).

III. Cost Analysis and Turnaround Time

The stable isotopic data from carbonate can now be obtained quickly and in many cases more cheaply than, for example, X-ray diffraction data. This is now possible because of technological improvements in the ability of isotope ratio mass spectrometers to analyze CO_2 gas from minute amounts of carbonate.

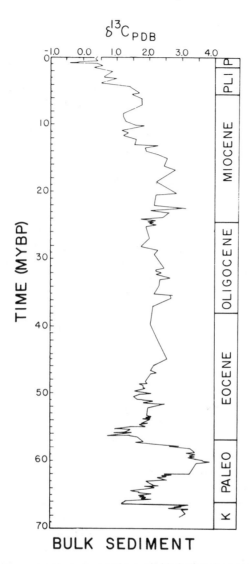

$\delta^{13}C_{PDB}$

BULK SEDIMENT

Fig. 24. The Tertiary carbon isotope record recorded in bulk carbonate from South Atlantic DSDP sites 525 and 528 (Shackleton, 1987). Reproduced with permission from Oxford University Press.

There are basically three options that a major or independent company has in developing an isotope chronostratigraphy program. The first approach would be to send samples of carbonate (either whole rock or of a particular $CaCO_3$ phase) to a commercial laboratory for an oxygen and carbon isotopic determination. Advantages of this approach are that it is not necessary to have persons on the staff trained in generating the isotopic data and maintaining a mass spectrometer and turnaround time on samples can be anywhere from 3–6 months, depending on the number of samples and commitments at the particular commercial laboratory. One of the disadvantages of this approach is that the stratigrapher or petroleum geologist interested in the data has no control over how the data are generated. Sometimes it is necessary to go to extraordinary lengths to determine that data produced from different commercial laboratories are comparable (the reader is referred to the interesting discussion on this subject in Scholle and Arthur, 1980).

The second approach would be to subcontract the work as a project with a stable isotope laboratory at a university or other research institution. This approach shares some of the same advantages and disadvantages as the first approach. Turnaround time may be a little longer than at a commercial laboratory but, in general, the level of commitment to the project and quality control are quite high, because academic stable isotope laboratories are usually interested in being able to detect very small isotopic changes. The major disadvantage of both approaches, however, is that personnel at both the commercial and academic laboratories are often not properly trained in the use of isotope chronostratigraphy to assist in the interpretation of the data. The second approach has the advantages that the data can usually be purchased fairly inexpensively and obtained fairly rapidly, and the funds are used to train graduate students, thereby benefiting the university and petroleum industry together. In addition, the funds very often help support research in other areas relevant to petroleum exploration. Unfortunately, many stable isotope laboratories and universities are not presently interested in this type of research with relevance to petroleum exploration problems. In many cases, the prevalent attitude is that diagenesis and stratigraphic complexities will be insurmountable problems.

The third approach would be to establish an isotope chronostratigraphy program within the company by hiring trained personnnel and setting up a stable isotope laboratory for generating the isotopic data. Many companies have laboratories with mass spectrometers capable of doing the basic analyses, but very often these mass spectrometers are heavily committed to other types of analyses, either in organic geochemical analyses of hydrocarbons or in diagenetic investigations. Very often company personnel are not trained in the nuances of isotope chronostratigraphy. This technique is new and innovative enough that not many graduate programs are producing geologists well versed in the application of isotope chronostratigraphy.

A disadvantage of this approach is that it is expensive to purchase a mass spectrometer and hire the personnel to maintain the mass spectrometer, perform the analyses, and interpret the data. The costs of purchasing the analytical equipment are not cheap. The state-of-the-art isotope ratio mass spectrometer ranges in price from $100,000 to $250,000 depending on the level of precision needed, computer automation, and other modifications. Initial establishment of the laboratory to house the mass spectrometer and extraction lines would probably involve an additional $15,000–$20,000, and maintenance would run on the order of $10,000 yearly at least. An advantage of this approach is that the company would have direct control over (1) what samples are analyzed, (2) the priority of those samples, and (3) how the data are generated and disseminated within the chronostratigraphy group and other segments of the company.

IV. New Technological Developments

In recent years, new technological developments have occurred in the computer automation of isotope ratio mass spectrometers. Computer automation offers many exciting possibilities including the potential for generating stable isotope data in a manner as routine to petroleum exploration as electric logs, well logging, and biostratigraphic workups. There are mass spectrometers currently on the market that are capable of analyzing 40–60 CO_2 samples a day via a computer-automated system totally independent of an operator.

Significant breakthroughs are being made in computer-automated carbonate extraction systems. The mating of an automated carbonate extraction system to an automated mass spectrometer has obvious advantages. Productivity is increased as the level of effort per unit of data production is decreased significantly. With such a system, it should be possible to perform 50–100 oxygen and carbon isotopic analyses a day. It is quite possible to install such a system on a drilling rig, offshore platform, or drill ship occupying a particularly important site. The system could then be used to provide isotopic data for a chronostratigraphic workup nearly as quickly as the samples are taken and processed, i.e., within 24–48 hr the data could be produced. If samples were being taken at approximately 60-ft intervals, it would be possible to generate isotopic data on as much as 3000 ft of well section per day. Such an approach would be particularly possible if the desired chronostratigraphic framework were based on either whole rock analyses or on a fraction of the carbonate present within the sediment (i.e., a calcareous nannofossil or detrital carbonate fraction <63 μm).

Chapter 4 | The Tertiary Oxygen Isotope Record

The major objectives of this chapter are to provide an up-to-date synthesis of the oxygen and carbon isotope records of the Tertiary. First, we examine examples for the entire Tertiary, and then we look at details of the Tertiary records by epoch. We also demonstrate the potential of using the Tertiary isotope signals for stratigraphic correlations and interpretations of exploration wells.

I. Development of the Tertiary Isotope Record

The historical development of our understanding of the Tertiary isotope record is closely tied to the Deep Sea Drilling Project and to efforts to understand the paleoceanographic history of the ocean. Well prior to the DSDP, Harold Urey and his colleagues developed empirical paleotemperature equations relating the $^{18}O/^{16}O$ ratio of mollusc carbonate to the temperature and isotopic composition of the water in which they grew (McCrea, 1950; Urey *et al.*, 1951; Epstein *et al.*, 1953). The experiments were based on Urey's (1947) theoretical work on the thermodynamic properties of oxygen isotopes as potential geothermometers. Dorman (1966) and Devereaux (1966) showed that the $^{18}O/^{16}O$ ratio in fossil molluscs and foraminifera appeared to provide a record of the climatic deterioration in temperature throughout the Tertiary. Cesare Emiliani was the first to apply the isotopic paleotemperature technique to fossil foraminifera from deep-sea sediment cores (Emiliani, 1954a, b, 1961a, b). Most of this early work on marine sediments was restricted to the Pleistocene and isolated sections of

41

the Tertiary because of the general inaccessibility of long sedimentary sequences from older parts of the Tertiary using conventional coring devices.

The advent of floating drilling platforms and the Deep Sea Drilling Project changed this situation dramatically in the late 1960s. For the first time, long and nearly continuous sequences of marine sediments could be recovered from most parts of the deep ocean basins. These new opportunities made it possible for micropaleontologists and geochemists like Douglas and Savin to report their stable isotopic studies for foraminifera from DSDP sites in a series of papers (1971, 1973, 1976; Fig. 25). Although their early work lacked great detail and was often based on analyses of mixed assemblages of foraminifera, it still showed that the Tertiary oxygen isotopic record is characterized by a series of major isotopic changes often exceeding 1‰ [i.e., changes in temperature or sea level equivalent to approximately 5°C or 90 m (>280 ft), respectively]. Some parts of the record showed what appeared to be very systematic and almost monotonic isotopic changes as a function

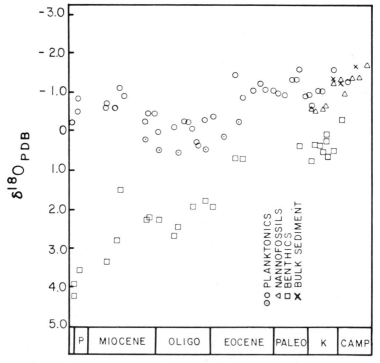

Fig. 25. The Tertiary oxygen isotopic record for the Tertiary and late Cretaceous based on oxygen isotopic analyses of planktonic foraminifera, calcareous nannofossils, and benthic foraminifera from Pacific DSDP sites 44, 47, 48, 49, 51, 67, 171, 305, and 306 (modified from Douglas and Savin, 1976).

of time (Fig. 25). Much of this work, however, was limited to sediments that contained huge abundances of calcareous microfossils, mainly foraminiferal or coccolith oozes, because of the large amounts of carbonate required to produce sufficient quantities of CO_2 for mass spectrometer analysis.

II. Global $\delta^{18}O$ Isotopic Changes in Tertiary Marine Carbonates

Improvements in the stability of head amplifiers and the development of mass spectrometers with inlets able to handle minute quantities of CO_2 (Shackleton, 1965) were significant steps forward. It was then possible to recover sufficient numbers of specimens of single foraminiferal species or of enough nannofossil carbonate in a restricted size fraction to make very precise oxygen and carbon isotopic analyses in most mid- to high-latitude DSDP sections (Savin, 1977). As a result, the scientific community now has access to detailed stable isotope records for various parts of the Tertiary from every major ocean basin and marginal sea (Fig. 26).

Shackleton and Kennett (1975a) and Savin (1977) significantly improved the early Tertiary record of Douglas and Savin by being able to perform isotopic analyses of small monospecific samples of planktonic and benthic foraminifera separately (<0.5 mg $CaCO_3$ by weight) (Figs. 17 and 18). By choosing sections from three overlapping DSDP sites from the South Pacific, Shackleton and Kennett (1975b) were able to enhance the resolution and timing of the major features shown in the early oxygen isotopic record of Douglas and Savin (1971) (Fig. 25). For example, beginning in the middle Miocene and extending into the Pleistocene, the $\delta^{18}O$ records of both the deep-water (benthics) and surface-water (planktonics) foraminifera are characterized by 2.5–3‰ changes. Superimposed on this systematic change in the isotopic composition of the foraminifera are some significant isotopic events that will be examined in more detail later. The Oligocene and early Miocene sections contain numerous isotopic events on the order of 0.5 to nearly 1‰ but no systematic change with time. A significant isotopic enrichment of nearly 1‰ was found to occur at or near the Eocene–Oligocene boundary recorded in both benthic and planktonic foraminifera (Figs. 17 and 18). This event was heralded as a major oceanographic event signaling the development of cold bottom water formation and vigorous circulation in the oceans. Preceding this major positive shift, the 10–12 MY time period from the early to the late Eocene is characterized by an isotopic deterioration of nearly 2‰, very similar in magnitude to the 2.5–3‰ enrichment in $\delta^{18}O$ values that occurs in the Neogene (mid-Miocene to Pleistocene). Many

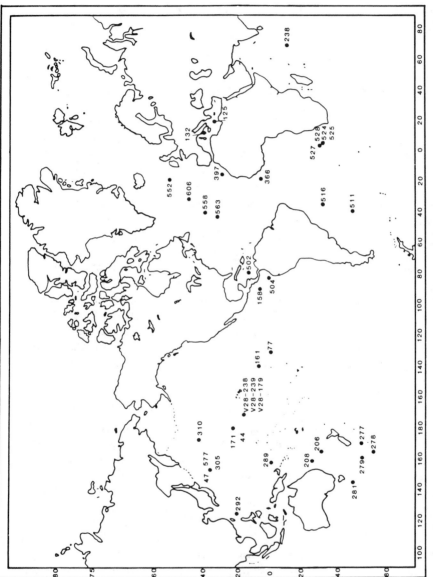

Fig. 26. Location map showing the approximate location of the DSDP sites, paleomagnetically dated piston cores, and other sediment sections from which detailed stable isotopic data are available.

other significant 0.5‰ isotopic events are visible on the overall Tertiary trend toward positive $\delta^{18}O$ values.

From these studies, we can see several significant relationships. First, all of the isotopic records for the Tertiary are characterized by a number of major shifts in absolute values. Second, the timing of many of these shifts appears to be globally isochronous, suggesting they may provide global tie lines. These shifts represent climatic and paleoceanographic events of global significance, as shown by the fact that an isotopic shift recorded in the South Pacific is also recorded in every other ocean basin of the same time period. This point is significant and will become clearer as we examine records from the other ocean basins and from other carbonate components of marine sediments. The third important point is that periods of time occur within the Tertiary that are characterized by unique isotopic values. For example, isotopic values between 3–2‰ are only found during one interval of the Plio-Pleistocene, represented by the section shown in Fig. 17. From the mid-Miocene to the Miocene–Pliocene boundary, isotopic values range between 2 and 0.5‰. Values during the middle Tertiary (i.e., early Miocene through early Oligocene) are between 1.5–0.5‰. This interval represents the largest period of time during the Tertiary in which oxygen isotopic values do not appear to vary in a unique, nearly monotonic function with time. To a first approximation, the time period from the last Eocene to the early Eocene is characterized by isotopic values of 0– −1‰ (Fig. 17). Of course, these specific values will vary as a function of the carbonate fraction analyzed (benthic foraminifera, planktonic foraminifera, or calcareous nannofossils, etc.) and the water temperature conditions at the latitudinal position of the site.

The importance of these observations for stratigraphic purposes and possible chronostratigraphy cannot be overemphasized because the records just described, even ones determined in the 1960s at very rough time intervals, strongly indicated that (1) the isotope signals represent global isotopic changes in ocean chemistry, (2) many of these changes occur synchronously on a global basis, (3) these records can be resolved into time units of millions or thousands of years before present (MYBP, KYBP), and (4) these records can be used as a new chemical stratigraphic tool to provide detailed tie lines independent of the regional changes in the microfossil assemblages and perhaps water depth.

III. Comparisons between the Nannofossil and Foraminiferal $\delta^{18}O$ Records

Margolis et al. (1975) demonstrated that the characteristic patterns found in the Tertiary oxygen isotopic record of foraminifera were also mirrored quite closely in the isotopic changes of the calcareous nannofossil fraction of marine sediments (Fig. 19). This study was important in demonstrating that the oxygen isotopic signal in the biogenic carbonate fraction was indeed reflecting major changes in the physiochemical properties of the oceans and not reflecting the changes in species composition of either the nannofossils or foraminifera. Bear in mind that evolutionary changes in foraminifera prevented a single foraminiferal species to be analyzed in the Shackleton-Kennett and Savin-Douglas studies. For any given interval of time, mono-specific analyses were performed (when possible). Both the nannofossil and foraminiferal assemblages vary significantly in their species composition throughout the Tertiary. Since Margolis *et al.* (1975) were unable to separate calcareous nannofossil species to achieve monospecific samples, they adopted the technique of sieving the sediment to achieve as consistent a size fraction as possible. Their record therefore shows that changes in species composition and microfloral changes throughout the Tertiary did not obscure the isotopic signal. They also showed that many if not all of the significant isotopic events are recorded in both the foraminiferal and nannofossil fractions. These results are significant from an exploration view-point because it may be eventually desirable to streamline the production of isotopic data from exploration wells by looking at either the total carbonate fraction or carbonate within a specific size fraction of the sediments.

IV. Whole Rock (Bulk Sediment) Analyses and the Tertiary $\delta^{18}O$ Record

The latest available $\delta^{18}O$ isotopic record for the Tertiary is from Shackleton (1987) (Fig. 23). This particular record is from analyses of bulk carbonate from the whole rock in overlapping sections of DSDP sites 525 and 528 from the South Atlantic (Fig. 26). Several significant points can be observed immediately. First, the overall pattern of the bulk $CaCO_3$ isotopic record is nearly identical to that observed in the foraminifera (benthic and plank-tonic) and calcareous nannofossils from the other DSDP sites discussed previously (Figs. 17–19). Second, the major isotopic events in the middle Miocene and Eocene–Oligocene boundary can be delineated. Third, the systematic changes in isotopic composition with time during the Neogene and early to late Eocene are readily observable, and fourth, this study was based on *whole rock* analyses from this site. No attempt was made to separate out individual foraminifera or to derive a pure fraction of the $<63-\mu m$ function for the calcareous nannofossils. Although the sediments at sites 525 and 528 are composed primarily of nannofossil oozes, the fact that the analyses are based on samples without any regard for microfossil content,

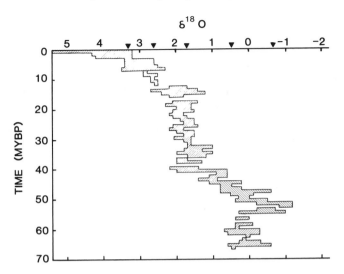

Fig. 27. A schematic interpretation of the stepwise shifts and long-term trends in the Tertiary $\delta^{18}O$ record from the Atlantic and Pacific oceans (modified from Shackleton, 1982).

varying proportions of different species, differing ratios of benthics to planktonics, or dissolution effects renders this record and its strong correlation with the other records (Figs. 17–19) a significant basis for the idea of developing an isotope chronostratigraphy for the Tertiary. The record shown in Fig. 23 can be considered to represent a whole signal of the $\delta^{18}O$ changes of the Tertiary oceans.

In his summary of oxygen isotopic data from benthic foraminifera from the last 70 MY (Fig. 27), Shackleton (1982) interpreted the long trends and stepwise shifts in the oxygen isotopic deep-water record to reflect a combination of the global change from a preglacial to a glacial world and major changes in the temperature and bottom water production in the world's oceans. He also predicted that interpolation from a given isotopic value in the record to the calibrated biostratigraphic zonation from the Tertiary (Berggren and Vancouvering, 1974) would lead to an uncertainty in resolution no better than +1 MY. In his compilation, however, Shackleton made the assumption that only the isotopic signal from biogenically precipitated calcareous microfossils would be suitable for stratigraphic correlations of marine sections. We believe that analyses of whole rock or a particular fraction of the detrital carbonate will extend the applicability of isotope chronostratigraphy to sediments with low carbonate content and into many clastic depositional settings. This approach could provide at least regional tie lines for depositional modeling in both temporal and spatial domains.

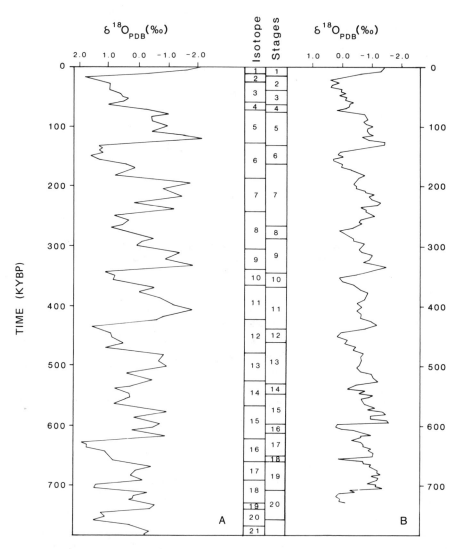

Fig. 28. A composite oxygen isotopic curve for the late Pleistocene based on spectral analyses of detailed $\delta^{18}O$ records (A) (Imbrie *et al.*, 1984) compared with the $\delta^{18}O$ record of core V28-238 (B) (Shackleton and Opdyke, 1973).

In summary, the Tertiary oxygen isotope records from different types of marine carbonate show clear evidence for global isotopic changes in ocean chemistry. These changes can be utilized for resolving stratigraphic correlations and calibrated to provide age estimates in Tertiary marine sediments. The degree of correspondence among the records is quite remarkable and serves as the principal basis for isotope chronostratigraphy. Details of the Tertiary oxygen and carbon records and the degree of stratigraphic resolution in parts of the Tertiary will be examined in succeeding chapters.

Chapter 5 | The Tertiary Carbon Isotope Record

I. The Foraminiferal $\delta^{13}C$ Record

The development of a carbon isotopic record for the Tertiary has had a curious history. For years it was assumed that the carbon isotopic composition of pelagic carbonates was very uniform and easily characterized by an isotopic value of approximately +2‰ (Craig, 1953). It is now recognized, however, that (1) the calcareous microfossils that contribute to marine sediments have widely different $\delta^{13}C$ values (Williams *et al.*, 1977; Berger *et al.*, 1978; Savin and Douglas, 1973; Duplessy *et al.*, 1970); (2) large changes have occurred in the mean $\delta^{13}C$ values for marine carbonate in the Tertiary; (3) many of these changes are recorded in both the benthic and planktonic carbonate fraction and reflect globally synchronous perturbations in the carbon reservoir of the oceans; (4) many of these carbon isotope changes can provide tie lines and a chronostratigraphy independent of and in some case superior to that provided by the Tertiary $\delta^{18}O$ record; and (5) changes in the storage of organic matter have had a significant impact on the carbon isotope record of marine carbonate.

It has not been easy arriving at this state of knowledge. For many years, geochemists and paleoceanographers did not understand the factors controlling the $\delta^{13}C$ of foraminifera, calcareous nannofossils, and marine carbonate in general. Therefore, very little of the $\delta^{13}C$ data was published. However, sufficient data now available in the literature show that significant patterns and age relationships are emerging that are analogous to, but not the same as, those seen for the oxygen isotopic record.

The early work by Shackleton and Kennett (1975b) provided an excellent indication of how the Tertiary carbon isotope record appears (Fig. 21). Comparison of the carbon isotopic composition of planktonic and benthic foraminiferal species from the late Paleocene into the Pleistocene with the $\delta^{13}C$ data in Fig. 17 shows that the oxygen and carbon isotopic records are entirely different in terms of overall character, the nature of the isotopic changes with time (i.e., in the negative or positive direction), and the magnitude of the $\delta^{13}C$ changes. For example, while there is an overall trend from the Paleocene to the Pleistocene of increasingly more positive values in the $\delta^{18}O$ record, no long-term trend exists in the $\delta^{13}C$ record. From the early Miocene to the Pleistocene, the carbon isotope records show systematic 2‰ trends in values ranging from +1.4 to −0.5‰ in the benthics and from > +2.5 to +0.7‰ in the planktonic foraminifera (Fig. 21). While both the $\delta^{18}O$ and $\delta^{13}C$ records for this time period of the Neogene exhibit systematic isotopic changes with time, the direction of the change in the $\delta^{13}C$ record is in the opposite direction to the $\delta^{18}O$ change. For example, the carbon isotopic values in the Neogene become more negative with time whereas the oxygen isotopic values become more positive. While the oxygen isotope trend in the Neogene has been explained as the result of the initiation of major glaciation on Antarctica (Shackleton and Kennett, 1975a), the exact reasons for the shift in the carbon isotope record are not explained entirely at this time.

Overall, the Tertiary carbon isotope records show evidence for two long-term cyclic changes. The Neogene change is the culmination of the last $\delta^{13}C$ peak in the early to middle Miocene (Fig. 21). For example, values in the early Eocene begin to become more positive in the middle to late Eocene, especially in the planktonic foraminiferal record. Superimposed on this trend are several major events with magnitudes on the order of 0.5-1‰. Beginning in the late Eocene to the late Oligocene, the carbon isotope records begin to shift back toward more negative values (Fig. 21). This shift is on the order of about 1.0-1.5‰. In the range of the middle to late Eocene boundary, the $\delta^{13}C$ signals recorded in the planktonic and benthic foraminifera appear to show a major divergence for some unexplained reason. Near the Oligocene–Miocene boundary, the carbon isotope values again show a systematic trend toward positive values in both the benthic and planktonic records, culminating at maximum values in the early to middle Miocene. Then in the early middle Miocene, the $\delta^{13}C$ values begin the Neogene trend toward negative (less positive) $\delta^{13}C$ values, with a major event in the late Miocene (Fig. 21).

II. Comparison of the Foraminiferal Nannofossil and Bulk Sediment $\delta^{13}C$ Records

The major features of the Tertiary carbon isotope records of foraminiferal carbonate are corroborated in the $\delta^{13}C$ signal for the calcareous nannofossil fraction (Margolis *et al.*, 1975; Fig. 20). The coccolith values are more positive on the whole than the foraminiferal values by as much as 1.0‰. This difference is approximately consistent with the nannoflora living in the more C-13 enriched upper waters of the euphotic zone than most foraminifera (Margolis *et al.*, 1975). Importantly, many of the same long-term trends are tracked by the nannofossil and foraminiferal fractions, i.e., from the middle Miocene to Pleistocene and from the early to middle Miocene positive peak. Large negative and positive isotopic events are superimposed on the long-term trends. From the early Oligocene to the late middle Eocene, the nannofossil values become more negative relative to the foraminiferal values. Margolis *et al.* (1975) used the caveat that isotope exchange with bottom waters may have been responsible for this effect, but a systematic follow-up has yet to be completed.

Many of the features seen in the carbon isotopic records from the foraminiferal and calcareous nannofossil fractions are also seen in the detailed carbon isotope record from whole rock analyses from the South Atlantic (Fig. 24) (Shackleton, 1985). The overall range in values is nearly 4‰, somewhat greater than the range observed in either of the biogenic $CaCO_3$ records individually.

A significant feature of the bulk sediment $\delta^{13}C$ record (Fig. 24) is a systematic decrease in $\delta^{13}C$ values of over 2.5‰ from the middle Miocene to the Pleistocene. Within this trend is a major positive carbon isotope event in the late Miocene. From the middle Eocene to the middle Miocene, the carbon isotope values remain within a range of 1.7-2.5‰. No long-term trend is discernible in this interval but, as will be seen when details of the Tertiary epochs are examined in later sections of this book, significant and resolvable cycles are present in the carbon isotope record and are potentially useful for stratigraphic determinations. The next major change in the bulk $\delta^{13}C$ record is a very negative carbon isotope event in the early Eocene. This change is preceded in the late Paleocene by the largest positive $\delta^{13}C$ event of the Tertiary, i.e., with values near +3.5‰.

It is widely recognized that the Maastrichtian–Paleocene carbon isotope record is characterized by a major negative carbon isotope shift across the Cretaceous–Tertiary boundary (Fischer and Arthur, 1977; Scholle and Arthur, 1980; Hsu *et al.*, 1982; Williams *et al.*, 1983, 1985; Zachos *et al.*, 1985). The characteristics of this boundary event will be discussed later. Following this Cretaceous–Tertiary carbon isotope shift, however, a trend exists in the late Paleocene toward the most heavy $\delta^{13}C$ values found in the entire Tertiary record (Fig. 24). A second major shift toward negative isotope values occurs near the Paleocene–Eocene boundary, achieving minimal $\delta^{13}C$ values in the early Eocene. The magnitude of this Paleocene–Eocene shift is larger than that across the Cretaceous–Tertiary boundary and nearly equal to that from the middle Miocene to the Pleistocene.

The large carbon isotope change across the Cretaceous–Tertiary boundary has been attributed to a dramatic decrease in marine productivity as a result of a catastrophic impact (Hsu *et al.*, 1982; Zachos *et al.*, 1985; Alvarez *et al.*, 1980). The very positive $\delta^{13}C$ values in the late Paleocene have been attributed to a very steep surface to deep-water $\delta^{13}C$ gradient as a result of enhanced productivity and photosynthetic removal of carbon (Shackleton and Hall, 1984b; Shackleton, 1985). The Paleocene–Eocene collapse in the $\delta^{13}C$ record may also signal a dramatic change in oceanic productivity. It will be demonstrated later that regardless of the exact nature of the cause, the magnitude of the carbon isotope shift across the Paleocene–Eocene boundary is an important event for precise, long-range stratigraphic correlation.

In summary, the $\delta^{13}C$ record for the Tertiary is characterized by at least 4 major $\delta^{13}C$ changes of potential importance for stratigraphic correlation of marine sections: (1) a negative event across the Cretaceous–Tertiary boundary; (2) a positive event from the early Paleocene into the late Paleocene; (3) the major $\delta^{13}C$ decrease across the Paleocene–Eocene boundary; and (4) the gradual but significant $\delta^{13}C$ deterioration from the mid-Miocene to the Pleistocene.

The absolute values of the $\delta^{13}C$ record are also important elements in isotope chronostratigraphy. For example, the lightest carbon isotope values in the Tertiary (Fig. 24) reside uniquely in the Pliocene–Pleistocene section (0–+1‰). The heaviest $\delta^{13}C$ values (i.e., >+3.5‰) are found in the late Paleocene. In addition, the systematic $\delta^{13}C$ change from +2.5‰ to near 0‰ from the mid-Miocene to the Pleistocene affords resolution in a chronostratigraphic framework, just as the oxygen isotope increase during much of the same time period. As will be shown in later sections, the cycles observed in the mid-Miocene through Oligocene section can also be resolved for stratigraphic correlation. The degree to which resolution of high-frequency events can be achieved (<100,000 yr in duration) will be examined in the later sections detailing the isotope records of each epoch.

Chapter 6 | Detailed Studies of the Tertiary $\delta^{18}O$ and $\delta^{13}C$ Records by Epoch

I. Pleistocene $\delta^{18}O$ Records

A. The Late Pleistocene $\delta^{18}O$ Record (0-1.0 MYBP)

The oxygen isotopic record for the Pleistocene is probably the best known part of the overall record in terms of detail for several reasons. Prior to the Deep Sea Drilling Project, most deep sea sediment cores available to the academic community were obtained with the Kullenberg piston corer [i.e., recovery of 10-20 m (approximately 30-60 ft) of sediment]. This sediment was represented predominantly by the upper Pleistocene except in areas of older subsurface exposures. Therefore, most initial isotopic studies of marine sediments were restricted to the Pleistocene epoch because of these sample restrictions.

Emiliani (1955a, b, 1958, 1964, 1970) was the first to examine the character of the oxygen isotopic record for the Pleistocene. He was able to demonstrate that the Pleistocene isotopic record was characterized by quasi-periodic fluctuations in the $^{18}O/^{16}O$ ratio of foraminifera from Caribbean and equatorial Atlantic cores. He made important contributions to our understanding of the glacial Pleistocene epoch by correctly interpreting these variations to reflect changes between glacial and interglacial climatic modes. He recognized and established the concept of oxygen isotopic stages, i.e., periods of time characterized by negative values (interglacials, odd numbered stages) or positive values (glacials, even numbered stages). Of course, given stages are not uniformly characterized by an average isotopic composition. In many cases it is possible to identify structure in stages of the late Pleistocene as substages.

Emiliani's isotope stages, extended by Shackleton and Opdyke (1973), became the basis for the oxygen isotope stratigraphy in Pleistocene sediments when it was recognized that the δ^{18}O record of Pleistocene foraminifera primarily reflects changes in the ^{18}O/^{16}O composition of seawater as a function of changes in sea level (Craig, 1965; Olaussen, 1965; Shackleton, 1967; Dansgaard and Tauber, 1969). The magnitude of this glacio-eustatic/ice-volume effect from glacial–interglacial changes of the last 1.0 MY has been estimated to be between 1.2–1.8‰ or, in terms of sea level change, between 110–164 m (approximately 360–540 ft) (Williams, 1984; see Chapter 8 for a discussion of the relationship between the δ^{18}O record and sea levels of the Tertiary). Radiometric dating of sea level records from coral terraces and oxygen isotope records from deep sea cores demonstrated that as water is removed from the ocean and secured as ice in high latitude regions of the globe, the resulting changes in the isotopic composition of the oceans and in sea level are recorded by lowered coral terraces (Broecker et al., 1968; Mesolella et al., 1969; Broecker and Von Donk, 1970; Veeh and Chappell, 1970; Bloom et al., 1974; Shackleton and Matthews, 1977; Fairbanks and Matthews, 1978) and also by the isotopic composition of foraminifera (Shackleton, 1967, 1977; Shackleton and Matthews, 1977a; Shackleton and Opdyke, 1973; Williams et al., 1981a) (Fig. 28).

Shackleton and Opdyke (1973) firmly established oxygen isotope stratigraphy as a global, high-resolution correlation tool in Pleistocene marine sediments through the use of magneto-stratigraphy (Fig. 29). For sediments of the last 900,000 yr, they were able to identify and date the boundaries of 22 isotope stages. The glacial and interglacial stages are interpreted in terms of high and low volumes of northern hemisphere ice and low and high stands of sea level, respectively. The importance of the Shackleton and Opdyke (1973) paper cannot be overemphasized because, through their work, oxygen isotope records have become the most precise means of establishing: (1) global correlations in marine sediments; (2) the synchroneity of various micropaleontological events in terms of isochronous last and first appearance datums (Hays and Shackleton, 1976; Thierstein et al., 1977; Burckle et al., 1982; Morley and Shackleton, 1978); and (3) the interpretation of marine sedimentary sections in terms of glacio-eustatic changes in sea level. Oxygen isotope stratigraphy also provides the time frame for most modern paleoceanographic studies of the Pleistocene.

Although the first oxygen isotope work dealt with the last 300,000–200,000 yr of the Pleistocene, in the last 10 yr, our knowledge of the δ^{18}O record has been extended back through the entire Pleistocene with the acquisition of (1) long piston cores that have been stratigraphically dated with paleomagnetic properties of the sediment (Shackleton and Opdyke, 1976; Ledbetter, 1984a); (2) hydraulically piston-cored sections from the

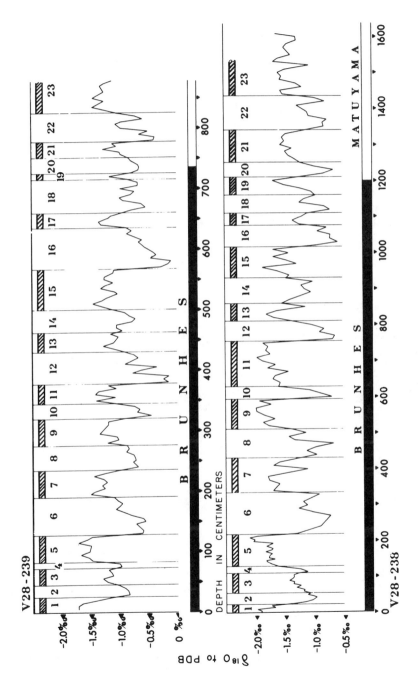

Fig. 29. A comparison of the timing and amplitude of the oxygen isotope stages in Pacific piston cores (bottom) V28-238 and (top) V28-239 along with their paleomagnetic stratigraphy (modified from Shackleton and Opdyke, 1973, 1976).

Deep Sea Drilling Project (Shackleton *et al.*, 1983a; Prell, 1982); and (3) continental margin exploration wells and boreholes from the Shell *Eureka* series (Williams, 1984).

The basic premises of oxygen isotope stratigraphy in the Pleistocene are described in greater detail in Williams (1984) but are briefly summarized here and illustrated in Figs. 30 and 31:

1. The isotope changes recorded in foraminifera and coccoliths in Pleistocene marine sediments are driven by glacio-eustatic changes in sea level. The removal of water from the oceanic reservoir, and the storage of this ^{18}O-depleted water as ice in polar regions during glacial periods, led to an increase in the ^{18}O of the remaining ocean reservoir.
2. The magnitude of the glacio-eustatic sea level changes is related to the volume of ice stored in high-latitude regions.

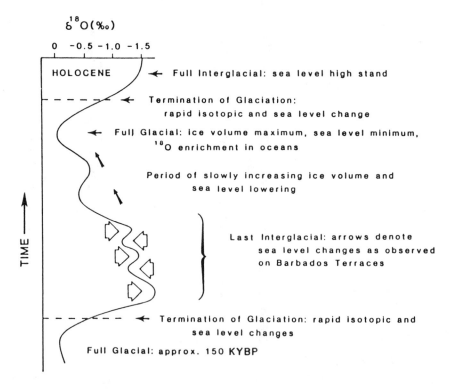

Fig. 30. Schematic representation of the oxygen isotope record for the last 150,000 yr and its approximate relationship with major climatic and sea level changes of the late Quaternary.

INTERGLACIAL CONDITIONS

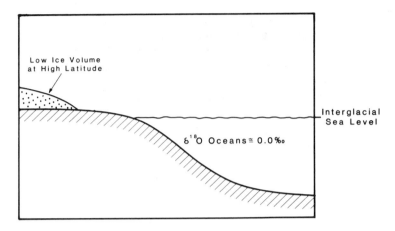

GLACIAL CONDITIONS

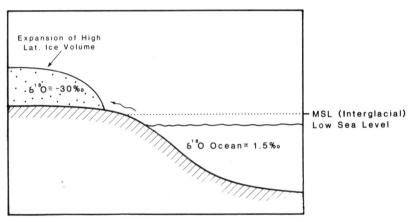

Fig. 31. A schematic representation of the relationship between changing oxygen isotopic composition of the oceans as a function of lowered sea levels during glacial periods.

3. The changes in glacio-eustatic sea level and continental ice volume affect the isotopic composition of the ocean rapidly and nearly synchronously within the mixing time of the oceans (much less than 1000 yr; Craig and Gordon, 1965).

4. Changes in the isotopic composition of the ocean occur quasiperiodically throughout the Pleistocene.

5. The isotope stages are recorded in biogenic carbonate in all the ocean basins regardless of latitude, location, or water temperature.

B. Extension of $\delta^{18}O$ Statigraphy into the Pliocene

Recently, Williams *et al.* (1988) have extended the concept of oxygen isotope stages, from the original 22 stages established by Shackleton and Opdyke (1973, 1976) (Fig. 29), into the middle and early Pleistocene (Fig. 32). Oxygen isotope records are now available for the late Pliocene–Pleistocene epoch from the equatorial Atlantic (Van Donk, 1976; Shackleton and Boersma, 1981), southwest Atlantic (Vergnaud-Grazzini *et al.*, 1983), equatorial Pacific (Shackleton and Opdyke, 1976), Gulf of Mexico (Williams, 1984), western Caribbean (Prell, 1982), and Mediterranean (Thunell

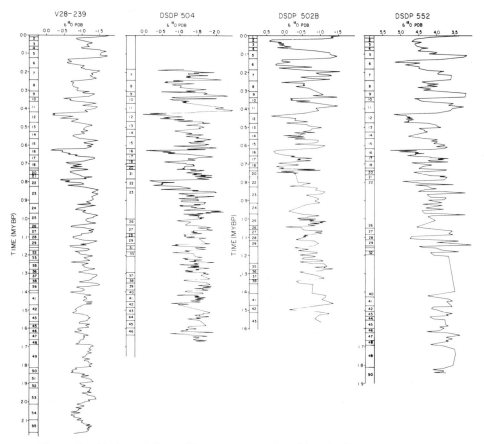

Fig. 32. Empirical correlation and isotope stage zonation of four detailed Pleistocene oxygen isotope records from the Pacific (core V28-239 and site 504), Caribbean (site 502B), and North Atlantic (site 552) (Williams and Trainor, 1986). The datum levels shown in Table I were used to convert the isotope recores to time.

Table I

Quaternary Datum Levels

Method[a]	Datum	Age (MY)	Reference
Isotope	Stage 5/6 boundary	0.128	Imbrie *et al.*, 1984
Foram	X/Y boundary	0.128	van Donk, 1976
Isotope	Stage 6/7 boundary	0.188	Morley and Hays, 1981
Isotope	Stage 7/8 boundary	0.245	Imbie *et al.*, 1984
Nanno	*Emiliania husleyi* FAD	0.275	Thierstein *et al.*, 1977
Isotope	Stage 9/10 boundary	0.339	Imbrie *et al.*, 1984
Isotope	Stage 10/11 boundary	0.362	Imbrie *et al.*, 1984
Isotope	Stage 11/12 boundary	0.423	Imbrie *et al.*, 1984
Nanno	*Pseudoemiliania lacunosa* LAD	0.474	Thierstein *et al.*, 1977
Isotope	Stage 12/13 boundary	0.478	Imbrie *et al.*, 1984
Foram	U/V boundary	0.490	van Donk, 1976
Isotope	Stage 13/14 boundary	0.524	Imbrie *et al.*, 1984
Isotope	Stage 17/18 boundary	0.689	Imbrie *et al.*, 1984
Isotope	Stage 18/19 boundary	0.726	Imbrie *et al.*, 1984
Paleomag	Brunhes/Matuyama boundary	0.730	Mankinen and Dalrymple, 1981
Isotope	Stage 19/20 boundary	0.736	Imbrie *et al.*, 1984
Isotope	Stage 20/21 boundary	0.763	Imbrie *et al.*, 1984
Isotope	Stage 21/22 boundary	0.790	Imbrie *et al.*, 1984
Isotope	Stage 22/24 boundary	0.828	Berggren *et al.*, 1980
Paleomag	Top of Jaramillo subchron	0.91	Mankinen and Dalrymple, 1981
Nanno	"Small *Gephyrocapsa*" LAD	0.92	Berggren *et al.*, 1980
Paleomag	Base of Jaramillo subchron	0.99	Mankinen and Dalrymple, 1981
Foram	S/T boundary	0.99	van Donk, 1976
Foram	R/S boundary	1.294	van Donk, 1976
Nanno	Large *Gephyrocapsa* GAD	1.29	Backman and Shackleton, 1983
Nanno	*Helicopontosphera sellii* LAD	1.37	Backman and Shackleton, 1983
Nanno	*Cyclococcolithina macintyrei* LAD	1.45	Backman and Shackleton, 1983
Foram	*G. fistulosus* LAD	1.60	Berggren *et al.*, 1980
Nanno	*G. oceanica* sl. FAD	1.62	Backman and Shackleton, 1983
Foram	Q/R boundary	1.656	van Donk, 1976
Paleomag	Top of Olduvai subchron	1.66	Mankinen and Dalrymple, 1981
Foram	P/Q boundary	1.729	van Donk, 1976
Nanno	*Discoaster brouweri* LAD	1.88	Backman and Shackleton, 1983
Paleomag	Base of Olduvai subchron	1.88	Mankinen and Dalrymple, 1981

[a] Foram, foraminiferal; Nanno, nannofossil; Paleomag, paleomagnetic.

and Williams, 1983). Using the Quaternary datums shown in Table I, each of the available Pleistocene isotope records has been plotted to a uniform Quaternary time scale (Fig. 32). Quasi-periodic isotopic cycles exist throughout the Pleistocene, providing evidence for two different climatic modes and enabling over 50 isotope stages to be defined for the last 2 MY. These features offer a high degree of stratigraphic resolution unparalleled in any biostratigraphic zonation presently available. It must be emphasized that in Plio-Pleistocene exploration wells it is important to integrate the oxygen isotope results with any or all of these: biostratigraphy (either foraminiferal or calcareous nannofossil), paleomagnetic stratigraphy (Ledbetter, 1984a), tephrochronology (Ledbetter, 1984b), and lithostratigraphy (i.e., seismic stratigraphy). However, we do not mean that the isotope chronostratigraphy in the Pleistocene is solely dependent upon any one of these other types of information. Obviously, the combination of these data bases will achieve the stratigraphic framework of highest resolution.

As shown in the synthesis with the Pleistocene oxygen isotope record (Fig. 32), this empirical approach, in conjuction with the foraminiferal and nannofossil datums, provides a stratigraphy capable of resolving stratigraphic correlations within 10,000–20,000 yr in the late Pleistocene and 40,000–50,000 yr in the early Pleistocene and late Pliocene. We are now in the process of quantifying and defining these stages using some of the digital processing techniques described in the latter part of this book.

In summary, the oxygen isotope record for the Plio-Pleistocene in DSDP core material is characterized by isotopic changes of an amplitude ranging between 1.0 to nearly 3.0‰, depending upon the sedimentary basin, the rate of sediment accumulation, and the thickness of the sedimentary section. For example, the relatively short (by industrial standards) piston core V28-239 has an average sediment accumulation rate of only approximately 1 cm/KY. The average isotopic change is <1.0 (Figs. 29 and 32). For the same time period, the oxygen isotope record in the Gulf of Mexico between 2.0 and 1.2 MYBP is characterized by changes <1.5–2‰ (Fig. 33). Note that the $\delta^{18}O$ record of *Eureka* E67-135 is quite comparable to those from other ocean basins and that the *timing* of the changes is the same between the Gulf of Mexico, Pacific, and Atlantic records. The quality of global correlations using $\delta^{18}O$ stratigraphy is also well illustrated in Fig. 34. This example shows very clearly that, despite the fact that the records are based on different foraminifera that inhabited different water masses (i.e., temperatures), the timing of the $\delta^{18}O$ events is synchronous.

We have attempted to extend the concept of isotope stages even further than stage 50, well into the Pliocene, using a combination of DSDP sites from the Pacific (V28-239) and Mediterranean (Site 125), proprietary evidence from exploration wells from the offshore northern Gulf of Mexico, and a tentative zonation of the paleomagnetically calibrated piston core V28-179 and site 572 (Fig. 35). While the particular stage designations shown are preliminary, they have already proven very useful in Plio-Pleistocene exploration wells from the Gulf of Mexico. We are presently in the process of using the signal processing techniques of cross-correlation and auto-correlation, which are described later in this book.

It is also important to note that the amplitudes of the $\delta^{18}O$ signal changes systematically during parts of the Plio-Pleistocene. Two reasons are principally involved. First, the Gulf of Mexico is a marginal ocean basin with a more restricted reservoir. Its circulation is therefore slightly more restricted than in the Pacific–Atlantic, meaning that the Gulf of Mexico oxygen isotope record will be subject to regional changes in the evaporation–precipitation–runoff balance. This is most easily seen in the influence of meltwater. Many times during the Pleistocene, the Gulf of Mexico oxygen isotope signal is much larger than the signal observed in the open ocean because of the input of meltwater from midcontinent ice sheets (Kennett and Shackleton, 1975; Emiliani et al., 1975; Emiliani, 1978; Falls, 1980; Leventer et al., 1982, 1983; Williams et al., 1983; Williams, 1984). This aspect will be discussed further in the Section I, C of this chapter because of its importance in resolving isotope chronostratigraphy in exploration wells.

Second, the isotopic differences reflected in the Plio-Pleistocene $\delta^{18}O$ records shown in Figs. 32–34 reflect the temperature differences in the waters in which benthic and planktonic foraminifera were residing during the time of their sedimentation. For example, the planktonic record in a low-latitude or equatorial location will have more negative $\delta^{18}O$ values because of warmer water temperatures than a surface planktonic record from a high-latitude region with colder water temperatures. The same temperature effect produces an isotopic offset between the benthic and planktonic signals from the same core because of the temperature differences that the benthic and planktonic foraminifera experience. Despite how temperature differences influence the records shown in Figs. 34 and 35, the timing of the isotopic changes is the same on a global basis, thereby providing the opportunity for high-resolution stratigraphy offered by few other methods and types of data (Williams, 1984).

C. Specific Application to the Gulf of Mexico

It has been clearly demonstrated that the Pleistocene $\delta^{18}O$ record of the Gulf of Mexico can be correlated to the global open-ocean record (Falls, 1980; Leventer et al., 1982; Williams, 1984, and unpublished data). The isotope record for the Gulf, however, has several unique characteristics because of its connection to the Laurentide ice sheet via the Mississippi

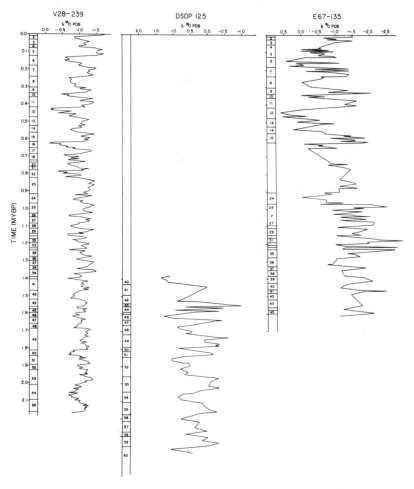

Fig. 33. Tentative definition of oxygen isotope stages for the late Pliocene–Pleistocene based on Pacific V28-239, Mediterranean site 125, and Gulf of Mexico borehole *Eureka* 67-135 (Williams and Trainor, 1986). All data are plotted to the same scale. Note the large difference in amplitude that exists in the semirestricted Mediterranean and Gulf of Mexico basins.

drainage system (Fillon and Williams, 1984). In addition to the global ice-volume signal, the Gulf record has been influenced by the periodic influx of isotopically light meltwater (^{16}O enriched) at the end of major continental glaciations (and possibly many of the minor glacial periods). The most recent meltwater signal, which occurred during the end of the last glacial maximum (11 KYBP), has been studied in detail (Fig. 36) (Falls, 1980; Leventer, 1981; Leventer *et al.*, 1982). Leventer's work has led to the best

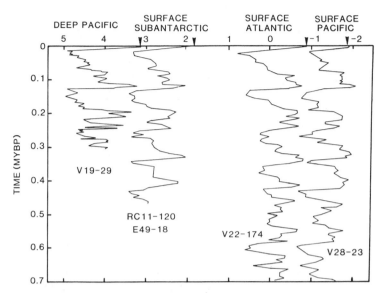

Fig. 34. A comparison of late Pleistocene oxygen isotope records based on benthic or planktonic foraminifera from the Pacific, Atlantic, and Antarctic. This is a good example of the temperature effect that determines the position of each record on the horizontal ($\delta^{18}O$) axis. Note, however, that the global glacial–interglacial signal is clearly visible in each basin, regardless of which species is used to generate the isotope signal (from Shackleton, 1982). Reproduced with permission from Pergamon Press.

understanding of the nature of this signal in the Gulf of Mexico and has enabled us to establish the maximum and minimum boundary conditions of the isotope signal in the Gulf of Mexico (Fig. 36). The absolute isotope values can be used to determine whether the sediment in a particular portion of a Pleistocene exploration section was deposited during a glacial or interglacial interval or a period of meltwater discharge. The major portion of the meltwater signal from the last discharge occurred near the isotope stage 2–1 boundary and the Y–Z biostratigraphic boundary of Ericson and Wollin (1968). Falls (1980) has shown the existence of a meltwater signal at the W–X biostratigraphic boundary and recognized the substages of isotope stage 5 (Fig. 37). Our studies of exploration wells from the Gulf show strong evidence of either meltwater signals or large fluvial events extending into the Pliocene. Isotope records from Pleistocene sections drilled in the Gulf suggest that these meltwater/fluvial events may be used as "fingerprints" with which to enhance our correlations.

D. Summary

The oxygen isotope record of Pleistocene deep-sea sediments has proven to be a very reliable tool for making global stratigraphic correlations.

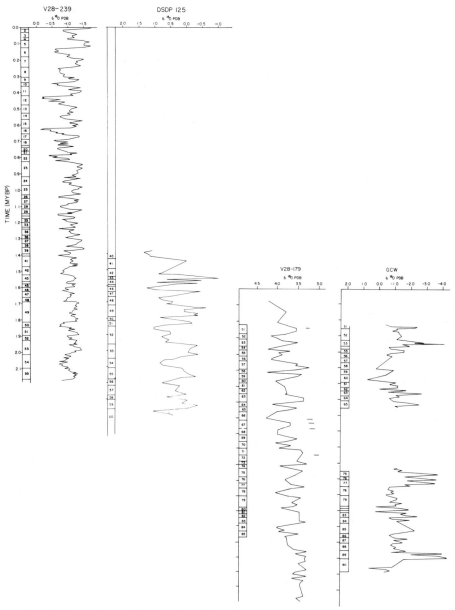

Fig. 35. A tentative empirical zonation of the Pliocene–Pleistocene using the oxygen isotope records from the Pacific (V28-239). Mediterranean (site 125), and Atlantic (V28-179) (Williams and Trainor, 1986).

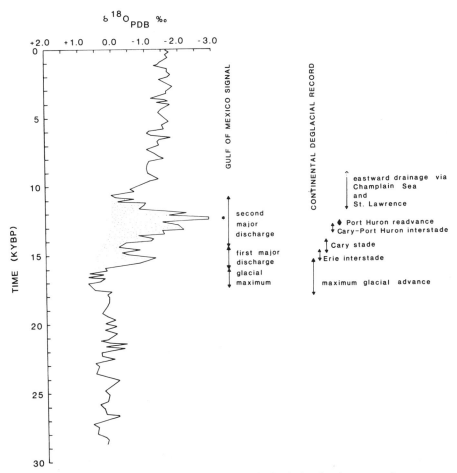

Fig. 36. A detailed study of the influence of isotopically depleted meltwater on the oxygen isotopic composition of planktonic foraminifera from the Gulf of Mexico. This record from the Orca Basin (Leventer *et al.*, 1982) enabled the establishment of specific delta values for distinguishing meltwater events from interglacial values from glacial–interglacial transition values and for glacial values representing different stages of glacio-eustatic sea levels.

Correlations can be made with minimal resolutions of 10,000–20,000 yr in the late Pleistocene, and 40,000–50,000 yr in the early Pleistocene (Figs. 32 and 35). Global correlations are possible because the isotopic changes in the oceans are driven by glacial–interglacial climatic changes in polar ice volumes. Because the oceanic reservoir is mixed almost instantaneously by geological standards, $\delta^{18}O$ changes controlled by glacial advances (increased ice volume) and periodic returns to interglacial conditions will

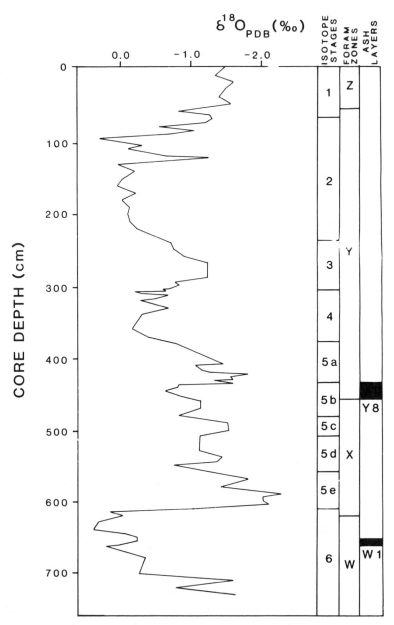

Fig. 37. A detailed oxygen isotope record for the last 140,000 yr for the Gulf of Mexico core TR126-23 in comparison with the Ericson biostratigraphic zones and tephrochronology (Williams, 1984).

be recorded worldwide by the $^{18}O/^{16}O$ ratio in the calcium carbonate tests of foraminifera and calcareous nannofossils. It has been demonstrated that such isotopic changes in marine sedimentary sequences can be unequivocally correlated and are synchronous.

Using $\delta^{18}O$ stratigraphy, it is possible to subdivide many of the commonly used biostratigraphic zonations of the Gulf of Mexico. For example, $\delta^{18}O$ stratigraphy can subdivide the Ericson and Wollin biostratigraphic X zone into five separate isotope substages of stage 5 (substages 5a, 5b, 5c, 5d, 5e) (Fig. 37). A potential of 14 stages and substages exists for the last 250,000 yr as compared with only 5 Ericson and Wollin biostratigraphic zones. The same holds true for the middle to late Pleistocene (Fig. 15). The potential exists to (1) enhance the correlation of Gulf Coast continental margin sequences; (2) identify horizons in the Plio-Pleistocene to within 20,000–50,000 yr within an absolute time scale; and (3) relate lithological and sedimentological changes to glacio-eustatic sea-level fluctuations and paleoclimatic changes.

II. Pliocene Isotope Records

Shackleton and Opdyke (1977) were the first to incorporate isotope and paleomagnetic stratigraphy for a detailed study of the Pliocene using a piston core from the Pacific Ocean (Fig. 38). This is one of the records that we have used to propose a tentative empirical isotope stage zonation back

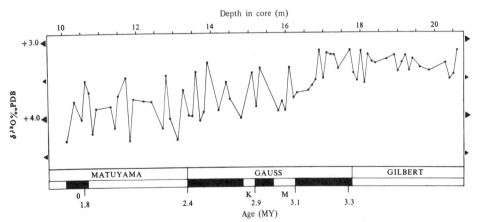

Fig. 38. Pliocene oxygen isotopic record based on detailed analyses of benthic foraminifera from paleomagnetically dated piston core V28-179 from the Pacific (Shackleton and Opdyke, 1977).

into the Pliocene (Fig. 35). This zonation will of course be revised as better records become available and the quantitative techniques described in the latter part of the book are utilized. Despite the low accumulation rate in V28-179 (an average of 0.55 cm/KY), significant Pleistocene-like isotopic variations approaching 0.7 to nearly 1‰ occur in this section of the Pliocene. At approximately 3.2 MYBP, Shackleton and Opdyke (1977) interpreted a particularly large enrichment in the $\delta^{18}O$ record to indicate the initiation of northern hemisphere glaciation. No carbon isotope data were published at that time.

As the result of this study, interest increased in reconstructing the deep-water history and paleoclimatic history of the Pliocene (Berggren, 1972), and a number of isotopic studies of Pliocene DSDP sections were reported (Leonard et al., 1983; Hodell et al., 1983, 1987; Thunell and Williams, 1983; Keigwin, 1979, 1982a, 1987; Shackleton and Hall, 1984c; Shackleton and Cita, 1979; Prell, 1984). A particularly significant enrichment in ^{18}O was found by Thunell and Williams (1983) to occur at approximately 2.5 MYBP in the middle part of the Pliocene faunal zone MPL 5 in several records from different ocean basins (Fig. 41): DSDP sites 132 and 125 from the Mediterranean (Fig. 39; Thunell and Williams, 1983); site 397 from the Atlantic off the northwestern African margin near Cape Bojador (Fig. 40, Shackleton and Cita, 1979); DSDP sites and piston cores from the Pacific (Site 310, Keigwin, 1979; V28-239 and V28-179, Shackleton and Opdyke, 1976, 1977); site 502 from the Caribbean (Keigwin, 1982b; Prell, 1982). Other long-term isotope shifts extend into the middle part of the Pleistocene and were interpreted by Thunell and Williams to represent a significant increase in northern hemisphere ice volume during glacial intervals. Cyclic glacial–interglacial fluctuations in $\delta^{18}O$ can be seen to extend into the early Pliocene. A second major climatic step is observed in the middle Pleistocene portion of the $\delta^{18}O$ records. This shift and the large change to heavier isotopic values during glacial intervals in the Pleistocene probably reflects further intensification of northern hemisphere glaciation.

Based on this accumulated work, it became clear that the significant but temporary change in $\delta^{18}O$ values observed between 3.2–3.0 MYBP in V28-179 (Fig. 38) (Shackleton and Opdyke, 1977) was related to a significant cooling of surface waters and not a permanent increase in ice volume (Prell, 1982; Thunell and Williams, 1983). Instead it appeared that the permanent shift at 2.5 MYBP (Fig. 41) was more likely acceptable evidence for the first major increase in northern hemisphere glaciation.

The second detailed study of the Pliocene comes from a hydraulically piston cored DSDP section from the north Atlantic (Shackleton et al., 1984a). Site 552A has, by deep-sea standards, a fairly high accumulation rate (1.7–1.8 cm/KY), especially when compared with the initial study by

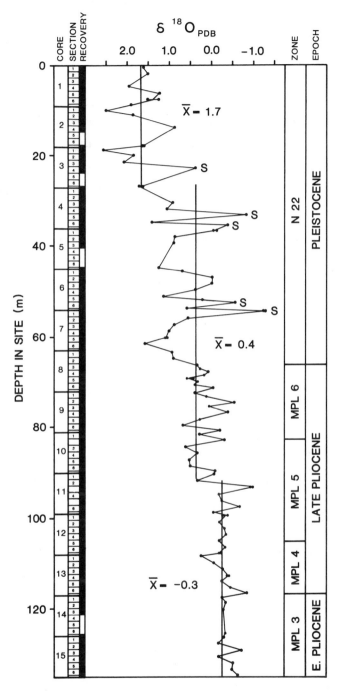

Fig. 39. A Plio-Pleistocene oxygen isotopic record based on analyses of planktonic foraminifera from Mediterranean DSDP site 132 (Thunell and Williams, 1983).

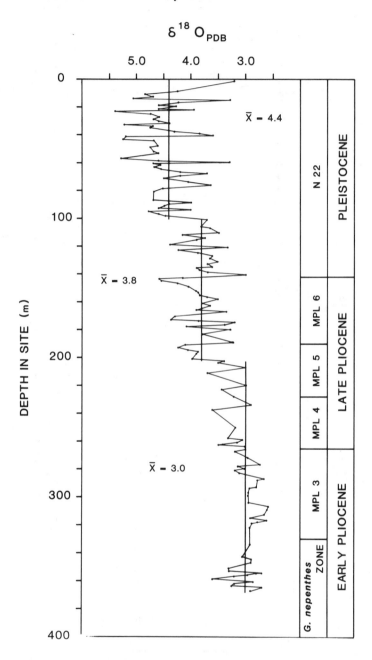

Fig. 40. A detailed Pliocene benthic oxygen isotopic record from northeastern Atlantic DSDP site 397 (modified from Shackleton and Cita, 1979; Thunell and Williams, 1983).

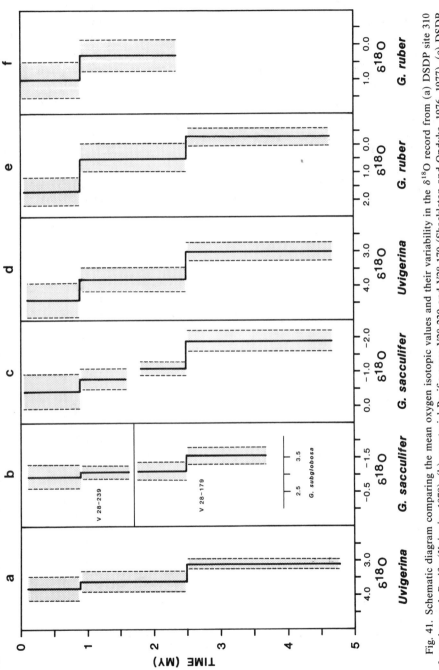

Fig. 41. Schematic diagram comparing the mean oxygen isotopic values and their variability in the $\delta^{18}O$ record from (a) DSDP site 310 from the north Pacific (Keigwin, 1979), (b) equatorial Pacific cores V28-239 and V28-179 (Shackleton and Opdyke, 1976, 1977), (c) DSDP site 502 from the Caribbean (Keigwin, 1982a; Prell, 1982), (d) DSDP site 397 off Cape Bojador (Shackleton and Cita, 1979), and DSDP sites (e) 132, and (f) 125 from the Mediterranean (Thunell and Williams, 1983). The vertical lines denote the mean $\delta^{18}O$ values for the time period representing significant shifts or steps in climatic conditions. The stippled areas represent +1 standard deviation.

Shackleton and Opdyke (1977) on piston core V28-179 (Fig. 42). Using paleomagnetic stratigraphy and calcareous nannofossil extinction horizons, Shackleton *et al.* (1984b) were able to cast the Pliocene oxygen isotopic data for site 552A and V28-179 into the approximate time scale shown in Fig. 42. Comparison of the 552A oxygen isotope record with carbonate and other indices of ice rafted debris in the north Atlantic indicated that the permanent shift in the oxygen isotopic record coincided with an increase in ice rafting indicators approximately 2.37 MYBP. This correspondence is now regarded as firm evidence for the first major northern hemisphere glacial event.

Keigwin (1983) examined the lower and upper Miocene portion of site 552A but at a sampling interval that is not very detailed. However, an examination of the data in Fig. 43 shows that an excellent correspondence exists between the oxygen isotopic results for the planktonic foraminiferal

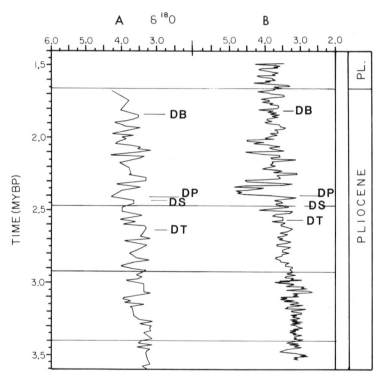

Fig. 42. A comparison of the Pliocene oxygen isotopic record of (A) site 552A (North Atlantic) with (B) Pacific core V28-179 along with nannofossil extinction horizons indicated by DT, *Discoaster tamalis*; DS, *D. surculus*; DP, *D. pentaradiatus*; DB, *D. brouweri* (modified from Shackleton *et al.*, 1984a).

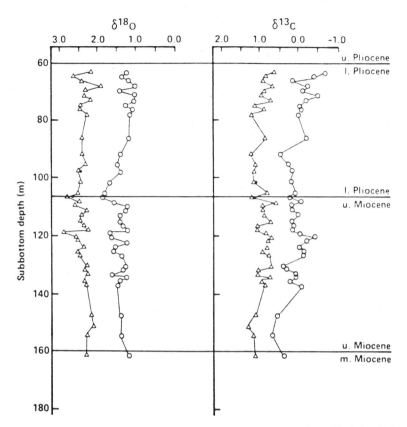

Fig. 43. Pliocene-Miocene oxygen and carbon isotope records of benthic (triangles) and planktonic foraminifera (circles) from North Atlantic DSDP site 552A (from Keigwin, 1983).

species *Globigerina bulloides* and the benthic species *Cibicides wueller-storfi*. The isotopic fluctuations in the lower Pliocene and upper Miocene portions of 552A are on the order of >0.5‰. Changes of this magnitude suggest that significant isotopic detail exists in this portion of the record, which should also be of stratigraphic use.

The concerns discussed previously regarding a permanent or nonpermanent shift in the Pliocene $\delta^{18}O$ record have been recently examined by Prell (1984) (Fig. 44). By a comparison of the covariance between the planktonic and benthic records from the time interval of 2.8–3.6 MYBP, Prell showed the lack of covariance between the planktonic and benthic $\delta^{18}O$ records around 3.2 MYBP, which has been previously described as the initiation of northern hemisphere glaciation (Berggren, 1972; Poore *et al.*, 1981; Shackleton and Opdyke, 1977; Hodell *et al.*, 1985; Weissert *et al.*,

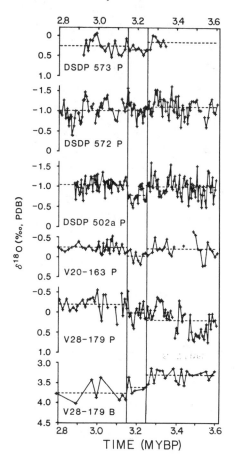

Fig. 44. A comparison of the covariance between detailed benthic and planktonic oxygen isotope records for the Pliocene interval between 3.6 and 2.8 MYBP (Prell, 1984). The interval highlighted with the horizontal lines centered around 3.2 MYBP has been interpreted to represent a bottom water cooling in the Pacific benthic record (V28-179B) and not the initiation of northern Hemisphere glaciation as previously postulated.

1984). Prell (1984) concluded that the mean volume of northern hemisphere ice did not substantially increase at 3.2 MYBP. The 3.2-MYBP shift in the benthic $\delta^{18}O$ values in V28-179 (Fig. 44) more likely represents a bottom water cooling in the Pacific or possibly new or increased production of cold bottom waters in high latitudes.

In summary, the Pliocene oxygen isotope record is characterized by significant shifts both in mean value and amplitude. Although much of this variation is less than that observed during glacial–interglacial fluctuations in the Pleistocene, sufficient change is recorded in biogenic carbonate to render the oxygen isotope record of use in stratigraphic correlations. This is particularly evident in the comparison between site 552A and core V28-179 in Fig. 42. If one takes the isotopic data between the Olduvai normal event in the Matuyama and the Matuyama–Gauss boundary at 1.66 and 2.47 million years, respectively (Fig. 42), a significant correlation can be seen between the oxygen isotope records. This correlation can be seen despite the fact that these records are from two different ocean basins and have very different accumulation rates (the average rate at site 552A is three times that of V28-179). New data from North Atlantic DSDP site 606 (Keigwin, 1987) also correlates well with the other Pliocene records.

Regarding the variability within the Pliocene isotopic record, Shackleton (1982) published detailed oxygen isotopic data from Atlantic and Pacific piston cores with excellent paleomagnetic stratigraphy and biostratigraphy. He was able to derive a detailed composite record for deep waters of the Pacific with excellent resolution from 2 to 6 MYBP (Fig. 45). Unfortunately, a direct comparison cannot be made to the Atlantic deep-water record at DSDP site 397 because of differences in the quality of the time control, sampling density, and the possibility of coring disturbance in site 397. Regardless, the composite Pacific record and the site 397 Atlantic record show that many of the same features occur in the records in terms of (1) changes in the isotopic amplitude and (2) long-term shifts toward more positive values from the early Pliocene into the Pleistocene (Fig. 45).

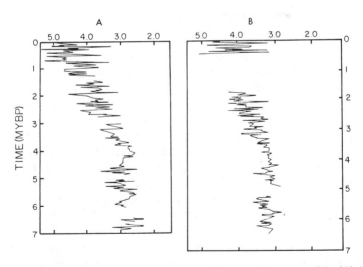

Fig. 45. Composite oxygen isotope records for the Pliocene-Pleistocene of the (A) Atlantic (DSDP site 397) and (B) Pacific (piston cores V19-29, V19-30, V28-179, V28-185, RC12-66) benthic foraminifera (modified from Shackleton, 1982).

An important point is that large isotopic variations of $>1\permil$ extend well beyond the currently accepted time for the large-scale glaciation of the northern hemisphere (2.5–3.2 million years) (Shackleton and Opdyke, 1977). Shackleton (1982) concluded from this analysis that a time resolution of approximately 100,000 yr was potentially achievable in this time interval of the Pliocene, depending on sediment disturbance and the degree of time control. It is also important to note that although a one-to-one correlation between the two records shown in Fig. 42 is not possible at this time they show evidence of divergence between the Atlantic and Pacific Ocean basins as a function of time (i.e., the records diverge prior to 3.0 MYBP (Shackleton, 1982).

It is now clear that northern hemisphere glaciation underwent a major and permanent advance approximately 2.4 MYBP and not at 3.2 MYBP as previously determined. For the purposes of isotope chronostratigraphy of the Pliocene, we have the possibility of using both the glacial–interglacial isotopic fluctuations, as in the Pleistocene, integrated with biostratigraphic and paleomagnetic control. In addition, we can utilize the large-scale shifts in the Plio-Pleistocene isotopic record to construct a chronostratigraphic framework. However, because much of the resolution of the isotopic record is dependent upon the accumulation rate, it is highly probable that we will find isotopic variations of greater amplitude as exploration wells containing Pliocene sections are examined in detail (i.e., at 30–60-foot intervals).

Pliocene Carbon Isotope Records

The reader is referred to Fig. 46 showing a comparison of the benthic carbon isotope records from Pacific V28-179 and north Atlantic DSDP 552A (Shackleton *et al.*, 1984c). Direct point-for-point comparisons are difficult because of the differences in sampling density and accumulation rate between the two cores. However, both records show that these two regions experienced $\delta^{13}C$ variations on the order of 0.5–1‰. No long-term trends that lend themselves to stratigraphic use are readily discernible in these records. Sections in exploration wells for this interval would need to be carefully chosen from reliable biostratigraphic tops or else one could possibly utilize paleomagnetic stratigraphy or tephrochronology (Ledbetter, 1984a, b) in conjunction with the isotopic records.

The $\delta^{13}C$ record of 552A (Fig. 46) shows possible evidence for a series of long-term cycles, separated by shifts or offsets, on the order of 200,000–300,000 yr in duration. These $\delta^{13}C$ shifts are observed to be on the order of 0.5–0.3‰ as indicated by the vertical lines in site 552A (Fig. 46). For example, from 3.5 to 3.3 MYBP, relatively positive values shift over to

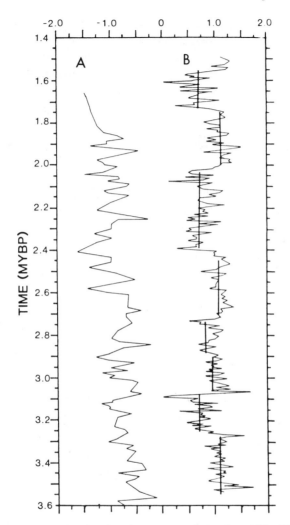

Fig. 46. A comparison of two carbon isotope records for the middle Pliocene to early Pleistocene from (A) a Pacific piston core, (V28-179) (Shackleton and Opdyke, 1977) and (B) an Atlantic DSDP site (552A) (Shackleton *et al.*, 1984a).

negative values between 3.25 and 3.1 MYBP. A broad positive interval exists from 3.05 to 2.45 MYBP, with another shift toward negative values. From 2.4 MYBP, the δ^{13}C values slowly become more positive until 2.05 MYBP. A positive interval exists from 1.95 to 1.75 MYBP followed by a negative interval from 1.7 to 1.55 MYBP with a return to more positive values of 1.25‰.

The mean isotopic composition of the two Pliocene records (Fig. 46) is different because of the well-documented 1‰ offset that exists between the deep waters of the north Atlantic and Pacific (Kroopnick, 1974, 1980). New $\delta^{13}C$ results from North Atlantic DSDP site 606 (Keigwin, 1987) corroborate the character of the $\delta^{13}C$ record shown in Fig. 46.

III. Miocene Isotope Records

The Miocene isotope record has received considerable attention from the academic community because of interest in reconstructing the paleoenvironmental conditions of the preglacial and postglacial Miocene oceans (Shackleton and Kennett, 1975a; Savin et al., 1981). On the basis of the oxygen isotopic records shown in Figs. 17, 25, and 47, Shackleton and Kennett (1975b) proposed that the large shift in the $\delta^{18}O$ values recorded in foraminifera from the late middle Miocene sections of southern Pacific DSDP sites reflected the formation of a permanent ice cap on Antarctica (14–11.5 MYBP). The positive $\delta^{18}O$ changes in the middle to late Miocene suggested large changes in the quantity of ice on Antarctica in the latest Miocene-Kapitean age, approaching volumes roughly equivalent to the present day volume (PDV). During the late Miocene-late Tongaporutuan

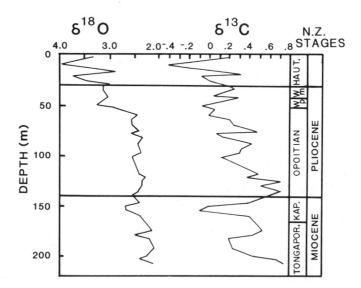

Fig. 47. One of the first detailed studies of the Neogene oxygen and carbon isotope records (modified from Shackleton and Kennett, 1975b).

age, one-third of the PDV was formed (Fig. 47). However, a more detailed study is critically needed for this interval of time to firmly test this hypothesis.

The important implications of the Shackleton and Kennett (1975a) study quickly led to a number of studies aimed at resolving the nature of the $\delta^{18}O$ signal in Miocene DSDP sections (Woodruff *et al.*, 1981; Savin *et al.*, 1981; Keigwin, 1979; Shackleton, 1982). The composite oxygen isotope record published by Barron and Keller (1982) serves as a good example of the broad overall changes that occurred during the Miocene (Fig. 48). The record is characterized by several long-term trends (i.e., values becoming more positive from the early middle Miocene boundary to the late middle Miocene). The values then undergo a steady enrichment from the early middle Miocene boundary to the Oligocene. Many short-term isotopic events equal to >0.5‰ also occur. These events would be equivalent to a stratigraphic resolution of ≤100,000 yr. This point of stratigraphic resolution will be discussed later in the context of detailed isotopic records from Shackleton (1982). Strong evidence for the initiation of Antarctic glaciation was invoked to explain the large and steady isotopic enrichment occurring just after 16 MYBP (Fig. 48). Keller and Barron (1981) also correlated the Miocene $\delta^{18}O$ shift to the widespread occurrence of deep-sea hiatuses and change in the Miocene onlap–offlap curve (Vail *et al.*, 1977).

The next attempt to document the nature of this major Miocene event was done by Woodruff *et al.* (1981) on DSDP site 289 from the western equatorial Pacific (see Fig. 26 for location map) (Fig. 49). The site 289 record shows that rapid (<0.1 MY) 0.5‰ $\delta^{18}O$ changes extend well back into the early Miocene. A long-term steady change of approximately 0.05‰ per million years occurs from planktonic foraminiferal zone N5 (21.5 MYBP), reaching the most negative isotopic values of +1‰ in zone N8 at approximately 16 MYBP (Fig. 49). Because of a high sample resolution from the period of 16–14 MYBP (Zones 9–11 in the early middle Miocene), the oxygen isotopic resolution shows that large (nearly 0.5–1‰) isotopic changes occur on a time scale approaching that observed in Plio-Pleistocene records (Fig. 50).

Woodruff *et al.* (1981) argued that this large expansion in the Antarctic ice cap during the middle Miocene perhaps caused, or was at least coincident with, an increase in the planetary temperature gradient from mid to high latitudes (Savin, 1977). The interpretations of these authors and those of Shackleton and Kennett (1975a, b) have been questioned recently by Matthews and Poore (1980) and Poore and Matthews (1984a). Matthews and Poore have proposed that the Miocene benthic $\delta^{18}O$ record reflects a global drop in deep bottom-water temperatures and that large volumes of continental ice existed throughout the Oligocene and Miocene. They suggest that the planktonic oxygen isotope record from tropical areas should be

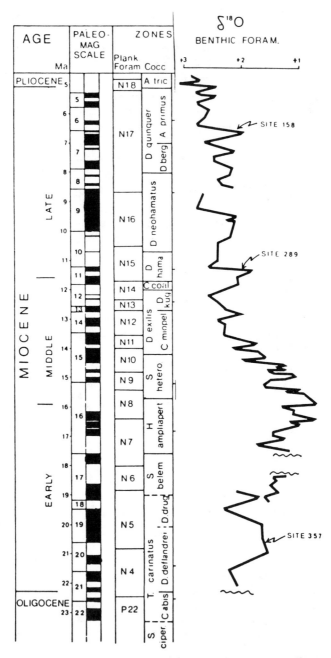

Fig. 48. Composite oxygen isotope record for the Miocene based on $\delta^{18}O$ records from three DSDP sites (357, 289, 158) compared to the absolute paleomagnetic time scale and biostratigraphic zones based on foraminifera and coccolithophorids (modified from Barron and Keller, 1982). The record is characterized by (1) several long-term trends (i.e., values becoming more positive from the early-middle Miocene boundary to the latest middle Miocene) and (2) many short-term isotopic events with a magnitude $\geq 0.5‰$ and with timing equal to a stratigraphic resolution equal to or less than 100,000 yr.

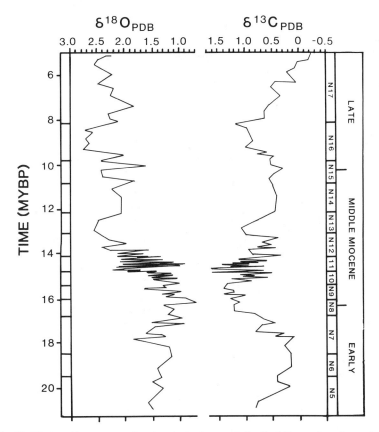

Fig. 49. The benthic oxygen and carbon isotopic records for the Miocene based on a detailed study of Pacific DSDP site 289 (modified from Woodruff *et al.*, 1981).

used as a template for comparison with other records. The matter is far from being resolved and is the subject of considerable debate and ongoing research.

In an effort to explain the initiation of large-scale Antarctic glaciation during the middle Miocene, Blanc *et al.* (1980) and Schnitker (1980) suggested that warm, saline, deep water, originating from the North Atlantic and upwelling in high southern latitudes, triggered the rapid growth of the Antarctic ice cap in the middle Miocene. Woodruff *et al.* (1981) have earlier documented that the largest change in the oxygen isotopic record occurred between approximately 14.8–14.0 MYBP.

Although all of the sites were not suitable for detailed stratigraphic analyses, Savin *et al.* (1981) synthesized the oxygen and carbon isotopic

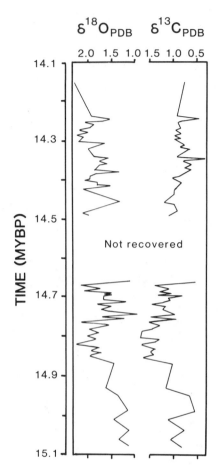

Fig. 50. A detailed examination of the middle Miocene interval (14.1–15.1 MYBP) in Pacific DSDP site 289 (modified from Woodruff *et al.*, 1981).

records from benthic foraminifera for the Miocene from 11 DSDP sites representing the Atlantic, Pacific, and Indian oceans. A summary diagram of data averaged for 0.5-million-year time intervals from three widely separated DSDP sites (Fig. 52) shows that the expansion of the Antarctic ice volume is recorded synchronously throughout the Pacific basin. They also suggested that the detail demonstrated in the site 289 record could provide a source for enhancing correlations within the Miocene.

By performing a very detailed study of two cores within the middle Miocene interval of 14.1–15.1 MYBP (Fig. 50), Woodruff *et al.* (1981) demonstrated that important features of the oxygen and carbon isotopic

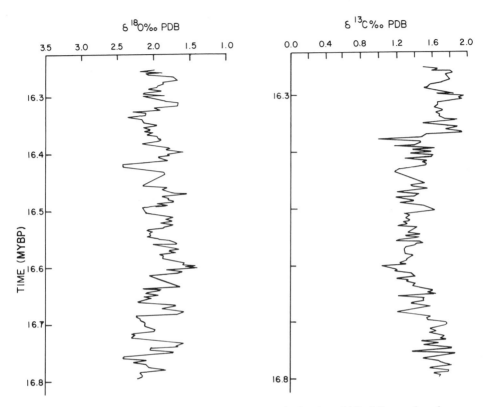

Fig. 51. High-resolution oxygen and carbon records for the middle Miocene based on benthic foraminifera from DSDP site 574 from the Pacific by Pisias *et al.* (1985).

record appear to reflect cyclical variations with periods on the order of those in the Pleistocene isotopic record.

Another example of the type of variability and time resolution that is possible to obtain using isotope records of the Miocene is well illustrated by the extremely high resolution study of the period from 16.2 to 16.8 MYBP in DSDP site 574 (Fig. 51) (Pisias *et al.*, 1985). The fact that these isotopic values are based on analyses of benthic foraminifera makes it highly unlikely that the rapid and large isotopic changes relate to changes in bottom water temperature. These data clearly show that a significant isotope signal exists into the Miocene and that this signal can be used to derive a high-resolution isotope chronostratigraphy.

In his evaluation of climatic variability and the deep-sea oxygen isotopic record, Shackleton (1982) also published detailed isotopic analyses of the planktonic foraminiferal species *Globoquadrina altispira* from the middle

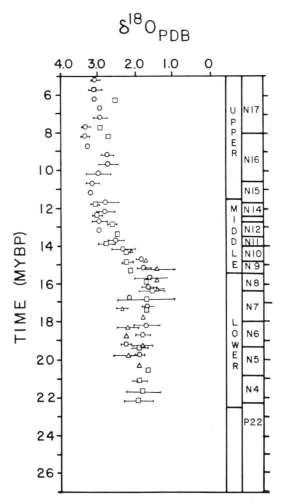

Fig. 52. A comparison of the mean oxygen isotopic value for Miocene benthic foraminifera from Pacific DSDP sites 71 (triangles), 77B (squares), and 289 (circles) averaged over 0.5-MY time intervals (Savin *et al.*, 1981). Reproduced with permission from Elsevier.

Miocene section of DSDP site 289. These data enabled us to examine and compare the character of the benthic and planktonic records previously published by Savin *et al.* (1981) and Woodruff *et al.* (1981). Figures 53 and 54 (redrafted from Shackleton, 1982) further demonstrate the nature and amplitude of the isotopic fluctuations in the Miocene. From these types of records Shackleton (1982) concluded that stratigraphic resolution of various parameters at a band width of ±100,000 years or worse was possible, depending on the quality of the recovered material under study.

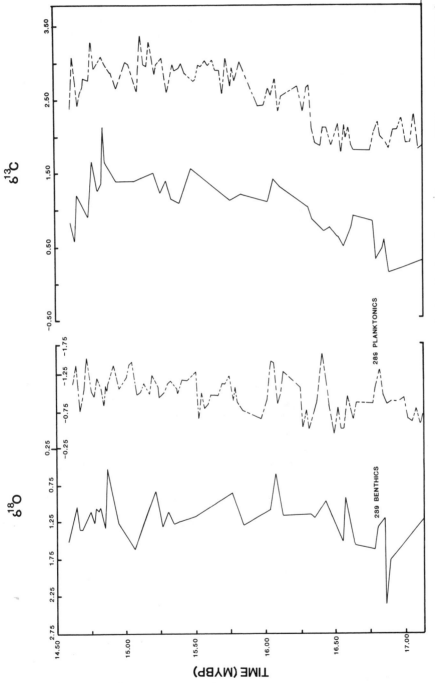

Fig. 53. A detailed examination of the Miocene carbon and oxygen isotopic signal in the surface and deep waters (*Globoquadrina altispira* and benthic foraminifera, respectively) (data from Shackleton, 1982, and Savin *et al.*, 1981).

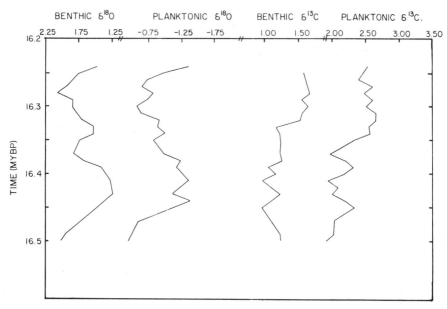

Fig. 54. Detailed comparison of benthic and planktonic oxygen and carbon isotope data from the middle Miocene portion of DSDP site 289 [modified from Savin *et al.* (1981) and Shackleton (1982)].

The fact that the frequency and amplitude of these isotopic fluctuations is very similar to those observed in the Pleistocene has important implications for understanding the driving mechanisms causing global climatic variability during the Miocene. Clearly, more work needs to be done, quite possibly with hydraulically piston-cored sequences in high accumulation rate areas. Alternatively, isotope studies of exploration wells from the Miocene of the Gulf of Mexico will prove to be very important. The purpose of this exercise, however, is to demonstrate that isotope variability in the Miocene exists at the correct amplitude and frequency to become an ideal chronostratigraphic tool for providing time lines in exploration areas.

Miocene Carbon Isotope Records

At the same time that the Miocene oxygen isotopic record was being resolved, increased attention was being directed toward understanding the carbon isotopic record. Keigwin (1979) was the first to suggest that a significant and permanent shift was present in the carbon isotopic record in the late Miocene. He dated the timing of this shift at various Pacific

DSDP sites to be approximately equivalent to the N17 faunal zone (Fig. 55). While some support for large carbon isotopic changes existed in the early work of Shackleton and Kennett (1975a) (Fig. 21), new evidence in support of the carbon shift in the late Miocene proposed by Keigwin came from the work of Woodruff *et al.* (1981). The δ^{13}C record of site 289 shows large cyclic changes within the late Miocene, as well as very rapid δ^{13}C fluctuations (Figs. 49 and 50). The 0.5–1‰ fluctuations in late Miocene δ^{13}C records are believed to reflect changes in the δ^{13}C of the marine bicarbonate reservoir (Bender and Keigwin, 1979; Keigwin, 1979; Haq *et al.*, 1980; Vincent *et al.*, 1980; Woodruff *et al.*, 1981).

In addition, Savin *et al.* (1981) demonstrated that the carbon isotopic records for benthic foraminifera varied "quasi-sympathetically" throughout the Miocene, most probably recording changes in the carbon isotopic composition of marine bicarbonate (Fig. 56). The time averaged data show that the cycle previously seen in site 289 (Woodruff *et al.*, 1981) is also recorded in other DSDP sites. Problems from sample spacing and rotary drilling prevented any further correlation at that time.

Keigwin (1979) was the first to publish relatively detailed oxygen and carbon isotopic analyses of both benthic and planktonic foraminifera from three east equatorial and central north Pacific DSDP sites (sites 157, 310, 158). Although analyzed in considerable detail, the Pliocene and Quaternary

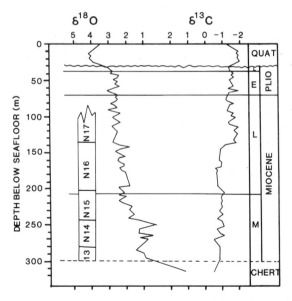

Fig. 55. The Miocene-Pliocene oxygen and carbon isotopic benthic records from eastern Pacific site 158 (modified from Keigwin, 1979).

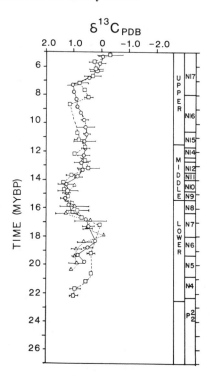

Fig. 56. A comparison of the mean carbon isotopic values for the Miocene records of benthic foraminifera Pacific DSDP sites 71 (triangles), 77B (squares), and 289 (circles) averaged over 0.5-MY time intervals (Savin et al., 1981). Reproduced with permission from Elsevier.

sections of sites 157 and 310 still do not have sufficient resolution to make long-range correlations possible. However, the expanded middle to late Miocene section of site 158 (Fig. 55) and the late Miocene section of site 310, enabled Keigwin to demonstrate that a permanent shift of approximately $-0.8‰$ occurred in the benthic $\delta^{13}C$ records at approximately 6.5 MYBP in the late Miocene (Fig. 55). The $\delta^{18}O$ records from both sites also show the characteristic enrichment in ^{18}O values from the late Miocene into the Quaternary as seen in other records (Fig. 48) discussed previously. Close on the heels of Keigwin's pronouncement of the late Miocene ^{13}C shift, Loutit and Kennett (1979) suggested that the late Miocene carbon shift recognized in Pacific DSDP sites was also recorded in a shallow marine sedimentary sequence of late Miocene age at Blind River, New Zealand. They proposed that correlations could be made from the land-based section to the deep marine DSDP section, and that carbon isotope stratigraphy could provide important datums for correlating late Miocene sequences.

Vincent *et al.* (1980) were able to show that this carbon isotope shift was also recorded in both the benthic and planktonic foraminiferal fractions of DSDP site 238 from the tropical central Indian Ocean (Fig. 57). The δ^{13}C shift in the Indian Ocean has the same magnitude and appears to be synchronous with the shift described in the Pacific. Vincent *et al.* (1980) thus suggested that the shift is a global datum at approximately 6.2 MYBP in magnetic Epoch 6. This late Miocene decrease in δ^{13}C has now been documented in other DSDP sites from the Pacific, Atlantic, and Indian oceans (Haq *et al.*, 1980; Savin *et al.*, 1981; Loutit *et al.*, 1983-84; McKenzie *et al.*, 1984). Except for still not having found the shift in the North Atlantic (Keigwin, 1987, results from site 552A), the shift is now presently considered

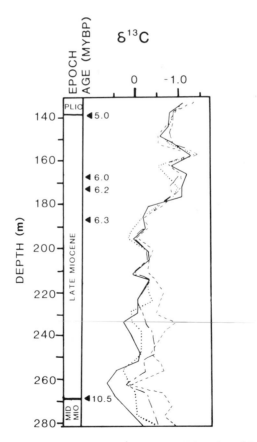

Fig. 57. A comparison of the Miocene carbon isotope records based on different planktonic and benthic species from Indian Ocean DSDP site 238 (modified from Vincent *et al.*, 1980). *O.* umbonatus, ——; *P. wvellerstorfi,* —·—; *C. subglobosa,* ····; *G. sacculifer,* - - -.

to be a global event possibly caused by an oceanwide decrease in upwelling rates and changes in circulation patterns (Bender and Keigwin, 1979). Vincent *et al.* (1980) alternatively suggested that the rate of supply of organic carbon from continental shelves and coastal regions underwent a rapid increase during a late Miocene regression. This change in organic supply, combined with changes in deep circulation patterns and ocean fertility, led to the negative carbon shift.

Haq *et al.* (1980) showed the relevance of using carbon isotope stratigraphy to refine and possibly calibrate calcareous nannofossil biostratigraphy. They showed that the FAD of the nannofossil genus *Amaurolithus* immediately precedes the major portion of the carbon isotopic shift (Fig. 13). Other important radiolarian and foraminiferal FADs are also known to be associated with the carbon shift in low- and mid-latitude Pacific DSDP sites (Vincent *et al.*, 1980). Extension of these results into piston cores with

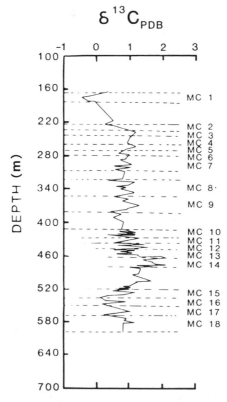

Fig. 58. Proposed definition of carbon isotope events in the Miocene based on a comparison of the $\delta^{13}C$ record from DSDP site 289 with those from other ocean basins (Loutit *et al.*, 1983). Reproduced with permission from Elsevier.

paleomagnetic stratigraphy enhanced the age resolution of the FAD for several calcareous nannofossil and diatom indicator species in the Miocene, and refined the age of the carbon shift to be 6.10–5.90 MYBP, with a duration of approximately 200,000 yr. As will be discussed in greater detail in another section, the combination of detailed biostratigraphy and isotope chronostratigraphy can lead to high-resolution datums and an accurate biochronology for the Miocene and other periods of the Tertiary.

Some of the general changes associated with the carbon isotope shift in deep- and bottom-water properties in the Miocene of the Pacific can be seen in Fig. 58 from Loutit *et al.* (1983–1984). Loutit *et al.* attempted to take the use of carbon isotopes in foraminiferal carbonate a step further by proposing a $\delta^{13}C$ zonation of the Miocene based on the $\delta^{13}C$ record from DSDP site 289. Although the effort shown in Fig. 58 should be regarded as preliminary, a description of the carbon isotope events has been given by Loutit *et al.* (1983). New data are continually being obtained to establish more firmly the character of the Miocene isotope record for stratigraphic correlations. This point becomes clear by examining the size and frequency of the Miocene carbon isotope signals shown in the new data presented in Figs. 53 and 54.

IV. Eocene and Oligocene Oxygen and Carbon Isotope Records

Currently, despite the efforts of Blanc *et al.* (1980), Miller and Fairbanks (1983), Shackleton *et al.* (1984a), Keigwin and Keller (1984), and Miller and Thomas (1985), the Oligocene oxygen isotope record appears to hold the least promise for detailed isotope chronostratigraphy of any interval of the Tertiary. We base this rather pessimistic statement on the fact that the character of the oxygen isotopic record for Oligocene foraminifera does not appear to exhibit any distinct steps or abrupt shifts in isotopic values that could be used for establishing tie lines between wells. For example, a comparison between benthic and planktonic foraminiferal $\delta^{18}O$ records for the Oligocene sections of North Atlantic DSDP sites 558 and 563 (Miller and Fairbanks, 1985a) shows that variations with 1‰ amplitudes can be seen in certain sections of the records (i.e., near the early Oligocene section of site 563 and the P22 section of late Oligocene time in site 558). However, the data are not sufficiently detailed to unequivocally determine the frequency of variation or the existence of long-term trends in the isotopic variability in the Oligocene. Further detailed work is therefore needed, perhaps in an expanded Oligocene section with a high sedimentation rate from a future hydraulically piston-cored site in the Ocean Drilling Program (ODP).

The Oligocene δ^{13}C record may hold more promise for isotope chrono-stratigraphy than the δ^{18}O record. For example, a comparison of the carbon isotopic values in the benthic and planktonic foraminifera from the same two sites (Fig. 59) shows part of the three long-term cycles with periods of 7–12 million yr and amplitudes of 1‰ (Miller and Fairbanks, 1983; 1985b; Shackleton *et al.*, 1984b). Small amplitude δ^{13}C fluctuations of <0.5‰ are

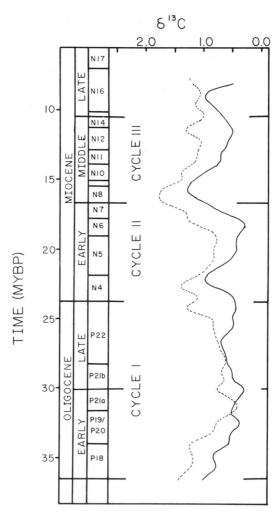

Fig. 59. Oligocene and Miocene global carbon isotope cycles as recorded in benthic foraminifera from Atlantic and Pacific DSDP sites (modified from Miller and Fairbanks, 1985a). The data are smoothed to illustrate the broad long-term trends only.

sparsely superimposed on rather long-term and otherwise smooth changes. From positive values near the Eocene-Ologocene boundary, minimal values appear at approximately 32-30 MYBP and extend to 28 MYBP. Gradually, the δ^{13}C values become more positive, reaching maximum positive values between 25 and 23 MYBP. By averaging the best available data for the Atlantic and Pacific to constant intervals of 0.2 million years and smoothing the data to eliminate events with periods less than approximately 1 million years, Miller and Fairbanks (1985a) present a case for what appear to be three carbon isotope cycles from the Oligocene to the middle Miocene. Each cycle is bounded by positive δ^{13}C values: Cycle I between 33 and 36.5 MYBP (early Oligocene); Cycle II between 22 and 25 MYBP (across the Oligocene-Miocene boundary); and Cycle III between 14 and 18 MYBP (across the early-middle Miocene boundary) (Fig. 59). The occurrence of similar cycles in foraminiferal δ^{13}C records from the Pacific, Atlantic, and Gulf of Mexico suggests that these cycles represent global changes in the relative amount of carbon buried as calcium carbonate versus the amount buried as organic carbon (Miller and Fairbanks, 1985b). The occurrence of these cycles in records from both benthic and planktonic foraminifera also lends support that these changes represent global δ^{13}C cycles on the order of 1‰. It may be possible therefore to use these cycles for stratigraphic correlations of Oligocene sections, but much more work needs to be done, particularly at sampling intervals designed to resolve isotopic changes occurring at frequencies shorter than 10^6 yr.

A. Eocene-Oligocene Boundary Event

The first indication of an abrupt climatic cooling associated with the Eocene-Oligocene boundary was based on stable isotopic analyses of microfossils by Dorman (1966) and Devereaux (1966). The Tertiary isotopic record published by Shackleton and Kennett (1975a) and Savin (1977) (Figs. 17 and 18) also provided an impetus to study the δ^{18}O changes associated with the Eocene-Oligocene boundary. Abrupt isotopic δ^{18}O changes occur during several intervals within the Tertiary. These changes are especially well defined near the Eocene-Oligocene boundary and within the middle Miocene (Fig. 17).

The events associated with the Eocene-Oligocene boundary are important because of coincident worldwide changes of several climatic and oceanic processes (Kennett and Shackleton, 1975; Keigwin, 1980). A later study of the latest Eocene to earliest Oligocene section of DSDP site 277 by Kennett and Shackleton (1976) showed that a change of 1‰ occurred within 0.1 MY in the lowermost Oligocene (Fig. 60). They interpreted this dramatic event as the development of the psychrosphere or the isotopic evidence for the

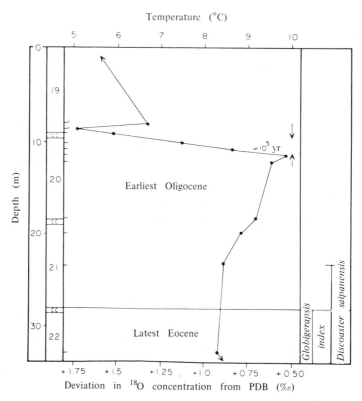

Fig. 60. The oxygen and carbon isotopic evidence across the Eocene–Oligocene boundary used to infer the development of cold, deep-water formation in the Paleogene (Kennett and Shackleton, 1976).

consistent production of cold, deep bottom waters in the oceans. Later work across the Eocene–Oligocene boundary in the Atlantic found similar ^{18}O enrichments (Boersma and Shackleton, 1977a; Vergnaud-Grazzini et al., 1978; Miller and Curry, 1982). The initial Kennett and Shackleton (1976) study was performed on mixed species of benthic foraminifera. The shift across the boundary was later confirmed by Keigwin (1980) using analyses of single species at DSDP sites 277 and 292 (Fig. 61) and by Murphy and Kennett (1985) at DSDP site 277 (Fig. 62). Keigwin (1980) and Murphy and Kennett (1987) were also able to show that in high-latitude regions (site 277) the magnitude of the $\delta^{18}O$ shift was the same in both benthic and planktonic foraminifera. In low latitudes near tropical regions (site 292), the benthic foraminifera recorded a $\delta^{18}O$ shift equal to that in the high-latitude regions, but the record from planktonic foraminiferal species shows a smaller ^{18}O enrichment (Fig. 61).

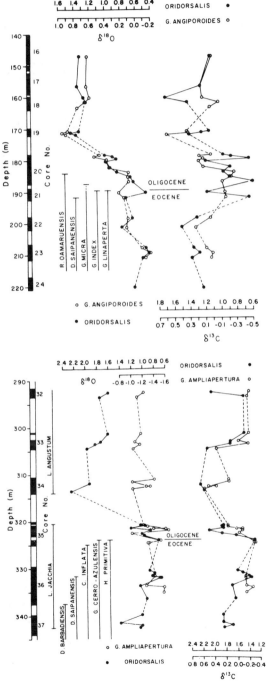

Fig. 61. A comparison of the planktonic and benthic foraminiferal oxygen and carbon isotopic responses across two Eocene–Oligocene sections from DSDP sites 277 and 292 (Keigwin, 1980).

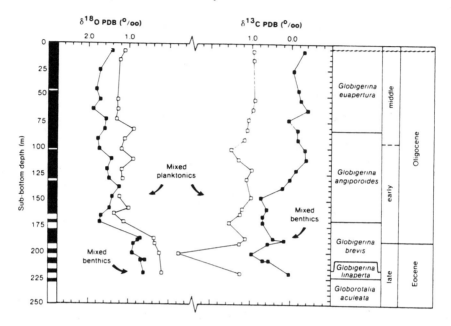

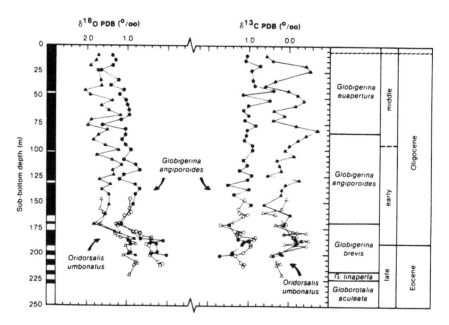

Fig. 62. Comparison of mixed foraminiferal and monospecific $\delta^{18}O$ records across the Eocene–Oligocene boundary in site 277 (modified from Murphy and Kennett, 1985).

The most extensive isotopic study of this important time interval involves detailed studies of (1) the middle Eocene to early Oligocene section in Atlantic DSDP sites 19 and 363; (2) Gulf of Mexico *Eureka* (Shell) section 67-128; (3) a Gulf Coast land based section at St. Stevens Quarry, Alabama; (4) Indian Ocean DSDP sites 291 and 253; and (5) Pacific Ocean DSDP sites 77, 277, and 292 (Keigwin and Corliss, 1986). From their synthesis, Keigwin and Corliss (1986) conclude that the 1‰ ^{18}O enrichment in benthic foraminifera associated with the Eocene–Oligocene boundary occurs nearly synchronously in all the ocean basins. Planktonic foraminiferal biostratigraphy provided the time control. The amplitude of the planktonic foraminiferal δ^{18}O change into the earliest Oligocene depends on whether they are from a low-latitude or high-latitude location. For example, at high latitudes, the δ^{18}O shift in planktonic foraminifera is approximately +1‰ (i.e., nearly the same enrichment as recorded by the benthic foraminifera); the planktonic ^{18}O enrichment is only 0.2–0.3‰ at low latitudes. From their study they conclude that no easily discernible pattern, long-term trends, or major shifts can be observed in the δ^{13}C record of the middle Eocene to Oligocene, but that there appears to be a significant increase in the surface water δ^{18}O gradient as a function of latitude (Keigwin and Corliss, 1986). They also find that the δ^{18}O gradient increases from 0.016 to 0.026‰ per degree latitude from Eocene to Oligocene time. Muza *et al.* (1983), Williams *et al.* (1985), and Vergnaud-Grazzini and Saliege (1987) also find similar results across the Eocene–Oligocene boundary in site 516 from the South Atlantic.

B. Results from the Gulf of Mexico

A comparison of the oxygen and carbon isotopic changes in benthic foraminifera from the two Gulf of Mexico sections analyzed by Keigwin and Corliss (1986) is shown in Fig. 63. One section from the St. Stevens Quarry, Alabama, represents a shallow-water sequence (Mancini, 1979), and the other section from Shell *Eureka* borehole E67-128 is presently located at a water depth of 4780 ft. The benthic foraminifera *Bulimina alazanensis* from E67-128 and *Uvigerina* species from St. Stevens Quarry both record a 1‰ shift toward positive values across the Eocene–Oligocene boundary (Fig. 63 modified from Keigwin and Corliss, 1986). This result is particularly interesting considering the differences in the depositional environments of the two areas. In addition, the ^{18}O enrichment is nearly the same magnitude as that observed in other benthic isotopic records from the Atlantic and Pacific (Keigwin and Corliss, 1986). It is also interesting to note that the average δ^{18}O value for *Uvigerina* in the upper Eocene sections of St. Stevens Quarry is approximately 0.7‰ heavier than that of

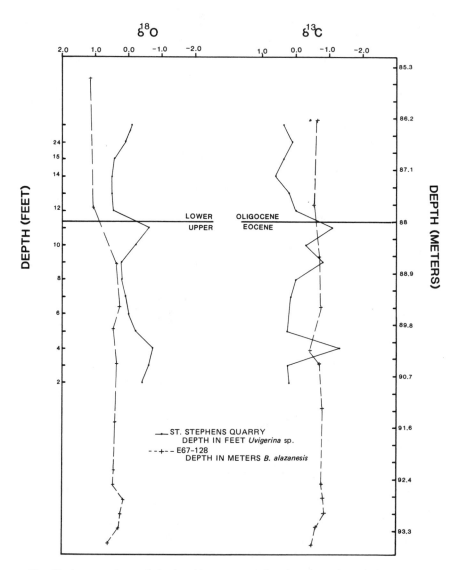

Fig. 63. A comparison of the benthic oxygen and carbon isotopic records across the Eocene–Oligocene boundary in offshore Gulf of Mexico borehole E67-128 and the land based section from St. Stevens Quarry, Alabama (data from Keigwin and Corliss, 1986).

that of *B. alazanensis* in the upper Eocene of E67-128. This 0.7‰ difference is approximately equivalent to a temperature difference of 3.50°C. In the lower Oligocene, this $\delta^{18}O$ difference between the benthic foraminifera remains $-0.7‰$, suggesting that the Eocene–Oligocene $\delta^{18}O$ shift recorded in the Gulf of Mexico may be more related to a change in the isotopic composition of the waters in the Gulf of Mexico rather than to a bottom-water temperature difference.

Although there are not many $\delta^{13}C$ data from E67-128 in the lower Oligocene, it is interesting to note that the mean $\delta^{13}C$ value of the St. Stevens Quarry section ($-0.8‰$) is equal to that of *B. alazanensis* in some portions of E67-128. It would be interesting to analyze more of the E67-128 section in order to see if the large 1.5‰ $\delta^{13}C$ change recorded from the late Eocene to early Oligocene in St. Stevens Quarry is also recorded in the deep-water section.

These Gulf of Mexico results have importance in demonstrating that stable isotopes can be used to provide tie lines between shallow- and deep-water environments of the Gulf of Mexico. As at most other sites, the benthic foraminifera from these Gulf of Mexico sections reach maximum $\delta^{18}O$ values within the lower to middle Oligocene. Keigwin and Corliss (1986) suggest that particularly abrupt $\delta^{18}O$ shifts near the Eocene–Oligocene boundary reflect unconformities as at Eureka 67-128. The magnitude of the shift in the St. Steven's Quarry is actually more rapid than in E67-128 (Fig. 63).

V. Isotope Records for the Paleocene and the Cretaceous–Tertiary Boundary

Except for the studies done at widely spaced intervals for the Tertiary (Shackleton and Kennett, 1975b; Savin *et al.*, 1975; Boersma and Shackleton, 1977b; Thierstein and Berger, 1978), the oxygen and carbon isotopic compositions of foraminifera from the Paleocene have received relatively little attention in the scientific literature. It is difficult to know the nature of the isotopic variability recorded by foraminifera in the Paleocene and whether the signal is global and potentially useful for isotope chronostratigraphy.

Recently, however, a new stratigraphic approach has been suggested by Shackleton and Hall (1984a). They demonstrate that excellent correlations using carbon isotope stratigraphy are possible with analyses of bulk carbonate (i.e., whole rock) from DSDP cores from the late Maestrichtian to Eocene interval (Fig. 64). Each of the records exhibits a dramatic change from relatively positive $\delta^{13}C$ values in the late Maestrichtian followed by a very abrupt shift in $\delta^{13}C$ values across the Cretaceous–Tertiary boundary.

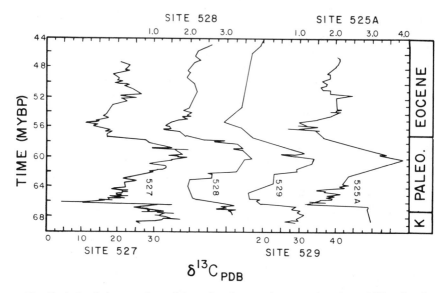

Fig. 64. A detailed comparison of the carbon isotope chronostratigraphy possible using the signals contained in the bulk carbonate from Atlantic DSDP sites 525, 528, 529, and 527 from the late Maestrictian into the middle Eocene (Shackleton and Hall, 1984a).

The magnitude of the shift toward light values varies from approximately 1.3‰ at site 528 to as large as 3‰ in site 527. Shackleton and Hall used these data to suggest that with increased sampling resolution it is possible to determine the completeness of the sections above and below the Cretaceous–Tertiary (K–T) boundary.

Following the dramatic K–T drop in $\delta^{13}C$ values, the $\delta^{13}C$ record undergoes a steady recovery toward positive values in a series of quasicyclic steps best illustrated in the site 527 record (Fig. 64). The most positive $\delta^{13}C$ values seen in any part of the 24-MY record (and indeed the most positive value indicated in all of the Tertiary as illustrated in Fig. 24) appear to be centered between 61 and 59 MYBP. From 59 to 55 MYBP, the carbon isotope record again undergoes a dramatic shift toward more negative $\delta^{13}C$ values, but more gradually than the K–T boundary event. Light $\delta^{13}C$ values are attained between 57 and 55 MYBP. After 55 MYBP, the $\delta^{13}C$ record begins another recovery producing another cycle in the $\delta^{13}C$ record of the Paleocene. The degree of correlation between the $\delta^{13}C$ variations at these sites (Fig. 64) is quite remarkable given the differences in sampling density and the possibility of coring gaps and hiatuses.

More detailed sampling in the late Maestrictian–late Paleocene section of site 527 and in the latest Paleocene section of site 525A (Fig. 64) indicates that rapid 0.5–1‰ $\delta^{13}C$ changes occur on time scales approaching 0.2–0.3 MY. These rapid changes are superimposed on cyclic long-term changes and abrupt boundary shifts (Fig. 64) (Shackleton *et al.*, 1984a).

The potential exists therefore of using the $\delta^{13}C$ record of bulk sediment to obtain detailed and accurate stratigraphic correlations of the early Paleogene. To illustrate the possibility of making long-range correlations using $\delta^{13}C$ isotope chronostratigraphy in the Paleocene, Shackleton (1985) compared the carbon isotope data for bulk sediment in DSDP site 527 from the South Atlantic with that from DSDP site 577 from the tropical Pacific on a common time scale (Fig. 65). Along the right vertical axis are the paleomagnetic chrons for this interval. The $\delta^{13}C$ scales for each site are

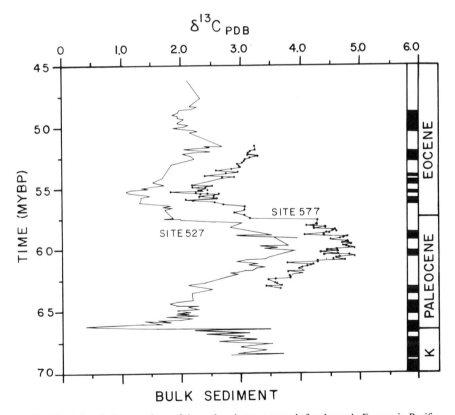

Fig. 65. A detailed comparison of the carbon isotope records for the early Eocene in Pacific DSDP site 577 and Atlantic DSDP site 527 (Shackleton, 1985). The records are based on whole rock analyses of the deep-water pelagic sediments. In both sites, paleomagnetic stratigraphy and biostratigraphy were used in the age model with adjustments made on the character of the $\delta^{13}C$ records.

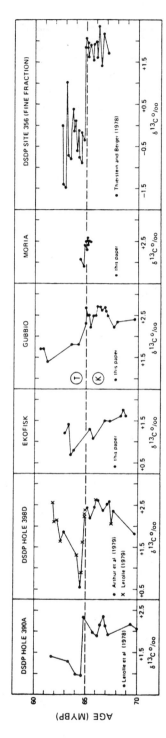

Fig. 66. Comparison of carbon isotope profiles across the Cretaceous–Tertiary boundary based on whole rock analyses of pelagic limestones (taken from Scholle and Arthur, 1980).

offset slightly so that one can see the excellent correlation between the Pacific and Atlantic records without the need for any type of cross-correlation analysis (of course this technique and others would certainly enable one to determine the degree of semblance and correlation more quantitatively; see chapters 12 and 14). The $\delta^{13}C$ recovery between 62.5 and 58 MYBP after the K–T boundary event shows up quite nicely in both records as well as the large $\delta^{13}C$ depletion between 58 and 55 MYBP. This mid-Paleocene $\delta^{13}C$ depletion of 1.65‰ equals a rate of change of approximately 0.5‰ per million years. Because it is possible to make $\delta^{13}C$ measurements to ±0.05‰, this rate of change lends itself to a correlation method with a high degree of resolution (~200,000 yr). Shackleton (1985) also proposed that the $\delta^{13}C$ record could be used to resolve missing sections and that the $\delta^{13}C$ changes represent important long-term changes in the global organic carbon reservoir. Scholle and Arthur (1980) had suggested previously that the carbon isotopic composition of oceanic carbonates could be used not only to define perturbations in the global carbon system, but also to provide stratigraphically meaningful signatures in the Mesozoic

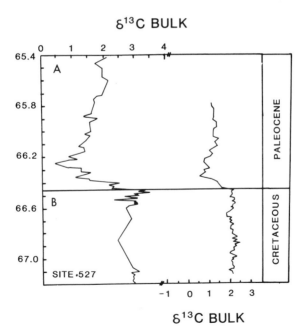

Fig. 67. A comparison of the carbon isotopic records across the Cretaceous–Tertiary boundary in late Maestrictian to early Paleocene time in South Atlantic DSDP sites 516F (A) and 527 (B) based on whole rock analyses (Shackleton and Hall, 1984a; Williams et al., 1983).

carbon isotope record. This subject will be discussed in an evaluation of the Mesozoic $\delta^{13}C$ record later in this book.

Regarding the Cretaceous–Tertiary boundary, while rock analyses of marine and land based carbonate sections (Figs. 65, 66, and 67) show a remarkable correlation of the K-T $\delta^{13}C$ shift from different ocean basins and localities (Scholle and Arthur, 1980; Boersma et al., 1979; Williams et al., 1983, 1985; Thierstein and Berger, 1978; Hsu et al., 1982; Letolle et al., 1978). Zachos et al. (1985) suggest that the magnitude of the $\delta^{13}C$ shift across the K-T boundary is related to the combined effects of a dramatic change in oceanic productivity and a decrease in the average sediment accumulation rates in the studied K-T sections. It will be interesting to test this hypothesis quantitatively.

Chapter 7 | Stable Isotopic Evidence for and against Sea Level Changes During the Cenozoic

I. Introduction

Sea level is one of the most important parameters influencing (1) past and present basin geometries; (2) the rate and mode of sediment input from continental margins into deep ocean basins; (3) the timing of sediment accommodation on continental margins; (4) facies relationships and stratal patterns in marine sedimentary sequences; (5) the deposition and preservation of marine and terrestrial organic materials; and (6) the subsequent occurrences of commercially important source rocks and hydrocarbon reservoirs. The sea level record is also important as a measure of large-scale tectonic and climatic processes on the earth's surface. Despite the considerable attention paid to deciphering this geological phenomenon since the early work of Suess, large uncertainties remain in our ability to determine, in a quantitative fashion, the *timing, magnitude, rates of change*, and causal mechanism(s) of sea level events in the Phanerozoic.

The integration of seismic stratigraphy and biostratigraphy (Fig. 68) has significantly advanced our knowledge of regional and perhaps global unconformity surfaces (Vail *et al.*, 1977; Pitman, 1978; Watts, 1982; Watts and Steckler, 1979). Translation of these seismically defined stratal patterns into regional unconformity surfaces, and then into a global record of eustatic sea level, is not without its controversies. The primary reasons for some of the controversy involves complex geoidal considerations, uncertainties in timing unconformity surfaces in a precise manner, difficulties in quantifying the magnitude and rates of sea level changes using the seismic approach, and last but not least, a degree of inertia in the industrial and academic

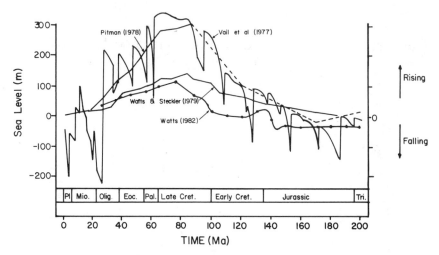

Fig. 68. Sea levels for the last 200 MY as inferred from seismic sequence interpretations (modified from Watts and Steckler, 1979).

communities toward rapid acceptance of unifying theories that proport to explain phenomena as complex as sea level, geohistories, etc.

We suggest that the stable oxygen isotope stratigraphies of exploration wells from passive margin settings can add an important new dimension concerning the timing, magnitude, and rates of change of Cenozoic, Mesozoic, and perhaps Paleozoic sea levels, particularly when integrated with seismic and biostratigraphic interpretations. Stable carbon isotopic records may provide important new information regarding the burial of organic carbon as a function of sea-level-influenced sedimentation. In addition, the combined $\delta^{18}O$ and $\delta^{13}C$ records can be used to provide a global stratigraphic framework that is independent of local or regional biostratigraphic zonations. This framework can be used to refine the chronostratigraphy of exploration wells and enhance the detection of unconformity surfaces that are transparent to biostratigraphic resolution. This enhanced precision will thereby improve the correlation of unconformity surfaces from section to section.

The purpose of this section, therefore, is to compare the information about Cenozoic sea levels contained in the oxygen isotope record of the Cenozoic with the relative sea level record determined from seismic stratal patterns (Vail et al., 1977; Vail and Hardenbol, 1979). As shown in Fig. 69, the Cenozoic oxygen isotope record should provide an independent evaluation of the seismically defined sea level record because the major variables controlling the sequence boundary patterns and the $\delta^{18}O$ signals of foraminifera are entirely different. The type of stratal patterns found on

SEQUENCE BOUNDARY PATTERNS	^{18}O SIGNAL
SEDIMENT SUPPLY	────
SUBSIDENCE RATE	────
TECTONISM	
	────
	TEMPERATURE
────────	DIAGENESIS
────────	
EUSTATICS	EUSTATICS

Fig. 69. A summary of the major variables controlling seismic sequence boundary patterns and the δ^{18}O signal in Tertiary marine carbonates.

continental margins are determined by the volume and supply rates of sediments, subsidence rates, tectonism, and eustatic sea level changes.

The δ^{18}O signal is determined by the water temperature in which the marine carbonates formed (in this case benthic foraminifera), post-depositional diagenesis, and the ^{18}O/^{16}O ratio of sea water, which has varied in the past primarily as a function of the glacio-eustatic sea level. The effects of temperature on the global δ^{18}O record of the Cenozoic can be minimized, but not totally eliminated, by utilizing benthic foraminifera from deep-sea cores recovered from water depths that experience less variable temperatures than surface waters. Matthews and Poore (1980) and Matthews (1984b) have attempted to model the sea level versus temperature components of the δ^{18}O record by using the signal in tropical planktonic foraminifera. The effects of diagenesis on the δ^{18}O record will vary as a function of time, temperature, and depth of burial. For the most part, these effects can be checked by using specimens that are well preserved and whose isotopic compositions are well understood. Certainly, the potential of diagenetic effects becomes more likely in the Mesozoic and Paleozoic.

Last, by establishing global patterns for the δ^{18}O and δ^{13}C records for various time periods and ocean basins, it should be possible to determine the eustatic component controlling the character and amplitude of the isotopic signal. To a first approximation, the amount of sea water removed from the oceanic reservoir and deposited in ice sheets during a sea level drop is translated into an increase in the ^{18}O/^{16}O ratio of the remaining sea water within the mixing time of the oceans (approximately 1000 yr today). This process is shown schematically in Figs. 30 and 31. Detailed studies of scores of deep-sea cores and several emergent coral terraces of the late Pleistocene have attempted to model how a given sea level change is reflected in an equivalent isotopic change. The best estimate currently available is that a 0.11‰ change in δ^{18}O results from every 10 m (approximately 30 ft) of sea water removed [Fairbanks and Matthews, 1978; see Matthews (1984b) for a review]. Mix and Ruddiman (1984) have attempted to model the nonlinear response of the ocean to the expansion of continental ice sheets of varying thicknesses and altitudes. While it is not possible to separate unequivocally the effects of temperature and sea water compositional changes in the past, the $\delta^{18}O$ signal is currently the best parameter available for estimating the magnitude and rate of sea level change independently of the seismic onlap records.

We begin, therefore, by examining the $\delta^{18}O$ evidence for the timing, magnitude, and rate of 2nd, 3rd, and 5th order sea level events in the Cenozoic. The definition of the 2nd, 3rd, and 5th order events is the same as used by the Exxon group, i.e., events that occur on the order of 36-9 MY, 5-1 MY, and 1,000,000 yr. We will try to address how the $\delta^{18}O$ record can be used as a chronostratigraphic tool for defining unconformity surfaces in offshore Plio-Pleistocene sections of the Gulf of Mexico.

II. Oxygen Isotopic Model for Cenozoic Sea Levels

Figure 70 presents a composite $\delta^{18}O$ record for deep-sea benthic foraminifera from Atlantic and Pacific Deep Sea Drilling Project sites. The data have been time averaged at one-million-year increments to filter out high-frequency events of the 4th order and higher (with periods <400 KYR) (Miller and Fairbanks, 1985a). In compiling Fig. 70, no attempts have been made to account for the effects of data from different species and biostratigraphic uncertainties of different sites.

A. Timing of Global $\delta^{18}O$ Events

The overall change in $\delta^{18}O$ through the Cenozoic is nearly 4‰ from the lightest $\delta^{18}O$ value in the early Eocene to the very positive values that characterize the Pleistocene (Fig. 70). Positive 1‰ $\delta^{18}O$ events are centered at approximately 3, 15, 38, 41-46, and 48-53 MYBP. Other significant events, defined as being >0.2‰, are present at 5.5, 7-9, 10, 12, 18, 24-26, 29.5-30.5, 41, 44, 58, and 61 MYBP (Fig. 70). Still more significant changes are present in individual records for individual sites but are unresolvable in this treatment.

Also shown in the left-hand margin of Fig. 70 is the position of unconformities of type 1 and 2 designated by seismically defined stratal patterns (Vail et al., 1977). Within the bias introduced by averaging the $\delta^{18}O$ data at 1 MYR increments, Fig. 70 illustrates the very good agreement that exists between the timing of most of the significant $\delta^{18}O$ events and unconformities defined by the Exxon group. More unconformities are listed in Table II than are presently seen in the composite Cenozoic $\delta^{18}O$ record because of the statistical bias of the latter. The agreement between these two

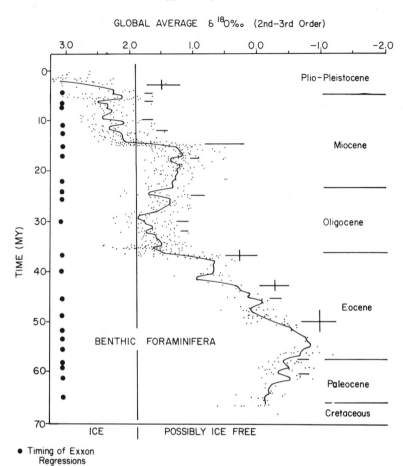

GLOBAL AVERAGE δ¹⁸O‰ (2nd-3rd Order)

Fig. 70. A composite benthic $\delta^{18}O$ record from Miller and Fairbanks (1985b) compared to the timing of major regressions inferred from the offlap/onlap record of Vail *et al.* (1977).

of the independent methods suggests that eustatic sea level is the primary control of the isotope signal and seismic patterns throughout much of the Cenozoic. While some temperature effect is present in the $\delta^{18}O$ signal, the agreement in timing also supports the basic arguments of Matthews and Poore (1980) for the existence of polar ice in the early Cenozoic prior to the major advance of ice in West Antarctica (Shackleton and Kennett, 1975a; Savin, 1977; Woodruff *et al.*, 1981).

B. Magnitude of $\delta^{18}O$ Inferred Sea Level Events

A strict interpretation of the $\delta^{18}O$ record solely in terms of sea level, using the Pleistocene calibration of 0.11‰ $\delta^{18}O$ per 10 m (33 ft) of sea level

Table II

Comparison of the Magnitude of Sea Level Events Based on the $\delta^{18}O$ and Onlap Records

Event[a]	Type	Timing (MYBP)	Agreement	Seismic (m)	$\delta^{18}O$ (m)
1	fall	15.5–6.6	−	~300	<50
2	fall	24	+	<50	<50
3	rise	30–15.5	−	>300	<100
4	fall	30	−	>400	<50
5	fall	52–37	−	<100	~250[b]
6	fall	40	+	~100	~100
7	fall	59	−	<150	<50
8	fall	62.5	−	~200	<50

[a] As shown in Fig. 71.
[b] A strong temperature component likely.

change (Fairbanks and Matthews, 1978), would predict an overall sea level drop through the Cenozoic of nearly 364 m (1200 ft). Sea level lowering at the 1‰ events would be equal to over 90 m (300 ft). These estimates compare with estimates of lowered sea levels during the late Wisconsin glacial maximum 18 KYBP of 80–163 m (Shackleton and Opdyke, 1973; Duplessy, 1978; Hecht, 1977; Williams, 1984; Matthews, 1984b). The average $\delta^{18}O$ change of 1.6‰ from deep-sea cores for this time period predicts a change of 145 m or 476 ft.

As seen in Fig. 71 and Table II, agreement on the magnitude of the sea level changes of the 3rd order is not good, particularly for the mid-Oligocene event 30 MYBP. In almost all cases, the inferred sea-level change using the sequence boundary patterns yields larger estimated changes than the $\delta^{18}O$ signal, except for the long-term fall spanning 52.5–37 MY (example number 5, Fig. 71). A temperature change is probably a part of the $\delta^{18}O$ signal in the early Eocene.

C. Rates of $\delta^{18}O$ and Sea Level Change

Applying the previous interpretation to the rapid 1‰ events of Fig. 70 implies that typical rates of sea level change (of the 2nd and 3rd orders) occur at an average rate of 10 m and 90–100 m per million years throughout the Cenozoic. These rate estimates represent gross long-term averages because of the broad sample intervals in the $\delta^{18}O$ data sets and the subsequent averaging of the data to construct Fig. 70 (Miller and Fairbanks, 1985b). This approach places constraints, however, on the estimated magnitude of the sea level changes from seismic stratigraphy (Vail et al., 1977).

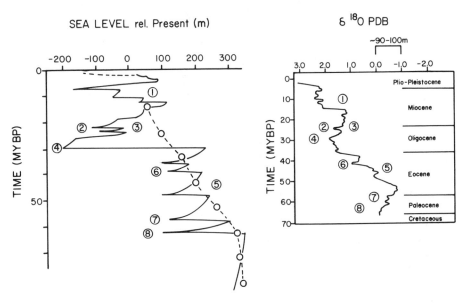

Fig. 71. Comparison of the magnitude of particular sea level events of the Tertiary as inferred from the seismic stratigraphy (Vail *et al.*, 1977) and the composite benthic $\delta^{18}O$ record (Miller and Fairbanks, 1985b). The encircled numbers refer to particular rises or falls examined in Table II.

We know from detailed $\delta^{18}O$ studies of high accumulation rate DSDP sequences that 50–100 m sea level changes of the 5th order can occur with frequencies of less than 100,000 yr throughout the Neogene (Figs. 32 and 51) (Woodruff *et al.*, 1981; Shackleton, 1982; Williams, 1984). The $\delta^{18}O$ records for the last 1.8 MY exhibit significant changes in the frequency and amplitude of the signal (Shackleton and Opdyke, 1976; Pisias and Moore, 1981; Prell, 1982; Thunell and Williams, 1983). Changes over the last 0.8–0.9 MY have a periodicity of 0.1 MY and an average amplitude of over 1.5‰ (a sea level equivalent of >130 m). The character of the $\delta^{18}O$ signal suggests that the rates of sea level rise and fall are not necessarily equal (Fig. 72). Broecker and Van Donk (1970) define the return from glacial low stand to interglacial high stands as *terminations* to describe the rapidity of the change, approaching 1–1.5 m/KY. The terminations are periods of rapid climatic, oceanographic, and biotic events. The subsequent return to full glacially induced lowered sea levels may take an order of magnitude longer (0.1–0.15 m/KY) (Fig. 72).

This pattern of unequal rates for sea level changes (rise and fall) of the 5th order suggests that the assumption that eustatic changes are curvilinear approaching sinusoidal may not be justified in all cases. The potential effects

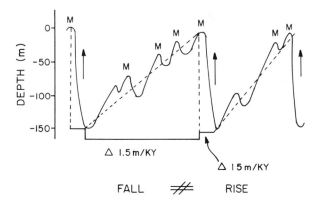

Fig. 72. A schematic representation of the asymmetry in the rate of sea level fall or rise during 5th (100 KY) and 6th (40 KY) order sea level $\delta^{18}O$ changes of the late Pleistocene. The M events represent the possible positions of meltwater events in a basin like the Gulf of Mexico superimposed on the global $\delta^{18}O$ change.

of this assumption on the stratal geometries of continental margins with differing subsidence rates and tectonic histories should be evaluated in models using sequence boundary patterns to estimate changes in accommodation of continental margins and eustatic sea levels. The $\delta^{18}O$ signal, even with its limitations regarding the relative temperature–ice volume effects, provides the only presently known independent check on this important parameter of eustatic sea level change.

III. $\delta^{18}O$ Chronostratigraphy, Gulf Coast Regional Unconformities, and Eustatic Sea Levels of the Plio-Pleistocene of Offshore Gulf of Mexico

The global nature of the $\delta^{18}O$ signal in the Plio-Pleistocene (Figs. 32 and 35) offers a unique opportunity to develop a chronostratigraphic framework of unprecedented resolution in offshore continental marginal sections. The chronology of isotope stages 1–23 are well known. Recently, Williams *et al.* (1987) compiled the most detailed $\delta^{18}O$ records for the late Pliocene–Pleistocene from Deep Sea Drilling Project cores and the Shell *Eureka* borehole 67-135 from the DeSoto Canyon (Figs. 32, 33, and 35). Radiometrically dated paleomagnetic chron boundaries and calcareous nannofossil datums shown in Table I were used to construct a chronostratigraphic

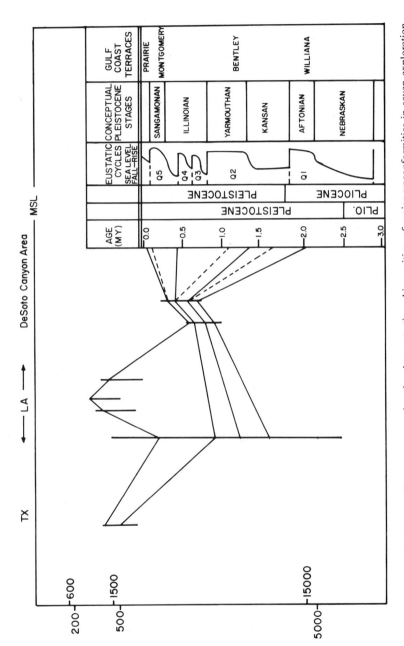

Fig. 73. A schematic fence diagram representing the chronostratigraphic position of major unconformities in seven exploration wells from offshore Gulf of Mexico. The timing of the continental slope unconformities is compared with eustatic cycles after Beard et al. (1982), the conceptual North American climate stages of the Pleistocene, and the Gulf Coast terraces (Fisk, 1944).

framework for each of the $\delta^{18}O$ records. A composite $\delta^{18}O$ record for the last 2 MYBP was derived from this empirical synthesis (Fig. 35), defining the boundaries of isotope stages 24–90. Quantitative techniques of spectral analysis, autocorrelation, cross-correlation, and semblance analysis are presently underway to refine this empirical record.

To test the applicability of this new type of chemical stratigraphy to offshore sections of the Gulf of Mexico, $\delta^{18}O$ records from the planktonic foraminifera, *Globigerinoides ruber*, were determined for a series of shallow and deep cored sections and exploration wells (cuttings). The sections were obtained from the upper and lower slope of the northern Gulf of Mexico (Fig. 73). From these $\delta^{18}O$ chronostratigraphic records, regional biostratigraphic tops were calibrated and several major unconformities were detected. Some of the unconformities involved missing time of several hundred thousand years and missing sediment spanning several thousand feet. Most of the unconformities were not indicated by the conventional biostratigraphic analysis. The regional significance of these unconformities and the power of this new chemical stratigraphy to exploration are readily apparent in Fig. 73.

The unconformities extend from offshore Texas to Louisiana east of the DeSoto Canyon. These unconformity surfaces have been informally designated LA–TX I, II, III, and IV with increasing age, and dated using the $\delta^{18}O$ chronostratigraphy. The subbottom depths over which these unconformities are present suggest that they are related to a widespread, eustatically triggered failure of the continental slope. Our knowledge of the $\delta^{18}O$ record contains no evidence for unusually large sea level drops during this time period to account for this regional failure nor does it provide a means of distinguishing the magnitude of one $\delta^{18}O$ sea level cycle from another. An apparent relationship exists between the timing of these regional deep-water unconformities and (1) the inferred paleobathymetric changes in the Gulf Coast, (2) the positions of the Gulf Coast Pleistocene terraces, and (3) the classical North American climatic stages (right column of Fig. 73). LA–TX IV appears to be equivalent to paleobathymetric event Q1 in the Nebraskan as defined by Beard (1969). LA–TX III suggests a correlation to Q2 in the Kansan. Unconformity II may actually be two events seen as Q3 and Q4 in the Illinoian. LA–TX unconformity I is recognized as consisting of paleobathymetric events Q5–Q6 associated with the Sangamonian (Fig. 73). Additional evidence suggests that several other minor hiatuses exist in individual sections, but more work is needed to clearly define their regional extent. Oxygen isotope chronostratigraphy has the time resolution not only to detect these regional events but also to better understand the relationships of these events to sea level changes and sediment depocenters in a calibrated absolute time framework.

IV. Conclusions and Recommendations

The stable oxygen isotope record for the Cenozoic is characterized by a series of large, third-order steps of +1‰ superimposed on a long-term second-order trend. This 2nd-order trend accounts for an $\delta^{18}O$ change of nearly +4‰ from the early Eocene into the Neogene. The 2nd- and 3rd-order changes in the $\delta^{18}O$ signal are driven primarily by a combination of global glacio-eustatic sea level changes and ocean paleotemperature responses to evolving circulation and climate patterns. Timing of the $\delta^{18}O$ events is in good agreement with the seismically defined changes in global coastal onlap (Vail *et al.*, 1977). Agreement is not good between the magnitudes of apparent sea level changes using the Exxon onlap record and oceanic $\delta^{18}O$ events. A consideration of the $\delta^{18}O$–ice volume–sea level relationship during the Pleistocene suggests that sinusoidal eustatics, i.e., the rise and fall of sea level being equal, are not a good assumption of 4th- and 5th-order sea level events. Agreement between the timing of $\delta^{18}O$ events and onlap/offlap inferred changes in relative sea level supports a common mechanism, perhaps that glaciation is apparent throughout much of the record. The $\delta^{18}O$ record certainly suggests that glaciation intensified in the Neogene. Although interpretation of the $\delta^{18}O$ record is not without its assumptions and limitations, it offers an independent geochemical check on seismically defined changes in stratal patterns. The $\delta^{18}O$ record also offers an independent chronostratigraphy, often with a temporal resolution exceeding many biostratigraphic zonations.

Seismic stratigraphy has become an inseparable part of exploration and has the ability, when used with proper ground truth control, to map large-scale depositional processes as a function of relative changes in eustatic sea level. Oxygen isotope stratigraphy offers an independent test of some of the sea level inferences made from onlap records and sequence boundary interpretations. The future integration of these geochemical and geophysical tools with improved biostratigraphic zonations and spectral processes of the $\delta^{18}O$ (and $\delta^{13}C$) signals offers exciting possibilities in terms of improved chronostratigraphy in exploration areas of passive continental margins. These results will in turn lead to a more accurate picture of global eustatic sea level records of the Cenozoic and possibly Mesozoic and older sequences. Such work is in progress at several academic and industrial laboratories. A paleobathymetric model for using the $\delta^{18}O$ composition of foraminifera to distinguish (1) eustatic from tectonic depth changes and (2) depth changes because of subsidence, uplift, and progradation, is described in Chapter 10.

Chapter 8 | Prospects for Applying Isotope Chronostratigraphy to Exploration Wells

I. Neogene Examples

A. The Gulf of Mexico

The character of the Neogene isotope record is sufficiently well established in terms of both long-term trends and short-term, high-resolution events to enable the application of isotope chronostratigraphy to continental margin sequences from the Gulf of Mexico without too much difficulty. This same technique can also be applied to other continental margins. Our experience indicates that small (<100 cc) samples of bulk sediment from either sidewall cores or cuttings typically contain sufficient foraminifera to make isotope chronostratigraphy a viable technique for time correlations between sections. For example, in a well with a 10,000 ft Pleistocene section, approximately 300 isotopic analyses would provide the resolution to define (1) the oxygen isotope stages, (2) missing sections, and (3) depositional modeling of the glacio-eustatic history at the well site. We are in the process of defining complimentary carbon isotope stages and determining the spectral character of carbonate contents for the Plio-Pleistocene using the quantitative techniques described in Chapters 11–20. High-resolution isotope records will concomitantly enhance the accuracy and reliability of the regional biostratigraphic zonations. A comprehensive isotope chronostratigraphy program in wells from shallow- to deep-water environments (i.e., zones 1–3) would pave the way for a very precise, standardized biostratigraphic and chronostratigraphic definition of drilled sequences from shelf to lower slope environments of the Gulf of Mexico. For example, we have already demonstrated that a detailed isotopic study in overlapping Pleistocene sections permits the establishment of a standard oxygen isotope "type section" for the upper slope (zone 4) of the Gulf of Mexico. A similar approach to that discussed throughout this book for exploration wells from shallow- and deep-water zones 1-3 and 5 could help relate the shallow shelf benthic foraminiferal zonations to zonations based on deep-water benthic foraminifera, planktonic foraminifera, and calcareous nannofossils.

Paleo-sea-level estimates for the Gulf of Mexico based on changes in the $\delta^{18}O$ record could be used to determine the linkage between augmented Mississippi River discharge and the lithology of drilled continental margin sections. Isotope chronostratigraphy would permit the construction of more accurate depositional models to permit an understanding of the relationships between such important phenomena as the glacio-eustatic sea level history of the region and the timing of major depocenter shifts. Isotope chronostratigraphy also provides the resolution to establish tie lines linking occurrences of source and reservoir rocks between small, isolated basins within offshore blocks. This information would of course lead to a better understanding of structural and stratigraphic effects of salt tectonics from one basin to another. Fine-scale changes in depocenters along the continental margin could be detected and related to high or low sea level stands.

To date, our work, using boreholes from the Gulf of Mexico, can be summarized with these points:

(1) With the perspective we have gained from our detailed work on several of the Shell *Eureka* boreholes and over 15 exploration wells from offshore Louisiana–Texas, we have been able to recognize the presence of major unconformities in specific parts of the Plio-Pleistocene of the northeastern Gulf continental slope (Fig. 73). For example, relying upon planktonic foraminiferal and nannofossil biostratigraphy alone (Brunner and Keigwin, 1981; Gartner et al., 1983-1984; Neff, 1983), the Pleistocene portion of E67-135 was thought to be complete. Although the Ericson foraminiferal zones and last appearances of *Discoaster brouweri, C. macintyrei* (*C. tropicas* of Gartner et al., 1983-1984), *Helicopontosphera selli,* and *Pseudoemiliania lacunosa* are all accounted for and occur in the proper sequence (Fig. 74), $\delta^{18}O$ analyses of the planktonic foraminifera *Globigerinoides ruber* (Williams, 1984) produced a detailed oxygen isotope record which, when integrated with the available biostratigraphic information, showed that many of the expected oxygen isotope stages were missing (Figs. 32, 33, and 35). We know that the record of E67-135 has at least three Pleistocene unconformities that were previously unrecognized. These unconformities are correlative to regional unconformities found in other Gulf of Mexico continental slope sections (Fig. 74) (Isotope Stratigraphy Group, unpublished).

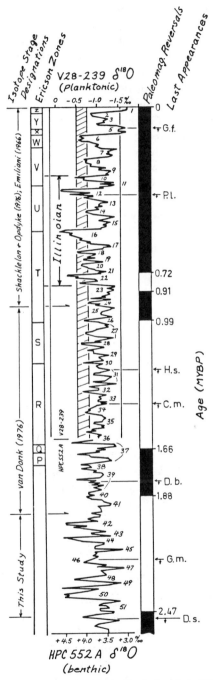

Fig. 74. Oxygen isotope chronostratigraphy in *Eureka* borehole E67-135 showing the isotope stages, positions of recognized unconformities, and positions of some commonly used biostratigraphic markets.

(2) Through isotope chronostratigraphy the regional unconformities of the continental slope can be incorporated into sedimentation rate diagrams that reveal that average accumulation rates are much lower than would be inferred on the basis of the biostratigraphic interpretation alone. Frequent unconformities in slope sediments obviously play an important role in exploration of deep-water Plio-Pleistocene tracts in the Gulf of Mexico. Shifting depocenters and large variations in the amount of marine versus terrestrial organic carbon delivered during low and high stands of sea level control the spatial and temporal distribution of source and reservoir rocks. At the same time, these depositional complexities often lead to large stratigraphic uncertainties.

(3) At least three and possibly four major regional unconformities exist in the Pleistocene sequences of the northern Gulf of Mexico; many more undoubtedly exist on finer time scales which require a resolution that only isotope chronostratigraphy can supply at this time. Applying our chronostratigraphic approaches described in this book, we have estimated the age of sediments below each unconformity. We can also show that many exploration wells are missing large portions of the Pleistocene in critical parts of the record, either through major intervals of erosion (slope failure?) or nondeposition (sediment bypassing?).

(4) A "type" oxygen isotope record has been established for the Pleistocene of the Gulf of Mexico continental margin. Isotope records from overlapping exploration wells cover the last 4 MY in sufficient detail to reveal the global isotope signals of the best deep-sea sections (Figs. 32, 33, and 35). In addition, this type record contains significant information that is not seen in the open ocean records (Shackleton et al., 1984a; Williams et al., 1984). We will soon have developed similar type records for the carbon isotope signal and carbonate content signal using the quantitative signal extraction techniques described in this book. One of the most exciting things about the concept of isotope chronostratigraphy is that the same approaches can be used in exploration wells from continental margins anywhere in the world.

(5) The composite oxygen isotopic record for the Gulf of Mexico Plio-Pleistocene provides a unique look at the history of glacial–interglacial fluctuations on the North American continent. The features of the composite record show that North American glaciers extended far enough south, as early as 2.5 MYBP, to drain substantial quantities of isotopically light meltwater through the Mississippi River system and into the Gulf of Mexico. These meltwater events play a large role in altering the type and quantity of sediment delivered to the Gulf of Mexico margin during the Plio-Pleistocene. With this composite record, it will be possible to determine the effects of meltwater events on the depositional system of the northern Gulf of Mexico.

(6) Nearly all of the last appearances of the microfauna and microflora currently used for biostratigraphic studies in the Gulf can be calibrated to an absolute time frame using isotope chronostratigraphy. Using the $\delta^{18}O$ type section, the last appearances of many microfossil events, which have not been previously correlated into an oxygen isotope stratigraphic framework, can be dated. For example, the extinction of the calcareous nannofossil *P. lacunosa* occurred in the Gulf of Mexico within 10,000 yr of the same event in open ocean sediments of Pleistocene age. This type of information is important for knowing how broadly a particular biostratigraphic zonation can be applied both within and between basins.

(7) We can correlate the several unconformities (and intervening packages of sediment) on the upper slope to the episodic transgressions and regressions on the continental shelf (Akers and Holck, 1957) and the Pleistocene terraces of the Gulf coastal plain (Fisk, 1944; Bernard and LeBlanc, 1965). Fillon and Full (1984) placed the Gulf Coast regressions corresponding to the bases of the Williana, Bentley, Montgomery, and Prairie formations on the continental shelf into a chronological context. The estimated ages for these regressions on the continental shelf agree quite closely with the timing of the major unconformities that we have now dated in exploration wells on the upper slope (i.e., Williana and unconformity IV, Bentley and unconformity II, Montgomery and Prairie regressions and unconformity I) (Fig. 73).

(8) Isotope chronostratigraphy can be used to recognize deep-water facies that are equivalent to those in shallow water. If our preliminary interpretations are correct, the task of correlating from the shallow-water environments of the shelf to the deep-water environments of the slope in the Gulf of Mexico may be considerably simplified. It may also be possible to extend the seismic stratigraphic framework(s) of the shelf formations to the slope where the benthic microfauna are of limited use because of the depth-related assemblage changes. We should also be able to determine which seismic horizons are synchronous as sequence boundaries are traced from shallow to deeper water, especially to parts of basins affected differentially by faulting. The ability to resolve regional unconformities in the Gulf Coast is important from an exploration standpoint because these regional unconformities can be used to intercorrelate thick but isolated reservoirs sandwiched between them.

B. Offshore California—The Miocene Monterey Formation

The Miocene Monterey Formation of California has received considerable attention for several reasons. The Monterey Formation attracts tremendous scientific interest as an example of the interaction between tectonic, climatic, and paleoceanographic events in the circum-Pacific region during the

Miocene. It also has significant economic importance for its major petroleum source beds in offshore and onshore California. The complicated tectonic, depositional, and diagenetic history of the Monterey Formation offers perhaps the consummate challenge to isotope chronostratigraphy, particularly because of postdepositional diagenetic transformations related to burial depth and temperature changes.

Very few data have been published on the stratigraphic oxygen and carbon records in the Monterey. Considering the importance of the Monterey Formation, a detailed isotope stratigraphic study of onshore and offshore Monterey sections could provide an interesting evaluation of the stratigraphic correlation schemes. For example, the biostratigraphic and chronologic scheme given in Fig. 75, adapted from the work of Barron

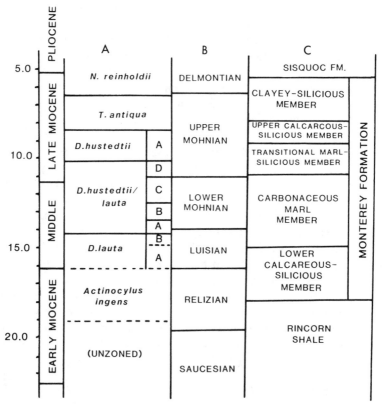

Fig. 75. A summary diagram showing the stratigraphic and chronologic relationships of (A) the Miocene diatom zones (Barron, 1980; Keller and Barron, 1981), (B) the California stages of the Miocene (Kleinpell, 1938), and (C) lithostratigraphic members of the Monterey Formation in the Santa Barbara coastal area (Issacs, 1984).

(1980) and Keller and Barron (1981), shows that it may be possible to utilize the same integrated approach using biostratigraphy and isotope stratigraphy to achieve a higher degree of stratigraphic correlation within the Monterey. As discussed in Chapter 6, Section III, large carbon and oxygen isotopic changes in marine carbonate are well documented throughout the Miocene (Figs. 48–52).

The occurrence of well-described onshore sections of the Monterey in the Santa Barbara and Santa Maria areas of California (Fig. 76) (i.e., Hondo Beach, Point Conception, Dos Pueblos, and Point Arquillo) provides easy access to Monterey rocks. Many of the outcropping sections are located

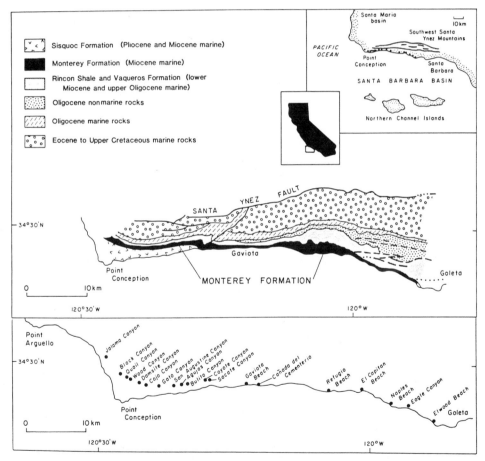

Fig. 76. Maps of southern California showing the excellent exposures of the rocks of the Monterey Formation in the Santa Barbara and Santa Maria areas (from Issacs, 1981b).

adjacent to important petroleum fields (Hondo Field, South Elwood Field, and Point Arquello–Hueso Field). Several COST wells (OCS-CAL78164 No. 1) and DSDP sites (173, 468B, and 470) offer an opportunity to determine whether stable isotopic studies of the carbonate fractions in the early to late Miocene sections of the Monterey can be used to correlate the onshore and offshore Monterey sections to the Miocene chronostratigraphic record. The biostratigraphic and diagenetic relationships in the Monterey Formation are sufficiently well described that a comprehensive attempt to use isotope stratigraphy appears to be warranted (Poore et al., 1981; Keller and Barron, 1981; Weaver et al., 1981; Pisciotto and Garrison, 1981; Isaacs, 1981a, b).

Stable isotopic data from the entire carbonate fraction, the biogenic carbonate fraction, or secondary carbonates (in the form of either disseminated dolomite or authigenic carbonates or in phosphatic rocks) may be able to detect the existence of important regional diagenetic trends affecting the porosity and permeability of different Monterey units. Previous isotopic work by Pisciotto (1981) shows that secondary carbonates in the Monterey have extremely variable oxygen and carbon isotopic ratios. These variations possibly relate to formation temperatures of the dolomites and interaction with microbial metabolic processes and organic carbon (Pisciotto, 1981). Ferm (1984) has shown that stable isotopic analyses in different dolomite phases can be used to determine important stratigraphic and diagenetic relationships, especially when the δ values are used in a stratigraphic context. Isotope chronostratigraphy may be most important in the lower carbonate bearing members of the Monterey, such as those at Elwood Beach and Goleta Slough. These sections appear to be the least altered diagenetically and to contain dispersed calcite (primarily) as foraminifera and coccolith debris (Isaacs, 1984). Preliminary isotope stratigraphic results on the Pedernales Point and Dos Pueblos outcrop sections are intriguing. The results suggest that some oxygen and carbon isotope events can be correlated from these two sections, even though the two sections differ significantly in lithology and diagenetic history.

II. Paleogene Examples

A. Offshore California

The borderland region of southern California is characterized by a series of northwest trending ridges and intervening sedimentary depressions with structural and stratigraphic complexities that reflect the Mesozoic and Cenozoic tectonic history of the region (Fig. 77; Crouch, 1981). The complex

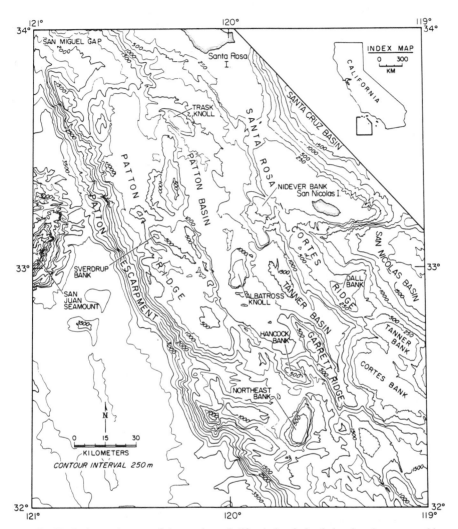

Fig. 77. Bathymetric map of the southern California borderland showing the topographic complexities of the offshore basins (from Crouch, 1981).

nature of this region makes it imperative to explore new techniques and strategies for achieving stratigraphic correlations, particularly of lower Tertiary rock units within borderland exploration wells. It is often difficult to accurately correlate wells in basins of the borderland because of sedimentological complexities, numerous unconformities, and frequent lack of abundant microfossils. Also, the available micropaleontological tops often appear to be diachronous. It is particularly difficult to correlate exploration sections from one sedimentary basin to another because of isolation by the numerous ridges, knolls, and escarpments in the region. Rocks from the Cretaceous to the Quaternary are known to be represented in the subsurface from exploratory COST wells and DSDP sites 467, 468, and 469. The lithology and thickness of the units in the basins are quite variable, but most units are marine siltstones, claystones, and sandstones. These lithological units are often separated by tuffs, igneous intrusives, volcani-clastic rocks, and occasionally nonmarine sandstones and shales (Crouch, 1981).

Again, as in the case of the Monterey Formation, this region was chosen as an area to test the use of oxygen and carbon isotopic records in a stratigraphic context as an innovative way of providing detailed correlations of lower Tertiary turbidites at a higher resolution than presently possible using conventional biostratigraphic techniques alone. In sections in which it is not practical (or possible) to determine the isotopic composition of foraminifera, it is often possible to develop a stratigraphic record using the isotopic composition of the carbonate found in a specific size fraction (i.e., < 63 μm or 63–90 μm; Fig. 12). Although no published data are currently available from such an approach, we have preliminary results that look quite promising; and sufficient exploration wells, COST wells, and DSDP sites are available from the southern California borderland to test this hypothesis.

B. The Gulf of Mexico

As an example of the potential use of isotope chronostratigraphy in the Paleogene of the Gulf of Mexico, the reader is referred to the comparison made in Fig. 62 of the oxygen and carbon isotopic change in benthic foraminifera from two Gulf of Mexico sections: one from the land-based St. Stephens Quarry, Alabama, and the other from Shell *Eureka* borehole E67-128. These sections were deposited in very different water depths, and yet the similarities in the oxygen and carbon isotopic signals across the Eocene–Oligocene boundary in these two sections suggest that stable isotopes may provide tie lines between shallow- and deep-water environments of the Gulf of Mexico in the Paleogene as well as the Neogene. Given the stratigraphic uncertainties in the sections reported by Keigwin and Corliss (1986) and the limited amount of data, the Paleogene results are encouraging (Fig. 62). Also, the fact that two different benthic species were analyzed in their study suggests that the stable isotope approach may be able to overcome the limitations imposed on biostratigraphic zonations that are based on benthic assemblages from shallow-water environments, which do not extend into deeper water offshore environments. We propose, therefore, that the adoption of isotope chronostratigraphy may overcome some of these limitations and be of use in Paleogene exploration sections at onshore drill sites. A blending of the empirical and signal processing approaches that we advocate should lead to an enhanced resolution of shallow- to deep-water biostratigraphic schemes that are presently being used by industrial micropaleontologists. This resolution will improve the understanding of the numerous wells with stratigraphic frameworks based on shallow-water biostratigraphic zonations.

III. Mesozoic and Paleozoic Examples

In discussing the use of stable isotope signals in stratigraphy, we are often asked if isotope chronostratigraphy has potential application in older sedimentary deposits. The potential exists, but the risks from diagenesis obviously increase. One of the first questions we asked ourselves, however, was, "Does evidence exist for $\delta^{18}O$ and $\delta^{13}C$ signals in older carbonate and dolomite rocks?" Examination of Fig. 78 shows that an ample signal is indeed present throughout the Phanerozoic, both in limestones and dolomites of any given time period. Studies of the stratigraphic signal in the oxygen and carbon isotopic composition of dolomite minerals by Ferm (1984) in the Knox Formation and Alsharham and Williams (unpublished manuscript) in the Jurassic Shuaiba Formation of Abu Dhabi show that useable signals are present in both isotope records. How correlative the signals are from section to section, even as regional diagenetic gradients, still remains to be determined.

A. Cretaceous $\delta^{13}C$ Events and Black Shales

In 1980, Scholle and Arthur proposed that the $\delta^{13}C$ records of Cretaceous limestones could be correlated using the character of the $\delta^{13}C$ records (Fig. 79). Indeed, the comparison is quite favorable within the constraints of the biostratigraphic control. Figure 80 shows the magnitude of the $\delta^{13}C$ signal in Cretaceous carbonates, especially in relation to the occurrence of black shales. Note that the $\delta^{13}C$ values of $CaCO_3$ become very positive during periods of enhanced organic carbon burial and removal of $\delta^{13}C$ from the ocean reservoir. It is also important to compare these Cretaceous black shale $\delta^{13}C$ values with the values of marine $CaCO_3$ during the late Paleocene (Figs. 24 and 64). No black shales are known for this interval of the Tertiary, but the nature of these $\delta^{13}C$ events suggests that the Cretaceous

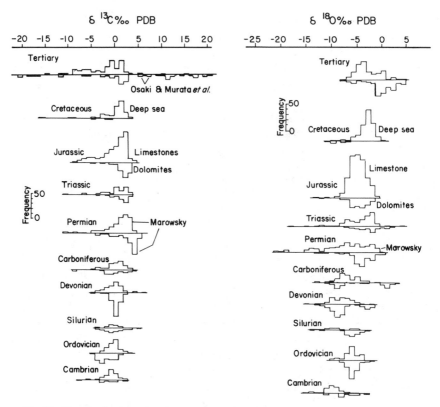

Fig. 78. The distribution of carbon and oxygen values of limestones (upper histograms) and dolomites (lower histograms) from the Ordovician to the Tertiary (modified from Veizer et al., 1980).

and Paleocene events are related to similar organic carbon burial rates. The present data, although limited, suggest that stratigraphic correlations using $\delta^{13}C$ are certainly feasible in some Cretaceous sections, as advocated earlier by Scholle and Arthur (1980). Exactly how useful these records will be requires more work.

B. Correlation of Basinal Clastics in Permian Strata, Delaware Basin, West Texas

Very few isotopic studies have been made of the carbonate matrix present in many clastic facies. A pilot study was conducted using subsurface cores

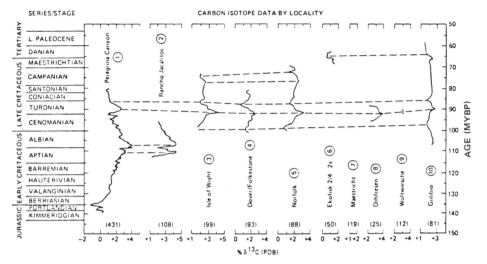

Fig. 79. Carbon isotopic ratios of whole rock analyses of pelagic limestone from 10 Cretaceous sections as functions of absolute age (from Scholle and Arthur, 1980). Each of the curves has been smoothed with a 5-point running average to filter data that fall more than 1‰ off average at that level. The data are from a combination of three commercial laboratories. Numbers in parentheses below each section indicate the total number of data points used to construct the record for that locality.

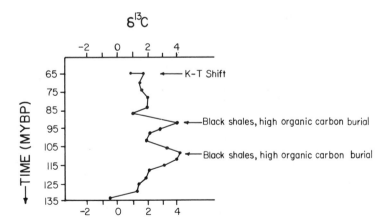

Fig. 80. An example of carbon isotope stratigraphy in Cretaceous limestones (M.A. Arthur, personal communication). Note the large positive $\delta^{13}C$ excursions associated with the time of high black shale deposition and the well-recognized negative $\delta^{13}C$ event associated with the Cretaceous–Tertiary boundary.

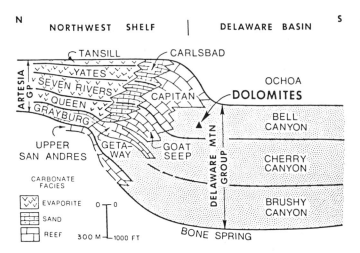

Fig. 81. Diagrammatic cross-section from the Northwest Shelf into the deep-water Delaware Basin showing the shelf and shelf–margin carbonate facies of the Permian Guadalupian Series and the basinal Delaware Mountain clastics. [The diagram is after Meissner (1972), as modified by Ahr and Berg (1982).]

representing the Brushy Canyon and Cherry Canyon formations of the Delaware Mountain Group (Fig. 81). The Delaware Mountain Group of the Delaware Basin is composed of three major clastic facies: the Bell Canyon, Cherry Canyon, and Brushy Canyon formations. During Guadalupian time, sediments composed mainly of thinly laminated, fine sandstones and siltstones with some interbedded dark limestones were deposited by both turbidity currents and low-density currents (Williamson, 1979). To the west of the Delaware Basin is the Guadalupian mountain range, which is a large reef complex divided into the Capitan Reef and Goat Seep Reef. Further to the west is the Northwest Shelf, which is composed of interbedded carbonate and evaporite facies. These facies are known as the Artesia Group and are subdivided into the Tansill, Yates, Seven Rivers, Queen, and Grayburg formations (King, 1942; Ward *et al.*, 1986). The Artesia Group represents a shallow-water, subtidal to supratidal environment in which carbonates were deposited during high stands of sea level. During low stands, carbonate deposition decreased and allowed large quantities of fine sandstone and coarse siltstones to deposit slowly (Ahr and Berg, 1982; Fig. 81).

The deep-water facies of the Brushy Canyon and Cherry Canyon formations were cored at sites located approximately nine miles from the ancient shelf margin. The Brushy Canyon sections are interpreted as a submarine channel site consisting of three major lithologic units (Stelting *et al.*, 1984). The lower unit consists of multiple cyclic events of upward thinning channel sequences. The middle unit represents a calcareous unit of widely different lithologies. The uppermost unit consists of interbedded laminated siltstones, sandstones, and limestones.

The other Brushy Canyon section is located approximately 0.25 mile west of the channel site and has facies indicative of an overbank depositional regime in a deep-water clastic system. Differences in the lithologies between the two sites suggest that they represent very differential depositional environments.

The entire Cherry Canyon section was cored for approximately 1000 ft at one location and consists of interbedded and finely laminated siltstones and fine to medium-grained sandstones. The Cherry Canyon Formation was also cored at two other sites, and equivalent 125-ft sections were sampled in each of the cores for a detailed isotopic and lithostratigraphic comparison. The lithologies of the Cherry Canyon sites are nearly identical except that the limestone interval in one section is slightly thicker than in the other two.

A detailed stable isotopic study of the carbonate matrix in the Permian basinal clastics of the Cherry and Brushy Canyon formations of the Delaware Mountain Group revealed these results:

1. Distinct isotopic differences exist between the two formations; $\delta^{18}O$ values in the Cherry Canyon Formation have a range of >30‰. Very positive $\delta^{18}O$ values indicate that the carbonate carrying the signal may have been precipitated in a highly evaporative environment, perhaps in the lagoonal back reef portion of the Northwest Shelf.

2. A significant relationship exists between these unusually positive $\delta^{18}O$ values and finely laminated, greyish to black siltstones. The exact mechanism by which these siltstones were deposited in the Cherry Canyon Formation is still not known but it may be associated with low stands of sea level.

3. Matrix $CaCO_3$ in the two Brushy Canyon sections has normal isotopic values for marine carbonate muds or recrystallized micritic calcite. The isotopic values are quite similar regardless of lithology and the inferred depositional environment of the Brushy Canyon clastics (i.e., overbank or channel parts of the deep-water fan complex; Stelting *et al.*, 1984).

4. The association of the highly positive $\delta^{18}O$ values with the tight siltstones suggests that diagenesis has not eliminated the stratigraphic signal.

5. As primary depositional signals tied to lithology, many of the positive $\delta^{18}O$ events in the Cherry Canyon section may be regionally correlative.

C. Summary

From this brief overview of Mesozoic and Paleozoic isotopic signals, we conclude that $\delta^{18}O$ and $\delta^{13}C$ signals exist in different carbonate phases in ancient rocks. Only time and work will tell what insights are possible by integrating geochemical and petrographic studies of these rocks in a stratigraphic context. As new data are published, the effects of diagenesis on chemostratigraphy will become better understood.

Chapter 9 | Paleobathymetric Models Using the $\delta^{18}O$ of Foraminifera

In this section we speculate on possible ways to use the $\delta^{18}O$ response of benthic and planktonic foraminifera to estimate paleobathymetry in terms of both absolute and relative water depth. We present very schematic and preliminary paleobathymetric models for a continental margin, such as the Gulf of Mexico, but the models, if valid, should work in other exploration regions. We illustrate how subbottom changes in water depth relate to (1) eustatic lowering of the sea surface (eustatic sea level), (2) subsidence, (3) abrupt or gradual tectonic movements upward or downward, or (4) progradation of the shelf–slope break, which may be deciphered from the $\delta^{18}O$ response of foraminifera.

I. Basis of the Models

The basis of the $\delta^{18}O$ paleobathymetric models is the fact that the water temperatures along the shelf and slope decrease in a systematic fashion with subbottom depth (Fig. 82). Along the Gulf Coast margin, for example, bottom water temperatures from the shallow shelf to beyond the shelf–slope break range from 23 to nearly 18°C. Benthic foraminifera living at the sea bottom record this temperature gradient in the $^{18}O/^{16}O$ ratio of their tests. The $\delta^{18}O$ difference between benthics living at different water depths can be predicted (Fig. 83) using the well-known temperature/^{18}O fractionation relationship (Epstein *et al.*, 1953). The $\delta^{18}O$ difference of benthic foraminifera will be the direct function of the temperature differences between their habitats (Fig. 82). For example, species A at a depth of 2000 ft (approximately 7°C) will have an $\delta^{18}O$ value that is approximately +1.3‰

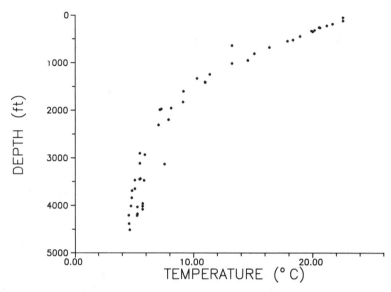

Fig. 82. Depth distribution of bottom water temperature taken along four transects from the offshore Louisiana–Texas shelf and slope (unpublished data, Morphometric Studies Group, University of South Carolina).

more positive than species A at 1000 ft (13°C) and +3.1‰ more positive than species A (or B) at 200 ft (21°C) (Fig. 84). We choose to ignore the salinity effects on the $\delta^{18}O$ of foraminifera for the moment, in order to keep the model as simple as possible initially.

If sea level is lowered 400 ft through glacio-eustatic processes, or the sea bottom is raised 400 ft by regional (salt or shale) tectonics, then the temperature experienced by the benthic species will change as a function of

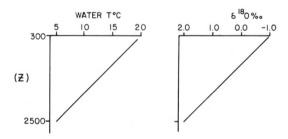

Fig. 83. Schematic representation of the predicted change in the $\delta^{18}O$ of benthic foraminifera on a continental margin, such as the Gulf of Mexico, as a function of the change in bottom temperature with water depth (left diagram).

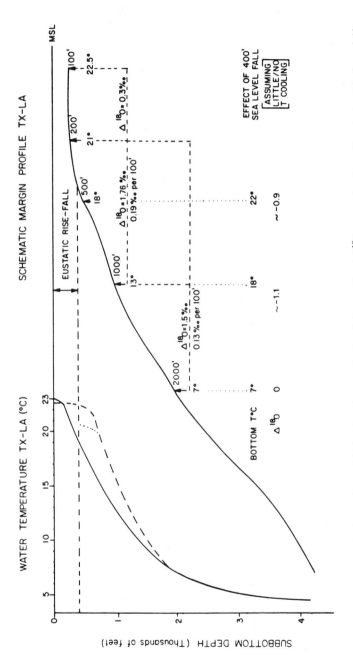

Fig. 84. A general paleobathymetric model showing the magnitude of the predicted $\delta^{18}O$–water depth gradient in benthic foraminifera along a continental slope, such as the Gulf of Mexico, during interglacial times (e.g., today) and during a eustatic lowering of sea level by 400 ft.

the depth change (Fig. 84). For example, the temperature profile in the water column to 2000 ft will become depressed with a 400 ft eustatic lowering of sea level. At 2000 ft (now 1600 ft), the temperature change is minimal (0–<1°C). At 1000 ft (now 600 ft) and 500 ft (now 100 ft), the actual water temperature would increase by 5°C and 4°C, respectively, thus decreasing the $\delta^{18}O$ difference in the benthic species relative to 2000 ft by as much as

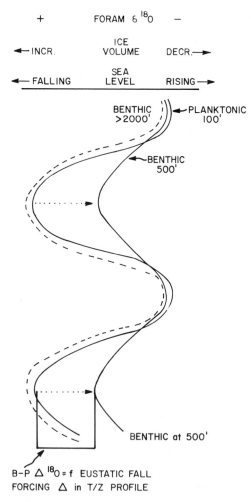

Fig. 85. A schematic representation of the $\delta^{18}O$ response of benthic and planktonic species inhabiting different parts of the water column during several glacio-eustatic changes in sea level. The planktonic record may be from the same well as the benthic records, form >2000 ft or from 500 ft, or all three records may be from three different wells from the same area.

-1.1 to $-0.9‰$. The change in the $\delta^{18}O$ difference will be a function of the depth change and the depth-temperature gradient at the respective depths of the foraminifera.

Eustatic rises and falls of the sea surface, holding subsidence, uplift, and progradation constant, produce predictable and measurable changes in the $\delta^{18}O$ differences between benthic species at different depths (Fig. 84). In addition, comparison of the benthic to planktonic $\delta^{18}O$ response to glacio-eustatic sea level changes may provide a means to estimate the magnitude of such changes. Figure 85 shows a schematic representation of the down-core $\delta^{18}O$ records for two benthic species living at different depths relative to the record of a planktonic foraminiferal species. The $\delta^{18}O$ records of the deep dwelling benthic species at 2000 ft and the planktonic species would have similar records in terms of the timing of the glacial-interglacial $\delta^{18}O$ changes. The two records would be offset, however, by the temperature difference between the two foraminiferal species. The $\delta^{18}O$ record for the benthic species living in water shallower than 2000 ft (but deeper than the mixed layer of the shelf) would have the same timing as the other two records, but the magnitude of the glacial-interglacial $\delta^{18}O$ changes would be less than the expected ice volume induced change in the $^{18}O/^{16}O$. The differences between the records would be due to the change in the thermo-cline as a function of the lowered sea level (Fig. 84). The magnitude of the $\delta^{18}O$ responses of the shallow benthic and planktonic $\delta^{18}O$ records would be a function of the degree of the eustatic fall forcing changes in the temperature-depth profile along the margin.

II. Distinguishing Eustatic from Tectonic Changes in Paleodepth

Figure 86 is constructed to illustrate how the $^{18}O/^{16}O$ records of benthic and planktonic foraminifera may change in one or more sections as a function of these situations: case 1—eustatic changes in sea level; case 2—abrupt tectonic uplift of 100 ft; case 3—abrupt tectonic collapse of 300 ft; case 4—gradual deepening because of subsidence or salt withdrawal; case 5—gradual shallowing because of uplift or progradation. All of the $\delta^{18}O$ paleobathymetric models are based on predictable changes in the bottom temperature regime experienced by the benthic foraminifera caused by changes in paleodepth.

In case 1, when only eustatics are invoked (i.e., of the 5th order, 330 ft every 100,000 yr, for example), the benthic and planktonic $\delta^{18}O$ response will generally be in-phase and the mean of the records offset by the tem-perature difference between the surface and bottom according to the con-straints discussed in Section I of this chapter (Figs. 84 and 85).

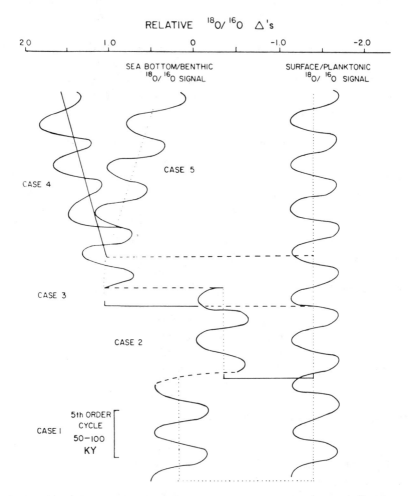

Fig. 86. A preliminary model showing the predicted response of time series $\delta^{18}O$ records for benthic foraminifera, relative to a planktonic foraminifera record that lived in the upper most water column, thereby being immune to changes in paleodepth. Case 1 represents the temperature-$\delta^{18}O$ offset between the benthic and planktonic record because of the temperature difference between the surface and bottom waters. Case 2 reflects the major offset that should occur because of a large, abrupt tectonic uplift of the sea floor. Case 3 represents the effect of an abrupt drop in the sea floor because of some form of tectonic collapse. Case 4 shows a gradual enrichment in benthic $\delta^{18}O$ values with time because of a gradual increase in paleodepth. Case 5 illustrates the exact opposite: $\delta^{18}O$ depletion as the site becomes more shallow and experiences warmer waters through time.

As a result of depth changes because of a rise or fall of the sea bottom, independent of eustatics (cases 2–5), each process will produce a significant, measureable, and predictable offset in the mean values or average character of the benthic and planktonic $\delta^{18}O$ records (Fig. 86). For example, a tectonic (salt) uplift of the sea bottom of 100 ft (case 2) could produce as much as a 10°C increase in the bottom temperature at that site, thus significantly decreasing the $\delta^{18}O$ difference between the benthic and planktonic records. An apparent bottom temperature decrease because of an abrupt tectonic collapse (case 3), or a gradual deepening because of the subsidence rate outpacing sediment input (case 4), would become manifested as an increase in the benthic to planktonic $\delta^{18}O$ difference. A gradual shallowing of the sea bottom because of progradation of the margin outpacing subsidence or regional tectonic uplift (case 5) would produce a convergence of the benthic and planktonic $\delta^{18}O$ records, independent of the eustatic component of sea level, depending on the initial depth of the sea floor (Figs. 84 and 86).

III. Distinguishing Uplift, Subsidence, and Progradation Changes in Paleodepth

To accurately predict the paleodepth changes modeled in Fig. 86, the benthic to planktonic response should be compared in several wells located up and down dip (Fig. 87). The schematic model presented in Fig. 87 could then be used to distinguish paleodepth changes related to salt or shale tectonics (case 2), growth faulting (case 3), or progradation through time (cases 4 and 5). Definition (calibration) of a reference surface (i.e., today; case 1 based on Figs. 82–84) could be used for comparison with changes in the $\delta^{18}O$ and temperature–depth gradients between well sites along subsurface reference surfaces. Of course, the reference surfaces should ideally be time horizons defined by biostratigraphic datums, isotope chronostratigraphy for higher resolution, or seismic stratigraphy.

IV. Summary

The paleobathymetric models presented here are entirely schematic. In their favor, however, is the fact that bottom temperature gradients exist today, and the gradients in the past are unlikely to have changed substantially enough to alter the basic components of the models. Depth–temperature gradients along continental margins produce predictable $\delta^{18}O$ differences between benthics living at different depths (temperatures). The magnitudes

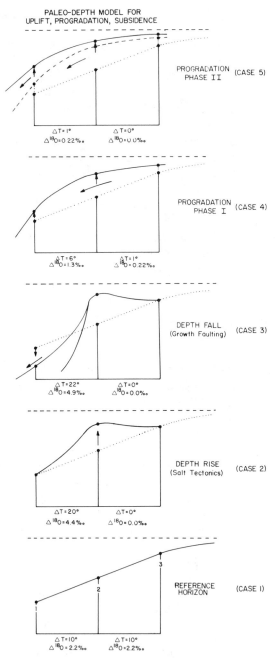

Fig. 87. Schematic representations of how $\delta^{18}O$ gradients with depth can be used to distinguish paleodepth changes because of salt (or shale) tectonics (case 2), a depth fall due to growth faulting (for example, case 3), and different phases of progradation, leading to overall shallowing in upward sequences at several depths along a continental slope, such as the Gulf of Mexico.

of the $\delta^{18}O$ differences produced by different types of paleodepth change are measurable. These facts yield predictable models that can potentially produce quantiative paleobathymetric reconstructions. These $\delta^{18}O$ paleodepth estimates could then be compared with paleodepth estimates from benthic assemblages, basin modeling studies, or morphometrically defined changes in the shape of benthic foraminiferal species (Gary, 1985). Results on several wells from the Gulf of Mexico slope and offshore Trinidad suggest that the models presented here have merit. Only more work will determine if these models have general applicability in exploration of continental margin wells.

Chapter 10 | General Overview of the Empirical Approaches to Isotope Chronostratigraphy

The carbon and oxygen isotopic records from marine carbonate fractions for the Tertiary and for specific epochs of the Tertiary have unique characteristics that provide high-resolution stratigraphic correlations and potential age dating frameworks. Figures 3 and 4 show that the Tertiary isotopic records can be modeled to determine the absolute age of marine sections using the broad long-terms features of each of the records. Once calibrated, such records could be used in a particular sedimentary basin without the need for a detailed biostratigraphic workup, particularly if some of the quantitative approaches described later in this book are employed as part of the isotope chronostratigraphy.

For a stratigraphy of the highest resolution, the best approach is to integrate isotope chronostratigraphy with other sources of stratigraphic information (i.e., biostratigraphic, seismic, and lithostratigraphic horizons) and with quantitative signal processing techniques. Using such an integrated approach it is possible to determine stratigraphic correlations and regional time lines in exploration wells to within ±20,000 yr in the mid to late Pleistocene; ±50,000 yr in the late Pliocene to early Pleistocene; ±100,000 yr in the Miocene to early Pliocene interval; ±300,000–500,000 yr in the late Eocene to Oligocene interval; and ±200,000 yr in the Paleocene to middle Eocene interval.

Our objectives in the first part of this volume are to synthesize what is presently known about the oxygen and carbon isotope signals of the Tertiary and to show how isotope chronostratigraphy may be applied to exploration wells from deep and shallow parts of the continental margin. One of the advantages of this technique is that it provides quantitative, globally synchronous correlations within an absolute time frame. Changes in sediment type and accumulation rates can be related directly to the Pleistocene glacio-eustatic sea level changes. The character and chronology of the isotope record are now established well enough to construct a standard isotope type section for continental margin boreholes of the Neogene and Paleogene. The empirical approaches of describing isotope signals are particularly promising when combined with existing quantitative techniques for extracting spectral information from the isotope signal. Predictive models are then possible for particular time periods and sedimentary basins. Having calibrated isotope chronostratigraphy models, standard $\delta^{18}O$ and $\delta^{13}C$ records can be used to:

1. Enhance the stratigraphic resolution of continental margin sections.
2. Determine regional to global time lines in exploration wells.
3. Provide a chronostratigraphic framework tying shallow- and deep-water wells into a common framework.
4. Detect significant but short duration hiatuses that are unresolvable ("transparent") using biostratigraphy, seismic stratigraphy, or magnetostratigraphy alone.
5. Determine which biostratigraphic markers are synchronous datums of chronostratigraphic significance.
6. Test the reliability of different biostratigraphic zonations within a basin and between widely spaced basins.
7. Place lithostratigraphic correlations within the framework of global sea level history (i.e., particular sediment packages, seismic boundary sequences, and stratal patterns). In the case of the Gulf of Mexico, lithological and depositional patterns can be related to important meltwater/fluvial discharges from the Mississippi River.
8. Build predictive models for use in exploration chronostratigraphy. The stable isotope signals from marine carbonate fractions are ideal for the signal extraction methods described in the latter portion of this book.

Chapter 11 | Quantitative Methods of Analysis: Theoretical Considerations

Many of the techniques used to analyze digital signals can be combined with the records of isotope chronostratigraphy to provide criteria concerning the identification of common events and the accuracy with which such events can be recognized and interpreted.

Two different types of records are available:

1. At a fixed location on the surface of the earth we record with depth (and so, by conversion, with respect to geologic time)[1] a set of different types of records (e.g., $\delta^{13}C, \delta^{18}O, \delta^{32}S$). Let $x(I, t)$ denote the measured record where we label the different types of records with the label I identifying the record type and the label t identifying the geologic time of the record.

2. At different locations on the surface of the earth we record with depth a set of similar or different records. Label the surface locations by the latitude and longitude pair $\mathbf{L} \equiv (\theta_\mathbf{L}, \phi_\mathbf{L})$ so that the record measured is $x(\mathbf{L}, I, t)$, where I and t have the same meaning and $\mathbf{L}$ denotes the locale. A given record will be sampled at a set of times $t_1, t_2, \ldots, t_N$. The N may vary from record to record.

Several questions are of interest. Illustrative examples of these questions are

1. Do different records at the same location show common events?

---

[1] The conversion of sampling with depth to sampling with time requires that we be given the rates of sedimentary deposition and compaction with respect to time. For the moment we shall assume that these are known but shall return later in this section to methods of determining the depositional and compactional velocities from the record itself.

2. Do records at different locations show common events, or events that have moved spatially in time?

3. What degree of accuracy can we ascribe to the common events?

4. The imprecision of knowledge on the paleoevolutionary behavior of foraminifera, the evolution of the oceans, isotopic diffusion with time, and diagenetic effects during and after sedimentary burial means that there is a certain amount of noise in each and every record. How do we extract the best signal from a record to be used in comparison against other records?

5. The sampling of a record is done at a set of discrete times over a finite span of geologic time. The record may be finely sampled in some time intervals (e.g., Holocene, Tertiary) and less finely sampled in other time intervals (e.g., Triassic). How do we allow for this variable time sampling in attempting to determine common events or spatially migrating events?

6. A record may have many "wiggles" on it but by sight we think that there might be an underlying, slowly varying behavior. How do we extract this slow behavior in the presence of noise?

For such problems of signal recognition and common event identification, we have at our disposal a large arsenal of methods based on information theory. In this chapter we outline (without too much rigorous detail) the more common methods used for quantitative analysis of signals. Details about the development of such methods can be found in modern texts on signal analysis (Papoulis, 1965; Arnold, 1974) and in the extended list of references at the end of this book. This chapter serves chiefly to outline the methods and introduce nomenclature, and it sets the stage for application of the methods to the isotopic chronostratigraphic records.

We shall discuss, in turn, several basic methods of data reduction, each of which performs a different type of operation on a data set, in order to explore the capability of extracting a signal from the data. The reader should not, however, form the impression that the methods to be discussed are either a universal panacea or form a complete system for analyzing data. They most certainly are neither. New, modern data analysis techniques are constantly being developed, researched, and improved, and it is fair to say that there is still a considerable degree of art in data analysis procedures. The techniques we discuss are proven "workhorses," which are usually used first in attempting to define signal behavior. Encouraging results from the application of any one technique are normally used to point the way to the use of more sophisticated methods of data analysis that try to improve upon the signal determination and resolution.

The processes we shall discuss are:

1. Semblance measures.

2. Filtering and deconvolution procedures.
3. Autocorrelation, cross-correlation, power spectral, and cross-spectral methods.
4. Maximum entropy and biased maximum entropy techniques and end member distribution methods (CABFAC, Q model, and "fuzzy" Q-model methods).

As we shall see, each method has its place, and each has something to tell us about the presence of signals in data.

Chapter 12 | Semblance Methods

The behavior of similar record types at different locations will record a common signal as the locations merge. Thus, the record types must reflect a degree of coherence. A measure of the coherence stored in the records can be obtained by constructing an "energy" for each time t:

$$E(I, t) \equiv \left[\sum_{\mathbf{L}} x(\mathbf{L}, I, t) \right]^2 \bigg/ \sum_{\mathbf{L}} [x(\mathbf{L}, I, t)^2], \qquad (1)$$

where $E(I, t)$ is the "energy" of coherence for a given record type I for time t, and Σ represents the summation of the given values $\mathbf{L}$ for all surface locations ($\mathbf{L}$). Also, E would presumably be large when the records of type I are in phase. Some difficulties with this approach are that not all of the type I records are necessarily sampled at the same time, and it is not clear that a common event at different spatial locations occurred at precisely the same time. Coherent events presumably are spread over some time range. In geophysics, Neidell and Taner (1971) introduced the concept of *semblance* to provide a more meaningful measure of coherent signals. The semblance at time t for records of type I is defined by

$$S_e(I, t, \Delta t) = \sum_{\tau = t - \Delta t}^{t + \Delta t} \left[\sum_{\mathbf{L}} x(\mathbf{L}, I, \tau) \right]^2 \bigg/ \sum_{\tau = t - \Delta t}^{t + \Delta t} \left\{ \sum_{\mathbf{L}} [x(\mathbf{L}, I, \tau)^2] \right\}, \qquad (2)$$

where $S_e(I, t, \Delta t)$ is the semblance at time t across all record types I for the time window $2 \Delta t$. The semblance will be large when a coherent event is present within the time window $2 \Delta t$ centered on t. The magnitude of the semblance will also be sensitive to the amplitude of the event.

One drawback to this semblance technique is that it does not recognize anti-coherent events. For instance, consider two records of type I at locations L and L'. If $x(L, I, t)$ is positive at L, but $x(L', I, t)$ is equally negative at L', then the semblance will be identically zero even though the two records are perfectly anticorrelated. Semblance dominantly recognizes events of the *same* signature, i.e., positively correlated.

The semblance behavior with increasing time window width Δt is of interest. If the window Δt is smaller than the time spacing between records of type I, and if, as is usually the case, the records are not all taken at precisely the same set of times $t_1, \ldots, t_N$, the semblance will pick up only a few record points for small Δt. A small semblance value will then be recorded. Equally, if the time window is large, say over the total length of the record, then again the semblance will be small, unless the records are perfectly in phase and are all taken at the same time set $t_1, \ldots, t_N$. Somewhere in between these two extremes there is a "best" range of time windows to choose so that the semblance is maximized. A sequence of common (coherent) events can then be measured by contouring constant semblance on a time, t, window, Δt, plot (see later). For time records of different types taken at the same or at different locations, a little more care has to be exercised since the records may have different dimensions. A way to avoid this technical problem is to make all records dimensionless by setting

$$X(L, I, t) = x(L, I, t) \Big/ \left\{ \sum_t [x(L, I, t)]^2 \right\}^{1/2}, \qquad (3)$$

where $X(L, I, t)$ represents the dimensionless record type at location L corresponding to $x(L, I, t)$, and Σ_t implies the summation of values over the entire length of the record. Note that Eq. (3) represents standard vector normalization commonly used in linear algebra. In most isotopic chronostratigraphic records a common scale is used and the problem of differing dimensionality does not arise in such cases. We can then use semblance as a measure of coherence for similar record types at different locations, and we can also use it at a given location for records of similar types. With the normalization given in Eq. (3), different records can be analyzed in similar fashion.

Thus, a local semblance measure is

$$S_f(L, t, \Delta t) = \sum_{\tau = t - \Delta t}^{t + \Delta t} \left[\sum_I X(L, I, \tau) \right]^2 \Big/ \sum_{\tau = t - t}^{t + \Delta t} \left\{ \sum_I [X(L, I, \tau)^2] \right\}, \qquad (4)$$

where Σ_I implies summation across the different record types for a given time t. This local semblance searches for coherence of common signal properties between records of different types at the same location, and yields a behavior for contours of constant local semblance in a time, t, window, Δt, coordinate space similar to that sketched previously.

A more ambitious project is to search in both record type I and coordinate $\mathbf{L}$, for semblance peaks, although this is not that often undertaken relative to the semblance methods previously recorded. If an event cannot be seen in semblance measures, either spatially or with respect to I types alone, it is not likely to be more easily identifiable by manipulating the $(\mathbf{L}, I)$ space of semblance measures. Some other procedure is needed. In addition, strongly anticorrelated events are minimized, leading to their lack of recognition on such semblance plots.

Chapter 13 | Filter and Deconvolution Techniques

The emphasis of nearly all modern techniques of data reduction is related to the separation of signal from noise in terms of a *desired* output. Presumably this is one of the reasons why the design and choice of filtering functions remains predominantly an art.

The input information for a record is $x(\mathbf{L}, I, t)$ at a sequence of times $t_1, \ldots, t_N$. We suppose that each record contains both a signal $S(\mathbf{L}, I, t)$ and a linear additive component $n(\mathbf{L}, I, t)$ that we call "noise." The task is to design some filtering operation on the data records $x(\mathbf{L}, I, t)$ that will extract the best signal while minimizing the noise content of the signal.

A common thread entangled in all such methods dates back to the work of Wiener (1949). Basically the problem is reduced to the form

$$x(\mathbf{L}, I, t) = S(\mathbf{L}, I, t) + n(\mathbf{L}, I, t), \tag{5}$$

with x measured k times. What is the linear filter function f that will enable us to produce a desired output $g(\mathbf{L}, I, t)$ from $x(\mathbf{L}, I, t)$? The underlying assumption is that with the noise minimized in some way the filtering operation will enable us to provide a measure g of the true signal S. Put another way, the problem is to define a filter f that will produce an output g that gives the best separation of signal and noise as reflected in Eq. (5).

Variations on the theme exist and are routinely available. Thus in place of minimizing noise to extract the best signal at a given location one could assume:

1. That a common signal is to be found at locations $\mathbf{L}$, $\mathbf{L}^1$;
2. That the shape of the signal is known but its magnitude is not;
3. That one knows the signal only up to some time t and wishes to predict the signal behavior at later times;

4. That the spectrum of the noise is known by running synthetically built samples and depositional and compactional models through the instrument, analyzing their output, constructing a filter function to reproduce the known input values, and then using this filter function on the field data samples one wishes to analyze for the presence of a signal, etc.

Each such assumption provides some constraints on the type of filter analysis one applies to the system, and each has its strengths and weaknesses. The previous four assumptions are those most commonly made. We discuss each of them in turn after setting up the analysis of signal extraction by noise minimization.

I. Least Squares Noise Minimization

Take a set of k measurements with time t at a given location.
We then apply to the raw data[2] $x(t)$ the linear filter $f(t)$ in the form

$$S(t) = \sum_{i=0}^{\infty} f(t_i)x(t - t_i), \tag{6}$$

and we require that $f(t)$ be such that $S(t)$ is the best, minimum noise estimate of the true signal $s(t)$.

For the set of k measurements of $x(t)$, Wiener (1949) showed that the best least squares solution for $f(t)$ is given by solving

$$\sum_{i=0}^{\infty} f(t_i)r(t_j - t_i) = g(t_j), \qquad j = 0, 1, 2, \ldots, \tag{7}$$

where

$$r(t) = \lim_{N \to \infty} N^{-1} \sum_{k=1}^{N} x^{(k)}(\tau)x^{(k)}(\tau - t), \tag{8a}$$

and

$$g(t) = \lim_{N \to \infty} N^{-1} \sum_{k=1}^{N} S^{(k)}(\tau)x^{(k)}(\tau - t), \tag{8b}$$

where $x^{(k)}(t)$ is the kth measurement of $x(t)$. Note that if the signal and noise are uncorrelated, then $r(t)$ will be the sum of signal and noise autocorrelations, and $g(t)$ will be just the signal autocorrelation. The frequency response of the Wiener filter then reduces to the classical form

---

[2] We have suppressed labeling the location **L** and measurement type **I** in the interests of clarity of presentation.

$P_s/(P_s + P_N)$, where P_s and P_N are the signal and noise spectral densities, respectively. Pragmatically, approximations to $r(t)$ and $g(t)$ are obtained by ignoring the limit statements. Fast numerical methods for solving the filter equation [Eq. (7)] using recursive algorithms have been developed by Levinson (1949) and have since been enhanced, elaborated, and generalized for multichannel filters.

II. The Common Signal Minimization Problem

In connection with similar types of data from different spatial locations yielding the *same* signal at an instant of time, and so representing a common event, we are interested in this problem: given measurements of $x(\mathbf{L}, I, t)$ at several locations $\mathbf{L}_1, \mathbf{L}_2, \ldots, \mathbf{L}_N$ at time t, how do we extract the best approximation to the signal $S(I, t)$ when each measurement is made in the presence of noise—be it of the instrument's response or because of local geologic conditions? That is, how do we extract $S(I, t)$ from

$$x(\mathbf{L}, I, t) = S(I, t) + n(\mathbf{L}, I, t), \qquad \mathbf{L} = \mathbf{L}_1, \mathbf{L}_2, \ldots, \tag{9}$$

where n represents the magnitude of the local noise at location $\mathbf{L}$ at time t?

This problem is treated in standard texts on noise analysis (e.g., Arnold, 1974; Papoulis, 1965) on the assumption that the noise obeys Gaussian statistics.

It can then be shown that the probability that a given set of data contains a signal S is maximized when the signal is chosen to be

$$S(I, t) = \sum_{i,j=1}^{N} \phi_{ij} x(\mathbf{L}_i, I, t) \Big/ \sum_{i,j=1}^{N} \phi_{ij}, \tag{10}$$

where ϕ_{ij} is the inverse matrix to the covariance matrix ρ_{ij}, and the covariance matrix istelf is related to the noise through

$$\rho_{ij} = \langle n(\mathbf{L}_i, I, t) n^*(\mathbf{L}_j, I, t) \rangle, \tag{11}$$

where angular brackets denote an ensemble average. Use of the ergodic hypothesis enables us to equate ensemble and arithmetic averages, thereby relating the covariance matrix to measurable quantities.

If the noises at the various spatial locations are uncorrelated, then ρ_{ij} will be a diagonal matrix with elements equal to the noise power. The inverse matrix applied to the data to produce the best estimate of the common signal will then also be diagonal with elements inversely proportional to the noise power. The filter then maximizes contributions to the common signal from locations with low noise and minimizes contributions to the common signal from locations with high noise.

III. The Location-Dependent Common Signal Problem

Instead of seeking a signal common to all locations at the same time, it is possible that the signal sought depends upon the location. (For instance, in the case of seismic waves, a normal move-out correction is applied to each seismic trace dependent upon its offset. The normal move-out and *stacking velocity* are then found by conventional semblance maximization methods.)

Thus, relative to some given location L_0, it may be that the signal sought arrives at a location L_1 at a different time than at location L_0 because of the type of transport of the cause of the signal with geologic time by some underlying time-dependent mechanism. The signal at L_1 then arrives at time t_1 with

$$L_1 - L_0 = \int_{t_0}^{t_1} dt' \, V(t'), \tag{12}$$

where V measures the velocity of transport in the direction $L_1 - L_0$ at time t'.

We seek the magnitude, time dependence, and direction of the velocity field that will maximize the correlation of a common signal from many locations. Many practical procedures are available for determining $V(t)$ based on ideas and concepts used in the petroleum industry to determine the subsurface medium velocities for seismic waves, that are directly applicable to the isotope chronostratigraphy case.

Perhaps the simplest theoretical concept to illustrate is that of a constant velocity in which case

$$t_1 = t_0 - V \cdot (L_1 - L_0)/|V|^2, \tag{13}$$

where modulus bars represent absolute value. Here, we are interested in finding the constant vector V which will maximize the coherence of a common signal from several locations.

Combining Eq. (13) with Eq. (9) yields

$$X(L_1, I, t_1) = S[I, t_0 - V \cdot (L_1 - L_0)/|V|^2] + n(L_1, I, t_1), \tag{14}$$

where the noise at time t_1 at location L_1 is *not* dependent on the signal propagation characteristics.

Again, this problem is merely a variation on the theme of Section II, this chapter. Basically, one applies some vector velocity to the data in the manner given in Eq. (14), then one constructs the best signal using the inverse of the noise covariance on the assumption of Gaussian statistics for the noise, and then one varies the vector velocity until the signal maximizes relative to the noise. This provides a measure of the best velocity vector controlling constant motion of the signal between spatial locations.

IV. The Magnitude of the Signal Problem

Quite often it happens that we know the *shape* of the signal but not its magnitude, or it happens that we have an estimate of the signal shape from previous data and we wish to ask whether that same shape is embedded with some magnitude in given sets of new data. In other words, we can write

$$x(\mathbf{L}, I, t) = CS(I, t) + n(\mathbf{L}, I, t), \qquad (15)$$

where we wish to determine the constant C given the shape of the signal. This method, developed by Capon (1969) for separating modes of teleseismic waves, is a variation of the previous methods. Again, assume Gaussian statistics for the noise. The best estimate of C is then given by

$$C = \sum_{i,j=1}^{N} x(t_j)\phi_{ij}S(t_i) \Big/ \sum_{i,j=1}^{N} S(t_j)\phi_{ij}s(t_i), \qquad (16)$$

and its variance is

$$(\delta C)^2 = \sum_{i_{ij}=1}^{N} S(t_j)\phi_{ij}S(t_i), \qquad (17)$$

where ϕ_{ij} is, once again, the inverse matrix of the autocovariance matrix ρ_{ij} of the noise.

V. Maximum Likelihood

All of the outlines of data reduction methods previously given are based in whole or major part on the assumption that the noise component follows Gaussian statistics. For a set of noise values $\mathbf{n} = [n_1, n_2, \ldots,]$, the Gaussian assumption of noise statistics means that the probability f of finding the noise vector $\mathbf{n}$ is given by

$$f \propto \exp[-\tfrac{1}{2}n_i\phi_{ij}n_j]. \qquad (18)$$

(The Einstein double summation convention over repeated indices is used.) The essence of the previous sections in this chapter is then contained in replacing n_i with $x_i - S_i$, where x_i are the measurements and S_i the signal, and then performing minimization operations to determine the best signal in the presence of noise.

An alternative way to investigate for signal behavior is hypothesis testing. Consider that a *known* signal is suspected of being embedded in noisy data. Two cases are possible. Suppose first that the data do *not* contain the signal. Then the observations x_i are normally distributed with zero mean (the noise has no steady value) and covariance matrix ϕ_{ij}. The probability density is

$$f_0 \propto \exp[-\tfrac{1}{2}x_i\phi_{ij}x_j], \qquad (19)$$

since x_i is then just the noise n_i. Suppose second that the data *do* contain the signal s, then x_i are again normally distributed but now with mean s_i and covaiance ϕ_{ij}. The probability density is now

$$f_1 \propto \exp[-\tfrac{1}{2}(x_i - s_i)\phi_{ij}(x_j - s_j)]. \tag{20}$$

The likelihood ratio $L(\mathbf{x})$ is formed by taking

$$L(\mathbf{x}) = f_1(\mathbf{x})/f_0(\mathbf{x})$$
$$= \exp[-\tfrac{1}{2}(-2x_i\phi_{ij}s_j + s_i\phi_{ij}s_j)]. \tag{21}$$

The presence of a known signal s in the data is then regarded as determined whenever $L > L_0$, where L_0 is some preassigned decision criterion; we regard the presence of the signal to be of maximum likelihood relative to the absence of signal.

Clearly, the separation of data into a signal and noise is determined by the selection (decision) criterion L_0. This is in line with the previous sections of this chapter where the autocorrelation of data with a predetermined signal behavior forms the basis for deciding whether the signal can be extracted from the data and at what level of significance. There must always be some decision rule applied, as has been so clearly demonstrated by Robinson (1967).

VI. Prediction Filters

Here the concern is to take a set of data with a signal known for all times less than t, and to use the data to predict the expected signal at later times or to determine a better approximation to the signal in regions where the signal might be missing or "confused" for some reason. This method basically designs a filter based on past data to apply to future data in order to extract an expected signal. The maximum entropy method of Burg (1967) is one such filter device. The Burg (1967) algorithm uses concepts from the bailiwick of frequency domain methods (i.e., Fourier transforms), which will be discussed in detail later. For the moment, only those portions of Fourier transforms central to the argument will be used in this discussion. It is not crucial to the argument that a thorough knowledge of Fourier transforms be at hand. In the isotope chronostratigraphic situation, our emphasis in respect to this problem can be simply stated—given the first $n + 1$ values, $\rho_0, \rho_1, \ldots, \rho_n$, of a uniformly sampled Fourier transform of a real positive function, what can we say about Fourier terms of higher order?

We first show that ρ_{n+1} must lie within a circle in the complex plane, which has a radius and centre that can be calculated from $\rho_0, \rho_1, \ldots, \rho_n$. Thus, ρ_{n+1} has known limits. For a value of ρ_{n+1} outside this circle, there is no positive function with a Fourier transform that passes through the values $\rho_0, \rho_1, \ldots, \rho_{n+1}$.

The known values $\rho_0, \rho_1, \ldots, \rho_n$ constrain higher order terms to sucessively larger, calculable, areas in the complex plane. If these areas are small enough, we may decide that $\rho_{n+1}, \ldots, \rho_{n+m}$ are determined to sufficient accuracy, and it is therefore unnecessary to measure them.

Now, in view of the residual uncertainty in each ρ_{n+m} $(m \geqslant 1)$, how do we select a value for each within its "allowable" range? It will be shown that *any* choice of ρ_{n+1} places additional constraints on all higher order terms, but if we estimate ρ_{n+1} by selecting the center of its *circle of constraint*, this imposes the minimal additional constraint on ρ_{n+m} $(m \geqslant 2)$. The value of ρ_{n+2} then lies within a calculable circle, and if we select the centre of this circle as our estimate of ρ_{N+2}, once again the additional constraints on higher terms are minimized. The process can be iterated indefinitely. This prescription provides the greatest likelihood that a measurement of ρ_{n+m} (where m is an arbitrarily chosen integer $\geqslant 1$) will fit with lower order terms to form an acceptable autocorrelation sequence. Choosing the center of the circle of constraint for each successive higher order term leads to the maximum entropy solution.

A. The Fourier Transform of a Positive Function

If $F(x)$ is a real, positive function, we may write

$$F(x) = f(x) \cdot f^*(x) \equiv |f(x)|^2, \tag{22}$$

where $f(x)$ is any arbitrary complex function and an asterisk denotes complex conjugation. Then the Fourier transform of F is the autocorrelation of g, where g is the Fourier transform of f. If we are interested in $F(x)$ for only those values of x within the range $-x_0/2 \leqslant x \leqslant x_0/2$ (x_0 is a constant), the autocorrelation function need be specified only at discrete intervals yielding an autocorrelation sequence with mth term

$$\rho_m = \sum_{k=-\infty}^{\infty} g_k g_{k-m}^*, \tag{23}$$

where g_k is the discrete Fourier transform of $f(x)$. This expression can be regarded as the scalar product of two complex vectors. Consider a set of unit vectors such that the jth vector $\mathbf{V}_j$ has as its lth component g_{j+l}. It can be shown that the mth term of the autocorrelation sequence may be written as the scalar product of the jth and $(j-m)$th vectors

$$\rho_m = \mathbf{V}_j \cdot \mathbf{V}_{j-m}^*. \tag{24}$$

Since the vectors $\mathbf{V}_j$ are unit vectors, $\rho_0 = 1$. In fact it may be shown that the numbers $\rho_0, \rho_1, \ldots, \rho_n$ are successive members of a normalized autocorrelation sequence if and only if a set of $n+1$ unit vectors exists such that Eq. (24) is satisfied for all j and m.

Without loss of generality, we may write

$$\mathbf{V}_j = \sum_{k=1}^{n} \alpha_{k,n} \mathbf{V}_{j-k} + \beta_n \mathbf{\epsilon}_0, \tag{25}$$

where β_n and the $\alpha_{k,n}$ are constants and the unit vector $\mathbf{\epsilon}_0$ is such that

$$\mathbf{\epsilon}_0 \cdot \mathbf{V}_{j-p} = 0, \qquad 1 \leq p \leq n. \tag{26}$$

It then follows that

$$\mathbf{V}_j \cdot \mathbf{V}_{j-p}^* = \sum_{k=1}^{n} \alpha_{k,n} \mathbf{V}_{j-k} \cdot \mathbf{V}_{j-p}^* + \beta_n^2 \delta_{p,0}, \qquad 0 \leq p \leq n, \tag{27}$$

where $\delta_{a,b}$ is the standard Kronecker δ function, unity if $a = b$, zero otherwise. In view of Eq. (24) we may write

$$\rho_p - \sum_{k=1}^{n} \alpha_{k,n} \rho_{p-k} = \beta_n^2 \delta_{p,0}, \qquad 0 \leq p \leq n. \tag{28}$$

Knowing the values of $\rho_0 (\equiv 1), \rho_1, \rho_2, \ldots, \rho_n$, we may solve the $n+1$ equations represented by Eq. (28) for the $n+1$ unknowns β_n^2, and $\alpha_{k,n}$ ($1 \leq k \leq n$), yielding values that are independent of j. Accordingly, in place of Eqs. (25) and (26), we may write

$$\mathbf{V}_{j+1} - \sum_{k=1}^{n} \alpha_{k,n} \mathbf{V}_{j+1-k} = \beta_n \mathbf{\epsilon}_1', \tag{29}$$

where

$$\mathbf{\epsilon}_1' \cdot \mathbf{V}_{j+1-p}^* = 0, \qquad 1 \leq p \leq n. \tag{30}$$

The unit vector $\mathbf{\epsilon}_1'$ may be written quite generally as

$$\mathbf{\epsilon}_1' = \lambda_{1,1} \mathbf{\eta} + \lambda_{1,2} \mathbf{\epsilon}_1, \tag{31}$$

where the unit vector $\mathbf{\eta}$ is a linear combination of the unit vectors $\mathbf{V}_j, \mathbf{V}_{j-1} \ldots, \mathbf{V}_{j-n}$, such that

$$\mathbf{\eta} \cdot \mathbf{V}_{j+1-k}^* = 0, \qquad 1 \leq k \leq n; \tag{32}$$

and the λ's are arbitrary constants limited as given below.

This defines $\mathbf{\eta}$ uniquely with

$$\mathbf{\eta} = \beta_n^{-1} \left[\mathbf{V}_{j-n} - \sum_{k=1}^{n} \alpha_{k,n}^* \mathbf{V}_{j-n+k} \right]. \tag{33}$$

The unit vector $\boldsymbol{\varepsilon}_1$ satifies the relation

$$\boldsymbol{\varepsilon}_1 \cdot \mathbf{V}^*_{j+1-q} = 0, \qquad 1 \leq q \leq n+1; \tag{34}$$

and the numbers $\lambda_{1,1}$ and $\lambda_{1,2}$ satisfy $|\lambda_{1,1}|^2 + |\lambda_{1,2}|^2 = 1$. From Eqs. (29), (31), and (33) we have

$$\mathbf{V}_{j+1} - \sum_{k=1}^{n} \alpha_{k,n} \mathbf{V}_{j+1-k} = \lambda_{1,1} \left[\mathbf{V}_{j-n} - \sum_{k=1}^{n} \alpha^*_{k,n} \mathbf{V}_{j-n+k} \right]$$
$$+ \lambda_{1,2} \beta_n \boldsymbol{\varepsilon}_1. \tag{35}$$

Taking the scalar product of Eq. (35) with $\mathbf{V}^*_{j-n}$, we obtain

$$\rho_{n+1} - \sum_{k=1}^{n} \alpha_{k,n} \rho_{n+1-k} = \lambda_{1,1} \beta_n^2. \tag{36}$$

An essentially identical argument yields

$$\rho_{n+2} - \sum_{k=1}^{n} \alpha_{k,n} \rho_{n+2-k} - \lambda_{1,1} \left[\rho_1 - \sum_{k=1}^{n} \alpha^*_{k,n} \rho_k \right]$$
$$= \lambda_{2,1} (1 - |\lambda_{1,1}|^2) \beta_n^2, \qquad |\lambda_{2,1}| \leq 1. \tag{37}$$

Equations (36) and (37) define the constraints imposed on ρ_{n+1} and ρ_{n+2} by the known values of $\rho_0, \rho_1, \ldots, \rho_n$. Any value of $\lambda_{1,1}$ (such that $|\lambda_{1,1}| \leq 1$) substituted into Eq. (36) yields an acceptable value of ρ_{n+1}, and any combination of $\lambda_{1,1}$ and $\lambda_{2,1}$ such that $|\lambda_{1,1}| \leq 1$ and $|\lambda_{2,1}| \leq 1$ yields an acceptable value of ρ_{n+2} when substituted into Eq. (37).

B. Extrapolating the Fourier Transform

The foregoing argument has shown that, given $\rho_0, \rho_1 \cdots \rho_n$, and hence $\alpha_{k,n}$ and β_n from Eq. (28), we can write

$$\rho_{n+1} = \sum_{k=1}^{n} \alpha_{k,n} \rho_{n+1-k} + \lambda_{1,1} \beta_n^2, \tag{38}$$

$$\rho_{n+2} = \sum_{k=1}^{n} \alpha_{k,n} \rho_{n+2-k} + \lambda_{1,1} \left(\rho_1 - \sum_{k=1}^{n} \alpha^*_{k,n} \rho_{k+1} \right)$$
$$+ \lambda_{2,1} (1 - |\lambda_{1,1}|^2) \beta_n^2, \qquad |\lambda_{1,1}| \leq 1, |\lambda_{2,1}| \leq 1. \tag{39}$$

Equation (38) tells us that ρ_{n+1} lies somewhere within a circle in the complex plane, the center of the circle being at

$$R_{n+1} \equiv \sum_{k=1}^{n} \alpha_{k,n} \rho_{n+1-k}$$

amd its radius being β_n^2. The value of β_n^2 depends, of course, on the nature of the power spectrum, and it may be shown that β_n^2 either decreases or remains constant as n increases. Clearly, the smaller β_n^2, the more accurately we know ρ_{n+1}.

In the absence of further information, we have no reason to prefer any one point within the circle as representing the true value of ρ_{n+1}. However, consider the result of assigning to ρ_{n+1} its extreme value, that is, consider the result of setting $|\lambda_{1,1}| = 1$, with the phase of $\lambda_{1,1}$ fixed. This constrains ρ_{n+1} to a particular, but arbitrary, value, as can be seen from Eq. (38). Furthermore, from Eq. (39), ρ_{n+2} is also determined, as are all terms of higher order. If we had some independent way of determining the value of ρ_{n+10} (for example) we have no reason to expect agreement with the value obtained by extrapolating from n with $|\lambda_{1,1}| = 1$.

However, if we set $\lambda_{1,1} = 0$, ρ_{n+2} is no longer completely constrained and ranges within values determined from Eq. (39). The range of values within which ρ_{n+2} can lie is largest when $\lambda_{1,1} = 0$. Similarly, it can be shown that if $\lambda_{1,1} = 0$, and also $\lambda_{2,1} = 0$, the range of values within which ρ_{n+3} can lie is greatest. By sequentially setting each term to the middle of its permitted range, we maximize the possible range of all higher order terms. If we had some independent method of determining ρ_{n+10} (for example), this procedure provides our best chance that its value will lie within the range permitted by the lower terms.

Setting $\lambda_{1,1} = 0$ leads to the iterative algorithm

$$\rho_{n+m} = \sum_{k=1}^{n} \alpha_{k,n} \rho_{n+m-k}.$$ (40)

This procedure is illustrated later.

In general, the constraints on ρ_{n+m} weaken as m increases. Thus, it can be shown that

$$\rho_{n+m} - \sum_{k=1}^{n} \alpha_{k,n} \rho_{n+m-k} = |\beta_n|^2 S_{n+m}, \qquad m \geq 1,$$ (41a)

where S_{n+m} satisfies

$$S_{n+m} = \boldsymbol{\varepsilon}_n^* \cdot \boldsymbol{\varepsilon}_m + \sum_{k=1}^{n} \alpha_{k,n}^* S_{n+m+k}, \qquad m \geq 1,$$ (41b)

where $\boldsymbol{\varepsilon}_n$ is an arbitrary complex unit vector. It can also be shown that

$$\rho_{n+m} - \sum_{k=1}^{n} \alpha_{k,n}^* \rho_{n+m+k} = |\beta_n|^2 P_{n+m}, \qquad m \geq 1,$$ (42a)

where P_{n+m} satisfies

$$P_{n+m} = \boldsymbol{\varepsilon}_n^* \cdot \boldsymbol{\varepsilon}_m + \sum_{k=1}^{n} \alpha_{k,n} P_{n+m-k}, \qquad m \geq 1.$$ (42b)

Equations (41) and (42) can be solved leading to expressions for S_{n+m} and P_{n+m} in terms of sums of the $\alpha_{k,n}$ and products of two $\boldsymbol{\varepsilon}$ factors. Here, we note that Eqs. (41) and (42) can be used as the basis for recursive algorithms expressing the uncertainty in ρ_{n+m} in terms of the ρ_{n+m-k} ($n + m > k \geqslant 1$). This point has already been anticipated by the illustration involving ρ_{n+1} and ρ_{n+2}.

C. A Numerical Example

An application of the procedure is illustrated in Fig. 88. A nonnegative function, sampled at 512 points, is illustrated in Fig. 88(a). The Fourier transform has been calculated for $\rho_0, \ldots, \rho_{32}$ by the usual Fourier inversion technique, and the values of $\rho_{33}, \ldots, \rho_{512}$ have been estimated by the extrapolation method just outlined. Transforming the result yields

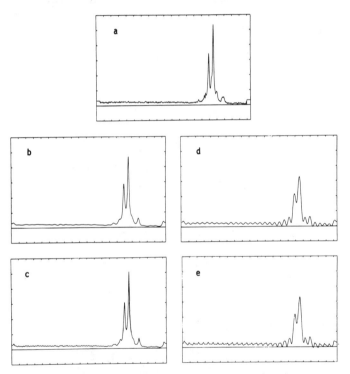

Fig. 88. (a) A 512-point measured spectrum. (b) For the data of (a), Fourier components $\rho_0, \ldots, \rho_{32}$ were calculated by Fourier inversion, and the missing higher order terms were estimated by the method outlined in the text. The result was then retransformed. (c) Fourier components $\rho_0, \ldots, \rho_{32}$ and $\rho_{64}, \rho_{96}, \rho_{128}, \rho_{160}, \rho_{192}, \rho_{224}$, and ρ_{256} were calculated by Fourier inversion, and the missing terms were estimated as in (b). (d) and (e) Fourier transform truncated as for (b) and (c), respectively, and the missing terms set to zero (after Komesaroff and Lerche, 1979).

Fig. 88(b). In the case of Fig. 88(c), the usual Fourier inversion technique has been used to calculate $\rho_0, \ldots, \rho_{32}$ and, in addition, $\rho_{64}, \rho_{96}, \rho_{128}, \rho_{160}, \rho_{192}, \rho_{224},$ and ρ_{256}. The missing values have been estimated from the extrapolation method. For comparison, the effects of merely truncating the transform to the values indicated are shown in Figs. 88(d) and (e).

D. Relation to Maximum Entropy

If we know the ρ_m (all m) *exactly*, then

$$Q(x) \equiv \sum_{m=-\infty}^{\infty} \rho_m \exp \frac{2\pi imx}{x_0}$$

is, of course, just proportional to $F(x)$. But if we use Eq. (40) to *estimate* $\rho_{m+n}(m > 1)$, the resulting estimate for $Q(x)$ is equivalent to the maximum entropy spectral estimate. Equation (40) is equivalent to demanding

$$\boldsymbol{\varepsilon}'_m \cdot \boldsymbol{\varepsilon}^*_0 = \delta_{m,0}\beta^2_n, \tag{43}$$

where

$$\boldsymbol{\varepsilon}'_m = \mathbf{V}_{j+m} - \sum_{k=1}^{n} \alpha_{k,n}\mathbf{V}_{j+m-k}. \tag{44}$$

By introducing the change in nomenclature,

$$\gamma_k = \delta_{k,0} - \alpha_{k,n}(1 - \delta_{k,0}), \qquad 0 \leq k \leq n, \tag{45}$$

then

$$\sum_{k=0}^{n} \gamma_k \mathbf{V}_{j+m-k} = \beta_n \boldsymbol{\varepsilon}'_m, \tag{46}$$

and thus from Eqs. (43) and (44),

$$\sum_{m=-\infty}^{\infty} \sum_{k=0}^{n} \gamma_k \sum_{l=0}^{n} \gamma^*_l \rho_{m-k+l} \exp \frac{2\pi imx}{x_0} = \beta^2_n. \tag{47}$$

With the replacement $M = m - k + l$, we have

$$\sum_{M=-\infty}^{\infty} \rho_M \exp \frac{2\pi iMx}{x_0} = \beta^2_n \left| \sum_{k=0}^{n} \gamma_k \exp \frac{2\pi ikx}{x_0} \right|^{-2}, \tag{48}$$

which is identical with the maximum entropy algorithm.

Many authors have written on the subject of maximum entropy. In particular Burg (1967) and Van den Bos (1971) have discussed its connection with the positivity constraint. The papers of Ables (1974) and Newman (1977) are also particularly clear expositions of the method.

VII. Relative Rates of Sedimentation

A variation of the procedure in Chapter 5 can be used to address the problem of converting isotope chronostratigraphic records, measured with respect to depth, into records measured with respect to time. If the sedimentation rate and compaction velocity were constant from site to site, then measurements at depth Z_j would correspond precisely to a fixed time t_j for each and every location. However, if the sedimentation or compaction rates vary from location to location, then an identification of events at the same geologic time in the same or different stratigraphic columns becomes difficult. The problem is that the same depth on spatially distinct records corresponds to different times. To determine the best variation in velocity becomes a question of obtaining the best semblance measure from the recorded sequence of isotope records.

To describe the sense of the argument, suppose a set of isotope records of the same type have been measured with respect to sedimentary depth. To convert the measured depth records into records with respect to sedimentary age we must either independently date the formations or determine the rate of sedimentation and compaction with time. Assuming that we do not have very many radiometrically determined ages, we wish to use the isotopic record itself to determine a depth-to-time conversion for the formations between known age horizons. Suppose then that we have a set $x(\mathbf{L}, Z_i)$ $(i = 1, 2, \ldots, N)$ of isotopic records at several neighboring spatial locations on the surface of the earth and at N equal depth intervals ΔZ, so that $Z_j = j\,\Delta Z$, where Z_j is the depth location of the jth sample at each site. The *relative* distortion of the different records can then be corrected. Let the rate of sedimentation with time at location $\mathbf{L}$ be $V(\mathbf{L}, t)$, so that a current depth $Z(\mathbf{L})$ corresponds to a depositional time T given by

$$Z(\mathbf{L}) = \int_0^T V(\mathbf{L}, t)\, dt. \tag{49}$$

For a record at location $\mathbf{L}^1$, with depositional rate $V(\mathbf{L}^1, t)$, the depositional time T corresponds to the current depth $Z(\mathbf{L}^1)$ given by

$$Z(\mathbf{L}^1) = \int_0^T V(\mathbf{L}^1, t)\, dt. \tag{50}$$

It follows that

$$Z(\mathbf{L}^1) = Z(\mathbf{L}) + \int_0^T [V(\mathbf{L}^1, t) - V(\mathbf{L}, t)]\, dt,$$

$$\equiv Z(\mathbf{L}) + \int_0^T \Delta V(\mathbf{L}, \mathbf{L}^1, t)\, dt. \tag{51}$$

Now, if the only difference in the isotopic records at L and L^1 is because of differences in rates of sedimentation with time at the two locations, we can obtain the relative rate of sedimentation between the two locations.

This is accomplished by first taking the discrete (spatial) Fourier transform (DFT) of each record (but see Section III of this chapter for definitions) so that

$$X[L, Z(L)] = \sum_{k=-N}^{N} x(L, k) \exp\left[\frac{i\pi z(L)k}{N \, \Delta Z}\right], \qquad (52a)$$

and

$$X[L^1, Z(L^1)] = \sum_{k=-N}^{N} x(L^1, k) \exp\left[\frac{i\pi Z(L^1)k}{N \, \Delta Z}\right], \qquad (52b)$$

where $x(L, k)$ $[x(L^1, k)]$ is the DFT of $x(L, Z)$ $[x(L^1, Z)]$ and $N \, \Delta Z$ is the record length. Then, if we replace $Z(L^1)$ by Eq. (51) in the expression for $x(L^1, Z)$, we produce

$$X[L^1, Z(L^1)] = \sum_{k=-N}^{N} x(L^1, k) \exp\left[\frac{i\pi Z(L)k}{N \, \Delta Z}\right]$$
$$\times \exp\left[i\pi k(N \, \Delta Z)^{-1} \int_0^T \Delta V(L, L^1, t) \, dt\right]. \qquad (53)$$

In order to extract $\Delta V(L, L^1, t)$, we then proceed in a conventional manner.

To construct the semblance of $x(L, Z)$ and $x[L^1, Z(L^1)]$ for short times T, replace $\int_0^T \Delta V \, dt$ by $\sum_{m=0}^{M} \Delta V_m \, \Delta t$ where the record has been split into equal time intervals Δt, with $M \, \Delta t = T$. Then, for a fixed value of M, by systematically increasing ΔV_m, we can find the value that will maximize the semblance. This procedure is repeated for each and every measurement point (corresponding to different depths and so different times). Gradually, we are able to construct, by this means, a sequence of values of ΔV_m, for increasing M as the depth sequence of points is increased. Then, the relative velocity of sedimentation at any time is determined. The *absolute* velocity of sedimentation cannot be determined by this method for we use one record (or some average over all records) to supply the template for comparison against which the other records at different spatial locations are tested for distortion. Thus, we need to obtain some measure of absolute age dating in order to convert the template record (measured with respect to depth) into an equivalent template record in time.

However, this method will (a) remove relative distortions between records and (b) provide a means of determining the relative rate of sedimentation and compaction with time between spatial locations.

An alternative method is available. The problem of correlating two or more isotopic records when one has been compressed or expanded is also amenable to inverse methods of data analysis using mapping function techniques. Indeed, Martinson *et al.* (1982) have followed precisely this route to relate coherence of two signals measured at different locations with different sedimentation rates. They were able to determine the distortion of one isotopic record relative to another and so obtained relative sedimentation rates.

In order to understand the sense of their argument, consider two records R_1 and R_2 measured with respect to sedimentary depth below the sediment-water interface. Let us use $R_1(Z)$ as a reference signal, so that we are interested in obtaining the best correlation of R_2 with respect to R_1. Let $R_1(Z_1)$ be related to R_2 through a mapping function $Z_2(Z_1)$, so that $R_2[Z_2(z_1)] = R_1(Z_1)$. The question is to obtain a method of finding the diffeomorphic map from Z_2 to Z_1, which maximizes the degree of agreement of the two records. Again the general underpinning logic of the method is least squares minimization. Martinson *et al.* (1982) considered a mapping function made up of a linear trend plus a truncated Fourier series in the form

$$Z_2(Z_1) = a_0 Z_1 + \sum_{k=1}^{n-1} a_k \sin\left(\frac{\pi k Z_1}{Z_{max}}\right), \tag{54}$$

where Z_{max} is the length of the record R_1. The problem is reduced to finding the coefficients $a_0, a_1, \ldots, a_{n-1}$, which will minimize the difference between the reference signal R_1 and the signal R_2 distorted relative to R_1.

First, the two signals are formally normalized so that

$$r_1(Z_1) = R_1(Z_1)\left[\int_0^{Z_{max}} R_1(Z)^2\, dZ\right]^{-1/2}, \tag{55a}$$

$$r_2(Z_2) = R_2(Z_2)\left[\int_0^{Z_{max}} R_2(Z_2(Z_1))^2\, dZ_1\right]^{-1/2}. \tag{55b}$$

Then, a least squared error is defined by

$$E(\mathbf{a}) = \int_0^{Z_{max}} \{r_1(Z_1) - r_2[Z_2(Z_1, \mathbf{a})]\}^2\, dZ_1, \tag{56}$$

where $\mathbf{a} = (a_0, a_1, \ldots, a_{n-1})$. This error is independent of magnitude differences between the records but is sensitive to the distortion of shape between the records. The clever observation made by Martinson *et al.* (1982) is that E is directly related to the coherence C between the two records by

$$E = 2(1 - C). \tag{57}$$

Thus, the shape error is minimized when the coherence is maximized with

$$C = \left\{ \int_0^{Z_{max}} R_1(Z_1) R_2[Z_2(Z_1)] \, dZ_1 \right\}$$
$$\times \left\{ \left[\int_0^{Z_{max}} R_1(Z_1)^2 \, dZ_1 \right] \left[\int_0^{Z_{max}} R_2(Z_2(Z_1))^2 \, dZ_1 \right] \right\}^{-1/2}. \tag{58}$$

The determination of the coefficients in the mapping function now proceeds in a conventional manner. Pick an initial set of values a_i. Then, if $\mathbf{a}$ be perturbed from $\mathbf{a}_i$ by $\delta \mathbf{a}$, we have $C(\mathbf{a}_i + \delta \mathbf{a})$ replacing $C(\mathbf{a}_i)$. Then, with ΔC as the change in coherence, we have

$$\Delta C \simeq \delta \mathbf{a}_i \cdot \nabla_{\mathbf{a}} C, \tag{59}$$

where $\nabla_{\mathbf{a}} C \equiv (\partial C/\partial a_1, \partial C/\partial a_2, \ldots, \partial C/\partial a_n)$. But we want the increment in $\delta \mathbf{a}$ to be such that the coherence heads most rapidly to its maximum value. Thus, $\delta \mathbf{a}$ should be in the direction of $\nabla_{\mathbf{a}} C$, which is the normal direction to contours of constant coherence. Then,

$$\delta \mathbf{a} = \Delta C \, \nabla_{\mathbf{a}} C / |\nabla_{\mathbf{a}} C|^2. \tag{60}$$

But, as Martinson et al. (1982) noted, this form of $\delta \mathbf{a}$ is of limited use since the coefficients $\mathbf{a}$ can be changed iteratively to make $\nabla_{\mathbf{a}} C$ tend to zero and so drive $\delta_{\mathbf{a}}$ unstable. They overcome this problem by making the vector $\delta_{\mathbf{a}}$ of fixed magnitude k. Then,

$$\Delta C = k^2 |\nabla_{\mathbf{a}} C|^2 / [\nabla_{\mathbf{a}} C \cdot \delta_{\mathbf{a}}]. \tag{61}$$

This process also aids convergence since iterations are continued as long as C increases and ΔC is decreasing. Resolution is maximal where ΔC begins to decrease. Martinson et al. also showed that the error in the mapping function $Z_2(Z_1)$ at any depth z_i is given as follows. The mapping is linear in the $\mathbf{a}$'s in the sense that at any depth $Z_1 = Z_i$,

$$Z_2(Z_i) = a_j F_{ij},$$

where F_{ij} is the coefficient of a_j evaluated at depth $Z_1 = Z_i$ in the map. Then, the error σ_i in the map at depth Z_i is

$$\sigma_i = [(\text{cov } Z_2)_{ii}]^{1/2}, \tag{62}$$

where $(\text{cov } Z_2)$ is the covariance matrix of Z_2 at $Z_1 = Z_i$. Numerous synthetic examples and case histories have been provided by Martinson et al. (1982) to illustrate the strength and accuracy of the method.

The problem of determining the best n number of coefficients $a_0, \ldots, a_{n-1}$ to use is addressed in the conventional manner. As the number of coefficients is increased, there comes a point where the uncertainty in the determination of the nth coefficient is as large as the coefficient itself. At this point the noise limit has been reached and there is no reason to continue to add coefficients. In short, we have available a method of estimating a mapping function that will remove distortions of one isotopic record relative to another—which latter is regarded as a template. Absolute conversion of depth to geologic time then requires age dating of the template sedimentary column.

One point, however, should be mentioned. Since the distortion and mapping function enter nonlinearly, there is no guarantee that the mapping function is unique. The resolution and precision of given choices of mapping functions can be, and have been, addressed using the procedure advocated by Martinson *et al.* (1982), but the question of uniqueness remains an outstanding concern.

Indeed, to put the point differently, Matthews (1984a) has boldly stated, "To a first approximation, many deep sea cores can be regarded as having a constant sedimentation rate. To a first approximation this rate can be estimated from carbon-14 dating of the upper portion of the core." Extrapolation to lower parts of the core with the assumed constant sedimentation rate is then done to provide a depth-to-time conversion. But how do we *know*, and not assume, that we can regard the sedimentation rate as constant? What control do we have on the validity and accuracy of such extrapolations to lower parts of the core? Clearly, unless we have some form of age dating as a control at deeper depths, we have nothing except that which we believe to be true but cannot justify. Even when such age-dated depth horizons are available, we have no knowledge of variability of sedimentation rates on a scale finer than the sampled age-dated horizons, even though the isotopic records may be sampled on a much finer scale. What portion of the isotopic record we ascribe to intrinsic isotopic variations and what portion we ascribe to variations in sedimentation rate are then matters of emotion rather than, more properly, matters of quantitative science. It is all too easy to impose on the isotopic record, either consciously or subconsciously, a variation that we hope will agree with some preconceived notion, such as orbital cycle effects of the earth. One cannot resolve the signal from the noise (sedimentation variability) on a time scale finer than that of the age-dated horizon samples—playing in the cracks between the piano keys produces a tune only for the deaf.

VIII. Matched Filters

Here the problem is to design a filter that has a response that is the same shape as the signal $S(t)$ but reversed in time. The advantage to this sort of filter is that the resulting signal-to-noise ratio is massively enhanced (Tukey,

1959; Aki, 1960; Capon, *et al.*, 1969) since the filtered response to the input signal is

$$S_{\text{out}}(t) = \sum_{i=0}^{N-1} S(t-i)S(-i), \qquad (63)$$

while the filtered response to noise is

$$n_{\text{out}}(t) = \sum_{i=0}^{N-1} n(t-i)S(-i). \qquad (64)$$

The filtered noise relative to the input noise is then enhanced by

$$\frac{\langle n_{\text{out}}(t)^2 \rangle}{\langle n(t)^2 \rangle} = (2W)^{-1} \sum_{i=0}^{N-1} S(i)^2, \qquad (65)$$

while the output signal *amplitude* is enhanced by the value

$$\sum_{i=0}^{N-1} S(i)^2 / \max[S(i)^2]. \qquad (66)$$

thus the gain G in signal to noise *power* is

$$G = 2W \sum S(i)^2 / \max[S(i)^2] = O(NW) \gg 1, \qquad (67)$$

where W is the frequency bandwidth taken to cover the range where the Fourier transform of the time signal exists and $O(x)$ means "of the order of x." Gains of 6–20 db are not uncommon (Capon, 1969) using the matched filter approach.

Chapter 14 | Frequency Domain Methods

The earth's response to past isotopic variations with time, as measured by isotopic variations taken today with depth, is complicated by many effects, including biological selection and preference of organisms for particular isotopes, chemical differentiation, thermal variations, subsurface fluid transport, diagenetic effects, sedimentary motion and compaction, and bioturbation. Some of these effects are to be considered as signals, since they carry information that enables the potential recovery of intrinsic isotopic variations with time, and some are to be considered as noise since they either have no discernible relation to any basic process of interest or they present too complex a signature for any understanding of this complexity as a signal to be achieved.

Often, by eye, we recognize some of the simpler patterns of behavior, visible as correlatable wiggles, rises, periodic components, etc. But the identification of more subtle patterns of signals, and the quantification of the simpler patterns, requires that we develop schemes to enhance signal identification in the presence of noise.

Frequency domain methods are probably the most powerfully developed methods of analysis to date and offer the greatest chance of signal identification and resolution. With the advent of high-speed digital computers and fast Fourier transform (FFT) routines, filtering operations in wavenumber and frequency space became accessible and are now routinely performed. The underlying reason is that the calculation of a time autocorrelation is slow, whereas, since the basic response is usually taken to be of the form

$$x(t) = \int_{-\infty}^{t} s(\tau)f(t-\tau) \, d\tau, \tag{68}$$

173

the convolutional type of autocorrelation or cross-correlation, it follows that in Fourier space we have $x(\omega) = s(\omega)f(\omega)$, for any angular frequency ω. Thus, since multiplicative operations are easier and quicker to perform, considerable effort has been expended in improving frequency domain analysis methods over the last quarter century.

Perhaps the pioneering work of Blackman and Tukey (1959) set the stage for the most common methods of frequency analysis in use today. Similar to their approach we start with an overview of noise analysis and then consider autocorrelation, cross-correlation, and spectral ratio methods of signal analysis with filtering.

I. Noise Frequency Spectra

We deal throughout this chapter with the situation in which measurements are made at equal time intervals.[3] Isotope chronostratigraphic measurements rarely follow this behavior but can be incorporated within an equal time frame by filling in the "blank" areas with zeros (as is conventionally done with other data records in order to compute fast Fourier transforms), by appropriate filtering (see later), or within the framework of entropy resolution methods in the time domain (also see later).

To illustrate and exemplify the problems, we deal in this subsection with N measurements, $x(n)$, of noise in the absence of a signal, at times $n \, \Delta t \, (n = 0, 1, \ldots, N-1)$ extending out to a total time interval $T = N \, \Delta t$. Each measurement is considered to be statistically independent of every other measurement no matter how short the time interval Δt, i.e., no coherent noise trend is present even for a short time interval. It is considered that repetitive measurements of $x(n)$ would yield a mean of zero and a variance σ^2 (Gaussian noise).

The discrete Fourier transform (DFT) is constructed:

$$X(k) = \sum_{m=0}^{N-1} x(m) \exp(ikm \, \Delta t \, \Delta \omega), \qquad k = 0, 1, 2, \ldots, N-1, \quad (69)$$

where $\Delta \omega = 2\pi / T$, and $X(k)$ is the kth term of the DFT in frequency space. Note that the transform repeats (aliases) at the angular Nyquist frequency $N \, \Delta \omega$.

Then, on the approximations made, it can be shown that the average value

$$\langle |X(k)|^2 \rangle = N(\sigma \, \Delta t)^2. \tag{70}$$

[3] Isotopic records are rarely taken this way. Often, however, we do have equal depth intervals, and the arguments are then identical if for time we read depth. However, since the frequency domain methods have been developed for equal time intervals, we stay with this notation so that readers interested in a more detailed treatment can read the literature without having to perform a mental gymnastic conversion.

We define the discrete power spectrum as

$$P(k) = \lim_{N \to \infty} |X(k)|^2/(N \, \Delta t) = \sigma^2 \, \Delta t, \tag{71}$$

which is independent of frequency (white noise) and independent of the record length $N \, \Delta t$ as $N \to \infty$.

Suppose that we pass the noise $x(n \, \Delta t)$ through a filter $f(t)$ with

$$n(t) = \sum_{i=0}^{N-1} x(t-i) f(i) \, \Delta t, \tag{72}$$

then the filtered noise power spectrum is

$$P(k) = |F(k)|^2 \sigma^2 \, \Delta t, \tag{73}$$

where $F(k)$ is the discrete Fourier transform of the filter function $f(t)$. Hence, the power spectral density of filtered noise is the product of the unfiltered noise power spectrum and the power spectrum of the filter function. Models of noise power spectra and filtered noise power spectra are often used to examine discretely sampled measurements for the behavior of a suspected signal relative to the assumed noise spectrum.

For instance, in autoregressive noise estimation, one takes the "raw" noise measurements $x(t)$ and constructs a filtered noise $n(t)$ from

$$n(i) + \sum_{m=1}^{P} \alpha(m) n(i-m) = \sum_{l=0}^{Q} \beta(l) x(i-l), \tag{74}$$

where $\alpha(m)$ and $\beta(l)$ are preassigned, with P, Q integer constants, greater than zero. The power spectral density of the filtered noise n is

$$P(k) = \sigma^2 \, \Delta t \left| \sum_{l=0}^{Q} \beta(l) \exp(ikl \, \Delta t \, \Delta \omega) \right|^2$$

$$\times \left| 1 + \sum_{m=1}^{P} \alpha(m) \exp(ikm \, \Delta t \, \Delta \omega) \right|^{-2}. \tag{75}$$

The choice $\beta(l) = 0$ $(l \neq 0)$, $\beta(0) = 1$ yields autoregressive noise filtering corresponding to a minimum delay filter (Aki and Richards, 1980). However, determination of the best $\alpha(m)$ to use can be improved over arbitrary preassigned values, since, if $x(i)$ is indeed unrelated to its past or future values, then $\langle n(i \, \Delta t) x(j \, \Delta t) \rangle = 0$ for $i < j$.

With the autoregressive assumption,

$$n(i) + \sum_{m=1}^{P} \alpha(m) n(i-m) = x(i), \tag{76}$$

multiplying by $n(j)$ and averaging we obtain

$$\sum_{m=1}^{P} \alpha(m)r(k-m) = -r(k), \qquad k = 1, 2, \ldots, P, \tag{77}$$

where $r(k)$ is the autocorrelation function

$$r(k) = \sum_{j=1}^{P} \langle n(j)n(j+k)\rangle \tag{78}$$

with $r(P-j) = r(j)$ (see, e.g., Van Schoonfeld, 1979, and references therein).

Since the autocorrelations $r(k)$ are known, we can solve for the $\alpha(m)$ using Toeplitz banded matrix methods. The Levinson (1949) numerical routine, or one of its derivative methods, is a quick and efficient procedure.

The fundamental reason for even bothering to be concerned with the noise spectra is that the problem of determining of "good" noise spectral estimate plays an important role in deciding the type of filtering operation to use on a set of data suspected of containing a signal. Crudely, one trades reliability of an estimate of the noise power spectrum for frequency resolution. The sense of the argument operates as given next.

Start with M noise measurements $x(j)$ at the time intervals $j\,\Delta t$ ($j = 0, , \ldots, M-1$) over a total record length $T = M\,\Delta t$.

Apply some filter $f(t)$ to the data so that the DFT of the filtered noise $n(t)$ can be written

$$N(k) = X(k)F(k), \tag{79}$$

where $F(k)[x(k)]$ is the DFT of the filter (data). Since we assume $x(j)$ is uncorrelated Gaussian noise, the $N(k)$ is also a statistical Gaussian variable. Then,

$$|N(k)|^2 = |F(k)|^2|X(k)|^2. \tag{80}$$

But each element of $X(k)$ is made up from the DFT

$$X(k) = \Delta t \sum_{l=0}^{M-1} x(l)\cos\left(\frac{2\pi kl}{N}\right) + i\,\Delta t \sum_{l=0}^{M-1} x(l)\sin\left(\frac{2\pi kl}{N}\right), \tag{81}$$

and each linear combination of $x(i)x(j)$ in $|X(k)|^2$ will be zero except for $j = i$ by the independence of $x(i)$. Hence, the chi-squared (χ^2) value

$$\chi^2 = 2|N(k)|^2/[|F(k)|^2 M\sigma^2(\Delta t)^2], \tag{82}$$

obeys the chi-squared distribution with two degrees of freedom since $|N(k)|^2$ is the sum of two (different) squares of Gaussian variables.

Now, a χ^2 distribution with n degrees of freedom has a mean n and an rms (root mean square) value relative to the mean of $(2n)^{1/2}$ (Feller, 1966). Thus, the relative error in the mean is $(2/n)^{1/2}$. For $n = 2$, the relative error is 1, i.e., 100%. To reduce the error in estimating the noise power spectrum, we average the noise power over neighboring frequencies centered on k with the DFT of the filter function chosen to vary slowly over the neighborhood. Over $2p$ points around k we then obtain the χ^2 for $4p+2$ degrees of freedom:

$$\chi^2(2p+1) = 2 \sum_{s=-p}^{p} |N(k+s)|^2/[|F(k)|^2 M\sigma^2(\Delta t)^2]. \tag{83}$$

The fractional rms error in the mean χ^2 is now $(2p+1)^{-1/2}$. The same fractional rms error applies to the filtered noise power spectrum after averaging it over the same $2p$ points in the neighborhood of k, i.e., we obtain a smoothed noise power estimate of

$$P_{\text{smooth}}(k) = [M\,\Delta t(2p+1)]^{-1} \sum_{s=-p}^{p} |N(k+s)|^2. \tag{84}$$

For a fixed fractional error $\delta \equiv (2p+1)^{-1/2}$, we must then smooth the power spectrum over the frequency interval $2\,\Delta k$ around k with

$$\Delta k = (2T)^{-1}(2p+1) \equiv (2T\delta^2)^{-1}, \tag{85}$$

where $T = M\,\Delta T$. For a ruggedly stable noise power estimate p must be large (δ small) compared to unity, so that the noise estimate must be smoothed over a very wide frequency band. For a fixed fractional error in the estimate, the frequency resolution achievable increases with the length of the record, while for a fixed record length (the more common situation) and a fixed frequency resolution, the accuracy of the ruggedly stable noise estimate only improves proportionately to the square root of the record length. The ultimate resolvable frequency width is limited to $2\pi/T$ for a record of length T.

Without making any further assumptions about the noise behavior, no further increase in resolution can be achieved. Finer resolution is achieved only when other assumptions are made, and that increase in resolution hinges critically on the assumptions made. For example, the maximum entropy method (Burg, 1967) assumes an autoregressive nature for the noise and fills in the "missing" frequency spacings consistent with the autoregressive power spectrum determined from a portion of the data. Occasionally, this technique works well but it is known not to be a universal panacea—particularly if the data refuse to be autoregressive or if additional physical constraints have to be satisfied by the data (see, e.g., Van Schoonfeld, 1979, and references therein for examples of such constraints and their impact on the maximum entropy method). But the conclusion derivable from such entropy methods is that they do provide some idea of the maximum allowable variations on resolution consistent with recorded data and, as such, can often point the way to the extra crucial measurements needed to more sharply hone the record for extraction of a suspected signal.

II. Autocorrelation and Cross-Correlation Spectral Methods

A. Autocorrelation Power Spectra

Our concern here is with the identification of a signal $s(t)$ from a set of noisy measurements $x(t)$ of one type taken at one spatial location. [Cross-correlation spectral methods (to be addressed in Section II, B of this chapter) are concerned either with (1) the identification of a common signal at one location from two or more sets of measurements of different types taken at the same location or (2) with the identification of a common signal from two or more sets of measurements of the same or different types at spatially distinct locations.] In order to ease the development for the cross-correlation section, we start here by assuming that some intrinsic signal $s(t)$ has been modulated by an applied filter $w(I, \mathbf{L}, t)$, which is site and type specific. The filter can be a combination of: (1) bias introduced by the data processor, or (2) it may be intrinsically geological in origin (e.g., in the sense that compaction, diagenesis, fluid flow, and thermal effects have modified the otherwise pristine signal). In any event we take the filter to modulate the intrinsic signal $s(t)$ such that the measurements recorded, $x(I, \mathbf{L}, t)$, are a linear combination of the modulated signal plus noise:

$$x(k, t) = W(k, t) * s(t) + n(k, t), \tag{86}$$

where k stands for the site and type variables $\mathbf{L}$ and I, an asterisk denotes convolution, and $n(k, t)$ defines the noise component.

The power spectrum at angular frequency ω yields the connection

$$P_{xx}(\omega) = |W(\omega)|^2 |S(\omega)|^2 + N(\omega), \tag{87}$$

where, as usual, we have taken the noise to be incoherent and to produce the noise power $N(\omega)$. The subscript of P_{ab} denotes the power spectrum of the a record with the b record. The signal to noise ratio at frequency ω is

$$S(\omega)/N(\omega) = |W(\omega)|^2 |s(\omega)|^2 / N(\omega), \tag{88}$$

while an estimate of the signal to noise ratio of the whole record is obtained from

$$\frac{S}{N} = \frac{\int_{\omega_{\min}}^{\omega_{\max}} |W(\omega)|^2 |s(\omega)|^2 \, d\omega}{\int_{\omega_{\min}}^{\omega_{\max}} N(\omega) \, d\omega}. \tag{89}$$

The problem, as in the time domain analysis, is to provide an estimate of the signal $s(\omega)$. Once again some estimate of the noise behavior is required, or a model signal or filtering behavior need to be defined. Thus, a predictive noise–filter analysis must be invoked. The techniques of the time domain are directly translatable into the frequency domain and provide the required behavior. Since we shall meet such techniques in a more general form when we consider cross-correlation methods, we look here only at the simple situation where we ask for a filter $f(t)$ that, when convolved with the data $x(t)$, will yield the best representation of the signal $s(t)$ in the presence of noise. We assume the noise to be completely uncorrelated so that it has a constant spectral noise power (white noise[4]) over the frequency bands of interest. (This is often achieved by applying white noise filters to the data.) On the assumption of white noise, we have several ways to estimate the noise power: (1) look for the lowest value of $P_{xx}(\omega)$ in the frequency band; (2) multiply $P_{xx}(\omega)$ by known functions of ω and integrate in the frequency band, gradually extracting an estimate of N; or (3) resolve $P_{xx}(\omega)$ into a set of orthogonal functions over the frequency band. The coefficient of the lowest orthogonal function, which we take to be constant, represents an estimate of the noise power. From any of these approaches, one can estimate the noise power. It follows that

$$|S(\omega)|^2 = |W(\omega)|^{-2}[P_{xx}(\omega) - N(\omega)]. \tag{90}$$

A myriad of such methods exist for estimating the noise power. No matter which technique is used, the point is that once the noise power has been defined then the residual power *must*, by definition, represent the signal. Matters improve significantly when more than one record exists over the same time span, and we now consider this aspect.

B. Cross-Correlation Power Spectra

Consider a set of isotopic chronostratigraphic records $x(k, t)$ where k denotes either type or spatial location. We take the records to exist over the same time interval $0 < t < T$. We again take each record to be a linear combination of a filtered intrinsic signal plus noise:

$$x(k, t) = W(k, t) * S(t) + n(k, t). \tag{91}$$

Let $\Phi_{ij}(\omega)$ be the spectral matrix of record i with record j, i.e.,

$$\Phi_{ij}(\omega) = x(i, \omega)x(j, \omega)^*. \tag{92}$$

---

[4] With isotope data, this assumption is likely accurate when the sample size for each spectrum is large (for example, greater than 40) but should be seriously reconsidered when smaller samples are analyzed.

Then,

$$\Phi_{ij}(\omega) = \Psi_{ij}(\omega) + N_{ij}(\omega), \qquad (93)$$

where N_{ij} is the cross-correlated noise power so that Ψ_{ij} represents the cross-correlated power spectral tensor of the signal. On the assumption of *channel independent* noise, we have N_{ij} as a diagonal matrix with energies at $i = j$, so that $N_{ij} = 0$ for $i \neq j$. The spectral matrix of the signal producing part of the record is

$$\Psi_{ij}(\omega) = W_j(\omega)^* W_i(\omega) |S(\omega)|^2. \qquad (94)$$

The signal to noise ratio $\rho_k(\omega)$, of the kth record at angular frequency ω, is then

$$\rho_k(\omega) = |W(k, \omega)|^2 |S(\omega)|^2 / N_k(\omega). \qquad (95)$$

The estimates of cross-coherence of the signal from record i to record j depend upon the complex cross-correlation coherence tensor $\Gamma_{ij}(\omega)$, where

$$\Gamma_{ij}(\omega) = \Phi_{ij}(\omega) / [\Phi_{ii}(\omega)\Phi_{jj}(\omega)]^{1/2}. \qquad (96)$$

The coherence of record i relative to record j is usually thought to be measured by

$$\gamma_{ij}(\omega) = |\Gamma_{ij}(\omega)|^2. \qquad (97)$$

It can be shown that γ_{ij} measures the proportion of power in the vicinity of the angular frequency ω on one of the records that can be predicted by linearly filtering the second record to produce a least squares misfit between the ith and jth records.

It follows that the frequency response of the filter, which converts record j into the least squares best fit to record i, is

$$F(j \to i, \omega) = \Phi_{ji}(\omega) / \Phi_{jj}(\omega), \qquad (98)$$

i.e.,

$$F(j \to i, \omega) = W_i(\omega)[W_j(\omega)(1 + \rho_j(\omega)^{-1})]^{-1}. \qquad (99)$$

In the special case of two records that respond in precisely the same way, $W_i(\omega) = W_j(\omega)$, the frequency response of the filter that converts record j into record i with the least squares best fit reduces to the conventional Wiener suppression filter

$$F(j \to i, \omega) = \rho_j(\omega) / [1 + \rho_j(\omega)]. \qquad (100)$$

More generally we see that the filter [Eq. (99)] is basically a Wiener noise suppression filter in conjunction with a filter that allows for the different response of the ith and jth isotopic chronostratigraphic records to the intrinsic signal. The coherence at frequency ω is directly related to the signal-to-noise power ratios of each record by

$$\gamma_{ij}(\omega) = \{[1 + \rho_i(\omega)^{-1}][1 + \rho_j(\omega)^{-1}]\}^{-1}. \tag{101}$$

A useful measure of the degree of fit of the two records is then provided by the proportion of power on record i, which is predicted by the filtered record j over the whole band of interest, i.e., an integral coherence measure,

$$C_{ij} = \int_0^\infty \gamma_{ij}(\omega)\, d\omega, \tag{102}$$

is a measure of the commonality of signal in the two records.

C. Multiple Cross-Correlation Power Spectra

In the isotropic chronostratigraphic case we have, however, to deal with more than just a pair of records since different spatial sites are occupied and, at the sites, different types of isotopic records are taken with sedimentary depth (and so age).

The multiple coherence function, $\Gamma_k(\omega)$, at the frequency of the kth record in a set of $N + 1$ records can be similarly related to the filter function $F_k(\omega)$, which predicts the proportion of power in the kth record by linearly filtering the other N records to produce a least squares best fit.

It is easy to show that

$$F_k(\omega) = \sum_{j=1}^N \Phi_{jj}^{-1} \Lambda_{jk}(\omega), \tag{103}$$

where Φ_{jj}^{-1} is the inverse of the spectral matrix of the N records and $\Lambda_{jk}(\omega)$ is the column vector of the cross spectra of the jth record $x(j, t)$ against the kth record $x(k, t)$.

The power in the kth record *not* predicted by the N records is related to the estimate of noise in that record. This nonpredicted power is given by

$$N_{kk}(\omega) = \Phi_{kk}(\omega) - \Lambda_{kj}(\omega)\Phi_{jj}^{-1}\Lambda_{jk}(\omega), \tag{104}$$

where N_{kk} is the kth diagonal element in the *inverse* of the full $(N+l) \times (N+l)$ spectral matrix.

The fraction of power predicted in the kth record from the other N records is again given by the coherence with

$$\gamma_k(\omega) = 1 - \Phi_{kk}(\omega)/N_{kk}(\omega). \tag{105}$$

The coherence at angular frequency ω is related to the signal to noise power ratio on the kth record by

$$\gamma_k(\omega) = A_k(\omega)[1 + \rho_k(\omega)^{-1}]^{-1}, \tag{106}$$

where

$$A_k(\omega) = [\rho(\omega) - \rho_k(\omega)][1 + \rho(\omega) - \rho_k(\omega)]^{-1}, \qquad (107)$$

where $\rho(\omega)$ is just the sum of all signal-to-noise ratios on all $N+1$ records; i.e.,

$$\rho(\omega) = \sum_{j=1}^{N+1} \rho_j(\omega). \qquad (108)$$

If we have 3 or more records ($N \geq 2$), the signal-to-noise ratios at each angular frequency of each record can be determined directly from the coherences $\gamma_k(\omega)$, which depend only upon the directly measurable power and cross-correlated power spectra of the $(N+1)$ records. This is a very powerful method of determining the intrinsic signal present in a set of records, and it also enables a good estimate to be made of which records are "noisy" and at what frequencies.

Several "corrections" have to be made to this analysis. First, as we have seen earlier, we have to filter the noise across a broad bandwidth in order to have a statistically reliable estimate of the noise. Second, we have the problem that either (1) our estimate of the geological filter $W(I, t)$ will be uncertain relative to the precise behavior of the actually occurring geological filter or (2) the filter we choose to apply to the data to enhance the signal-to-noise ratio can itself be altered depending on how we choose to assess our predictive capability for modeling the signal structure on a given set of records. Third, we have the problem that in the records of length T we may wish to concentrate not on the total record length but on a windowed fraction of each record, and the records are discretely sampled in time any way, so that some form of frequency smoothing must be applied to the data records.

In terms of a filter "error" δW_k, we then have a noise power spectrum that consists of two parts, that because of the nonpredicted noise power on record k and that because of variations in the filter function. Therefore,

$$N_k(\omega) = N_{kk}(\omega) + \delta W_k^* \, \Phi_{kk} \, \delta W_k, \qquad (109)$$

represents the variance power due to both sources. Following a standard chi-squared approach, we recognize that this variance power consists of two factors: the first corresponding to a χ^2 with $(2T \, \Delta B - 2N)$ degrees of freedom, the second with $2N$ degrees of freedom. Here ΔB is the bandwidth of the spectral window being considered. Thus, an unbiased estimate of the noise power is

$$N_k^{(u)}(\omega) = T \, \Delta B(T \, \Delta B - N)^{-1} N_{kk}(\omega), \qquad (110)$$

while an unbiased estimate of the signal power is

$$\Phi_{kk}(\omega) - N_k^{(u)}(\omega). \qquad (111)$$

Note that we must choose $T \Delta B > N$ in order that the unbiased noise power estimate be real and positive.

The point is, that based on such manipulations with the data records, confidence intervals can be established for hypothesis testing of whether a common signal has been detected and whether this signal is recorded in both phase and magnitude with the same degree of confidence (see, e.g., White, 1984). Attempts to determine both the relative rate of sedimentation between different locations as well as examining the earth's orbital effects on the Pleistocene isotopic records have, over the last few years, become regarded as viable research topics using such quantitative methods of analysis. In particular using time series methods and autocorrelated power spectra, Pisias and Shackleton and their collaborators in a variety of publications (see extended reference list) have started to examine the effect of various assumptions and decisions on the problems of resolution, precision, and uniqueness of suspected signals in the isotopic record.

For instance, on the assumption that individual isotopic events can be identified and correlated from core to core, graphical methods have been used to determine maximal correlations, on the further assumption that the depth intervals between identified events in the cores can be connected by straight line segments. Clearly, we need to investigate this technique further, in a less subjective manner, by allowing for uncertainty in the individual events identified or misidentified; by increasing or decreasing the number of identified events to determine sensitivity; and by replacing the straight-line assumption by one of the more common autocorrelative approximations in order to determine the resolution, precision, accuracy, and uniqueness of the mapping relating sedimentary depth to time. It is not good enough to assume *a priori* that the sedimentation rate is constant between age-dated horizons without *a posteriori* justification.

Likewise, the accuracy of power spectra must be determined in order to compare the response to global climate changes of different oceanographic regions. It is important to quantify not only the degree of uncertainty in the power level (i.e., the noise power) but also to be aware that the Nyquist aliasing and sampling interval are formidable opponents not to be ignored with impunity. In this connection it is important to note that an obtained correlation varying on the order of the sample interval is surely a warning signal indicating that adequate resolution is not available. A smoothing, or bandwidth averaging, technique is needed to eliminate spurious signals arising from an unwarranted use of statistical methods of analysis in a regime beyond their realm of applicability.

These caveats must be borne more firmly in mind than seems to have been recognized to date in appllctions of quantitative methods of analysis to isotopic records. Without an understanding and appreciation of the degree of resolution, precision, and uniqueness of such tools applied to the records, it is all too easy to obtain a signal from a sequence of measurements that cannot be justified, resolved, or proven.

III. Spectral Ratio Methods

As we have seen, the general sense of either time series methods or spectral methods is to devise filtering schemes with the filtered output being the desired behavior. Clearly, by taking appropriate ratios it should be possible to eliminate the effect of common factors. The danger in the method of spectral ratios is that noise can sharply limit the capability of extracting meaningful information. Assume that the noise on each of two records obeys independent Gaussian statistics, then the cosine and sine transforms of the record will also obey the Gaussian distribution. With $C(k, \omega)$ and $S(k, \omega)$ the cosine and sine transforms at angular frequency of the kth record $x(k, t)$, with zero mean noise, and with the ratio

$$r \exp(i\theta) = [\langle C(k', \omega) \rangle + i\langle S(k', \omega) \rangle]$$
$$\times [\langle C(k, \omega) \rangle + i\langle S(k, \omega) \rangle]^{-1}, \qquad (112)$$

where r is the relative amplitude and θ the relative phase of the two records at angular frequency ω, and where the two records are labeled k, k', respectively, we have

$$\langle C(k', \omega) \rangle = r \cos \theta \langle C(k, \omega) \rangle - r \sin \theta \langle S(k, \omega) \rangle, \qquad (113a)$$

and

$$\langle S(k', \omega) \rangle = r \cos \theta \langle S(k, \omega) \rangle + r \sin \theta \langle C(k, \omega) \rangle, \qquad (113b)$$

where angular brackets denote mean values. With a Gaussian noise model, we follow Pisarenko (1970) and maximize the noise distribution function to obtain, from the maximum likelihood method, the estimates of θ and r as

$$\tan \theta = G/F, \qquad (114)$$

$$r = \tfrac{1}{2}(V - ZU)(F^2 + G^2)^{-1/2} + [Z + \tfrac{1}{4}(V - ZU)^2(F^2 + G^2)^{-1}], \qquad (115)$$

with

$$U = \langle C(k, \omega)^2 \rangle + \langle S(k, \omega)^2 \rangle,$$
$$V = \langle C(k', \omega)^2 \rangle + \langle S(k', \omega)^2 \rangle,$$
$$F = \langle C(k, \omega) C(k', \omega) \rangle + \langle S(k, \omega) S(k', \omega) \rangle,$$
$$G = \langle C(k, \omega) S(k', \omega) \rangle - \langle C(k', \omega) S(k, \omega) \rangle,$$

and

$$Z = \sigma_1^2 / \sigma_2^2,$$

where σ_1^2 and σ_2^2 are the variances of the cosine and sine transforms of the noise at sites k and k^1, respectively.

Again, with more than two records, a multiple set of ratios and angles can be defined—just as in the cross-correlation spectral methods. However, except for relative attenuation estimates of seismic waves, and very low frequency estimates of wave dispersion, the spectral ratio methods have not enjoyed a very popular reception, probably because the phase θ is just too sensitive to noise in the data to be of much use. In a similar manner, cepstral methods (see, e.g., Tribolet, 1979) for separating reflection sequences from source sequences in the seismic record also suffer from severe phase unwrapping problems.

IV. Homomorphic Deconvolution

Thus far we have assumed that the noise is additive to the signal. However, it is quite possible, perhaps even probable, that an intrinsic isotopic signal impressed at an instant of geologic time on the depositional surface is later altered as the depth of burial of the sedimentary material is increased. Such changes can be caused by fluid transport of different isotopes, diagenetic alteration, cementation, dissolution, temperature and pressure influences, chemical replacement, compaction, and biological effects, such as bioturbation, preferential isotopic selection, and varying isotopic equilibria. Thus, there is a modulation produced by earth processes acting on the original intrinsic isotopic signal. In order to determine the pristine signal, we must remove the earth modulation effect.

We shall assume that the modulational effect of the earth acts on all depositional intrinsic isotopic signals in a convolutional manner, so that the measured signal $x(t)$ is considered to be the convolution of an intrinsic isotope signal $w(t)$ and the earth's modulational response $r(t)$ with

$$x(t) = \int_{-\infty}^{t} r(t-\tau)w(\tau) \, d\tau. \tag{116}$$

Homomorphic deconvolution is a nonlinear technique of separating the components of a convolution. The technique was invented in the mid 1960s and first applied to reflection seismic data in 1971 (Ulrych, 1971) for the purposes of wavelet extraction and echo removal. The potential utility of homomorphic deconvolution for attenuation determination from reflection seismic data was recognized early in its development (Schafer, 1969), but homomorphic deconvolution is still a topic of active research (e.g., Lines and Clayton, 1977; Lines *et al.*, 1980).

In this section we have two goals: (1) to give a short synopsis of the mathematics involved in homomorphic deconvolution, and (2) to discuss the possible application of a homomorphic deconvolution technique to problems in isotope chronostratigraphy. We address each of these goals in turn.

A. Mathematical Framework

The mathematical formula forming the basis of homomorphic deconvolution is easily given. The basic quantity obtained in the analysis is called the *complex cepstrum* $x(t')$ (an anagram of "spectrum"), defined by (e.g., Lines and Clayton, 1977)

$$\hat{X}(t') = (2\pi)^{-1} \int_{-\infty}^{\infty} \ln X(\omega) \exp(i\omega t') \, d\omega, \tag{117}$$

where $X(\omega)$ is the Fourier transform of the record $x(t)$. The utility of this definition becomes obvious when we consider that the isotopic record can be expressed in terms of the convolution of the depth alterations $r(t)$ of the earth (due perhaps to diagenetic or other influences) with the spectrum of the initial signal $w(t)$:

$$X(t) = \int_{-\infty}^{\infty} R(\omega) W(\omega) \exp(i\omega t) \, d\omega, \tag{118}$$

where $R(\omega)$ and $W(\omega)$ are the Fourier transforms of $r(t)$ and $w(t)$, respectively. The convolution theorem (e.g., Morse and Feshbach, 1953) tells us that we can also represent the record by

$$X(t) = \int_{-\infty}^{t} r(t-\tau) W(\tau) \, d\tau. \tag{119}$$

Thus, with the time origin appropriately chosen, $X(\omega)$ in Eq. (117) is given by

$$X(\omega) = R(\omega) W(\omega). \tag{120}$$

Thus, the natural logarithm in Eq. (117) is given by

$$\ln X(\omega) = \ln R(\omega) + \ln W(\omega). \tag{121}$$

Hence, the cepstrum $x(t')$ of the record can be rewritten as

$$x(t') = r(t') + w(t'), \tag{122}$$

where $r(t')$ is the cepstrum of the earth's modulation and $w(t')$ is the cepstrum of the initial signal. Hence, sequences that are convolved in the time domain are transformed into cepstra that are additive in another domain (called the "quefrency" domain).

Homomorphic deconvolution consists of separating $r(t')$ and $w(t')$ by filtering the cepstrum function $x(t')$. While this cannot be done in general, it can be done in certain cases of physical interest. Under these conditions, to be listed below, the transform of the initial signal can be extracted in the cepstral space. The isotopic signal can then be recovered via the inverse of the transformation in Eq. (117). For a time interval on a record, the record's distortion Q can be obtained from the log spectral amplitudes of the recovered signal $w(t)$ at the bottom and top of the interval. The least squares slope of the log spectral ratio values (bottom/top) versus frequency is computed. The computed slope is b, and the computed standard deviation of b is s_b. The estimate of Q for the time interval is

$$Q = \pi \, \Delta t / b. \tag{123}$$

The uncertainty in Q is given by

$$\delta Q = Q^2 s_b / \pi \, \Delta t. \tag{124}$$

Thus, Q for an interval is easily determined from the difference in isotopic signatures between the top and bottom of the interval.

While the transformation from the isotope record to the cepstrum is a completely straightforward procedure, the partitioning of the cepstral space into a part corresponding to the intrinsic isotopic signature and a part corresponding to the earth's modulation is an art. The cepstral separation, and hence determination of Q, depends on clever use of the phase properties of the two components $r(t)$ and $w(t)$. Some of the relevant properties that can be used in separating $r(t')$ from $w(t')$ are tabulated here (see Tribolet, 1979):

Property (1): The complex cepstrum of a minimum phase sequence (see, e.g., Sherwood and Trorey, 1965), whether of $w(t)$ or $r(t)$, is zero for $t' < 0$. The complex cepstrum of a maximum phase sequence is zero for $t > 0$. The complex cepstrum of a zero phase sequence is centred on $t' = 0$. Mixed phase sequences are neither positive-definite not negative-definite.

Property (2): The complex cepstrum $w(t')$ of a signal $w(t)$, which has a smooth spectrum, values and low ω.

Property (3): If a minimum phase sequence $r(t)$ has a first interarrival time T, so that $t_1 - t_2 = T$, where t_1 and t_2 are the first two major event times, then the complex cepstrum is zero for $0 < t' < T$.

Simple examples of the use of the above properties in homomorphic deconvolution applied to seismic signals are presented in the classic paper by Ulrych (1971), but correct determination of the phase of the components of the convolution is still a major difficulty.

A *phase unwrapping* problem arises in evaluation of the complex logarithm in Eq. (121). In evaluating the logarithm we have

$$\ln X(\omega) = \ln|X(\omega)| + i \arg X(\omega). \tag{125}$$

We can write the phase term as

$$\arg X(\omega) = \text{Arg } X(\omega) + i2\pi m, \tag{126}$$

where $m = 0, 1, 2, \ldots$, and the principal value of the argument is defined as

$$-\pi < \text{Arg } X(\omega) < \pi. \tag{127}$$

Thus, the complex logarithm is multivalued. In deconvolution we require

$$\ln X(\omega) = \ln R(\omega) + \ln W(\omega), \tag{128}$$

and

$$\arg X(\omega) = \arg R(\omega) + \arg W(\omega). \tag{129}$$

Clearly, Eq. (129) will hold only if we choose the appropriate values for $\arg X(\omega)$, $\arg R(\omega)$, and $\arg W(\omega)$. Hence, we have the phase unwrapping problem.

Under certain circumstances the phase unwrapping problem in homomorphic deconvolution does not arise. Two cases are noteworthy: (1) the case of a zero phase intrinsic isotope signal and (2) the case of a minimum phase record. Let us consider each case separately:

1. In the case of zero phase signal, we have $\arg W(\omega) = 0$, and so $\arg X(\omega) = \arg R(\omega)$. In this case there is no phase unwrapping problem.
2. In the case of a minimum phase record $x(t)$, Schafer (1969) has shown that for positive t' the cepstrum is

$$\hat{x}(t') = \pi^{-1} \int_{-\pi}^{\pi} \ln|X(\omega)| \exp(i\omega t') \, d\omega, \tag{130}$$

   i.e., twice the cepstrum resulting from the corresponding zero phase case. In both of these cases, the determination of $\arg X(\omega)$ is irrelevant and disappears from the calculation.

In general, however, the phase unwrapping problem does not enable us to resolve separately the individual phases of *both* the earth's modulation and the intrinsic isotopic signal series, only their sum. Several options exist for choosing the phases of the separate components. Through the use of these options it is possible to perform analysis of the sensitivity of Q to various phase choices. The more commonly available phase options are listed here:

1. Unwrap the phase of $X(\omega)$. Set arg $X(\omega) = $ arg $W(\omega)$, arg $R(\omega) = 0$.
2. Unwrap the phase of $X(\omega)$. Set arg $X(\omega) = $ arg $R(\omega)$. Set $W(\omega) = $ minimum phase.
3. Set arg $X(\omega) = $ arg $R(\omega) = 0 = $ arg $W(\omega)$, i.e., all components assumed zero phase. This is the option applicable to a zero phase signal.
4. Set arg $W(\omega) = $ minimum phase.
5. Use exponential weighting to make $X(\omega)$, $R(\omega)$, $W(\omega)$ all minimum phase.

At the present time, a comprehensive sensitivity analysis of the effects of these choices on Q does not appear to have been carried out.

B. Intrinsic Isotopic Signal Extraction Methods

In principle optimal cepstral filtering depends on the details of the local geological events.

There are two distinct methods of extracting the intrinsic isotopic signature via filtering of the complex cepstrum. In the first, the signal is extracted by low-quefrency windowing of the cepstrum, i.e., keeping the cepstrum inside a quefrency window $-t'_- < t' < t'_+$ and discarding the rest. This makes use of the concentration at low t' from property (2). A modification of the quefrency windowing, a t' comb filter, could be designed using property (3), which puts a minimum-phase earth modulation sequence a distance T away from the signal cepstrum, where T is determined by the local geology. Buttkus (1975) has emphasized the use of comb filters for homomorphic deconvolution.

The second filtering method for intrinsic signal extraction has been discussed by Ulrych (1971). In his technique the intrinsic signal is extracted by low-quefrency filtering of the complex cepstrum. This makes use of the concentration at low ω from property (2). Optimum filtering in quefrency space is also influenced by the local geological structure through property (3). As discussed, in quefrency space the cepstrum in a quefrency region $0 < t' < T$ is determined by the intrinsic signal. The uncertainty principle implies a region $0 < \omega < 1/T$ should be primarily determined by the signal. Thus, both comb and low-pass quefrency filters tailored to the local geology structure should be of value. However, we must remember that the data are band limited, and discrete, introducing aliasing and other problems.

As discussed by Schafer (1969), there are competing effects because of the discrete and band-limited nature of data, which affects the accuracy of homomorphic deconvolution. Because of the discrete nature of data, it is necessary to sample at a high rate (i.e., small Δt) to avoid aliasing in the Fourier transform. Since, however, the high-frequency content of signals is quite low (i.e., data are band limited on the upper frequency end), there can be a significant interval of frequency over which $X(\omega)$ is very small. Since it is necessary to compute $\ln|X(\omega)|$, accuracy problems can arise in this high-frequency regime. Thus, a trade-off is possible between minimizing aliasing and minimizing the range over which $\ln|W(\omega)|$ is large and negative.

A suggestion by Schafer (1969) would minimize the range over which $\ln|X(\omega)|$ is large and negative by low-pass filtering and then resampling the data at a larger sampling time interval. This has the effect of lowering the effective Nyquist frequency. The effects of this technique on aliasing are not evident and should be studied. An alternative has been suggested by Tribolet (1979). Basically Tribolet (1979) argues that the pristine signal has a slow time variation. He then develops short time scale signal models that represent the pristine signal as a series of windowed time-invariant models. But the homomorphic analysis of such short time representations requires a complete understanding of the distortion effects caused by the windowing operation. While this method takes into account the low-frequency, band-limited nature of the data, as well as the high-frequency data, the detailed effects of this varied algorithm have not, to date, been studied further.

Another result of the finite sampling of $X(\omega)$ at N points is that the cepstrum $x(t')$ becomes a periodic function, with a period $N \Delta t$, where Δt is the sampling interval. Thus,

$$\hat{x}(t') = \sum_{r=-\infty}^{\infty} \hat{x}([n+rN] \Delta t). \tag{131}$$

A consequence of this periodicity is that high-quefrency components of the cepstrum are translated back and occur in the discrete cepstrum at small negative quefrencies. Thus, a window filter that passes $-t'_- < t' < t'_+$, with $t'_- = 0$, will include unwanted contributions.

As we have remarked already, an observed isotopic record $x(t)$ is considered to be the convolution of an intrinsic isotopic signal $w(t)$ and the modulation $r(t)$ of the earth so that

$$x(t) = \int_{-\infty}^{t} r(t - \tau)w(\tau) \, d\tau. \tag{132}$$

Note that causality implies the upper limit of t on the integral in order that the record at time t not be a result of processes taking place after time t.

Furthermore, the convolution theorem enables us to write Eq. (132) in the form

$$x(\omega) = r(\omega)w(\omega), \tag{133}$$

where $r(\omega)$ is the Fourier transform of $r(t)$.

Homomorphic deconvolution attempts to separate $w(\omega)$ and $r(\omega)$ by noting, from Eq. (133), that

$$\ln x(\omega) = \ln r(\omega) + \ln w(\omega). \tag{134}$$

A cepstrum $x(t)$ is then constructed from

$$\hat{x}(t) = (2\pi)^{-1} \int_{-\infty}^{\infty} \ln x(\omega) \exp(i\omega t) \, dw. \tag{135}$$

Various *filtering* operations are carried out on the cepstrum $\hat{x}(t)$ in attempts to remove the contributions produced by the earth's modulation since, from Eq. (134),

$$\hat{x}(t) = \hat{X}_w(t) + \hat{x}_r(t), \tag{136}$$

so that the complex cepstrum is the additive sum of the individual cepstra from $w(t)(\hat{X}_w)$ and from the modulation $r(t)(\hat{x}_r)$.

After the filtering operations, the result is a filtered cepstrum $\hat{x}_f(t)$, which is then inverted to recover a filtered estimate, $w_f(t)$, of the intrinsic isotopic signal from

$$x_f(\omega) = \exp\left[\int_{-\infty}^{\infty} dt\, \hat{x}_f(t) \exp(i\omega t)\right], \tag{137}$$

and

$$w_f(t) = \int_{-\infty}^{\infty} x_f(\omega) \exp(i\omega t) \, d\omega. \tag{138}$$

The requirement on the ω integral in Eqs. (137) and (138) is that $w_f(t)$ and $x(t)$ both be zero for $t < 0$ in order to satisfy physical causality for an intrinsic signal that is "switched on" at $t = 0+$.

The success, or lack of, in cepstral filtering depends upon the frequency separation of the intrinsic signal and the earth's modulation. If it is assumed that the cepstral signal is basically a low-frequency quantity and that the cepstral earth modulation is mainly a high-frequency quantity, then window filters can be introduced to null either the intrinsic signal or the modulation as appropriate (Tribolet, 1979; Stoffa *et al.*, 1974).

In practice, however, a continuous Fourier representation is not available because of finite sampling. A DFT is then necessary. The DFT results corresponding to the continuum representation for a unit sampling interval are

$$X(Z) = \sum_{t=-\infty}^{\infty} x(t)Z^{-t}, \qquad Z = \exp(i\omega), \tag{139}$$

$$\hat{X}(T) = (2\pi i)^{-1} \int_c \ln X(Z)Z^{T-1}\, dZ, \qquad T = 0, \pm 1, \pm 2, \ldots . \tag{140}$$

The inversion, after filtering has converted $x(T)$ to $x_f(T)$, is

$$x_f(z) = \exp\left[\sum_{T=-\infty}^{\infty} \hat{x}_f(T)Z^{-T} \right], \tag{141}$$

and

$$x_f(t) = (2\pi i)^{-1} \int_{c'} x_f(z)z^{t-1}\, dz, \qquad t = 0, \pm 1, \pm 2, \ldots \tag{142}$$

The contours of integration in the complex z plane, c and c', have to be chosen so that physical causality is still satisfied. Thus, we require $x_f(t) = 0$ in $t < 0$, and $x(T) = 0$ for $T < 0$. Since a DFT has $Z = e^{i\omega}$, so that $|Z| = 1$, the contours of integration c and c' in the complex Z plane *have* to be unit circles. The physical causality requirements then *demand* that all poles and branch cuts of $\ln X(Z)$ and of $X_f(Z)$ be inside or on the unit circle, in order that the integrals around the unit circles be zero for $T < 0$ and $t < 0$, respectively. Schafer (1969) calls such functions "minimum phase" functions. Conversely functions with *all* of their poles and branch cuts *outside* the corresponding unit circles are called "maximum phase" functions (and they have $x_f(t) = 0$ in $t > 0$ and $x(T) = 0$ for $T > 0$). Functions with all of their poles or branch cuts *on* the unit circle are called "zero phase" functions while functions with poles or branch cuts both inside *and* outside the unit circle are called "mixed phase" functions.

Additionally, weighting can be attached to the original time series to modify the cepstral decomposition. For example, Schafer (1969) pointed out that by multiplying $x(t)$ by α^t (with α a real positive constant and less that unity) the poles of the resulting $X_a(Z)(X_a(Z) \equiv \sum_{t=-\infty}^{\infty} x(t)a^t Z^{-t})$ can be moved inside the unit circle to make the function minimum phase. Such a maneuver also aids with the aliasing problem caused by the DFT as pointed out by Schafer (1969) and further expounded upon by Stoffa *et al.* (1974).

Problems arise, however, since the act of taking the complex logarithm of $x(\omega)$ leads to a multivalued function for the phase of $x(\omega)$. The phase unwrapping problem [i.e., the determination of the continuous phase of $x(\omega)$] is a severe problem because of cycle skipping (because of the phase varying by one complete cycle between sampling intervals in the DFT) and aliasing and noise in the data. Two basic algorithms are available for phase unwrapping. Schafer (1969) presented an iterative algorithm for unwrapping the principal value of the phase spectrum, while Stoffa *et al.* (1974) presented an algorithm that integrates the derivative of the phase in frequency space to produce a continuously varying phase. Note that the weighting one ascribes to $x(t)$ does not have to be restricted to exponential multiplication as done by Schafer (1969).[5] One can take $x(t)$ and weight it convolutionally with an arbitrary function of time $R(t)$.

Several questions demand consideration:

1. What is the best weighting function, $R(t)$, to use on the isotopic record and how does one define best?
2. What is the best functional form to use in order to efficiently extract the relevant quantities?

The answers to these questions are not uniquely determined but depend on some defining prescriptions for "best," "efficient," etc., and on the degree to which one believes, or asserts, that ancillary constraints are playing a dominant role.

For instance, consider the determination of the weighting function $R(t)$. The resulting DFT, $X_R(Z)$, is

$$X_R(Z) = \sum_{t=-\infty}^{\infty} X(t)R(t)Z^{-t}. \tag{143}$$

Many choices of $R(t)$ will move the poles of $X_R(Z)$ inside the unit circle. Schafer (1969) showed that a special choice of the functional form of $R(t)$ with only one parameter (namely, α^t) enabled the parameter α to be bounded by $\alpha < a_0$, so that the resulting $x_\alpha(Z)$ was minimum phase for $\alpha < a_0$. However, different functional forms with more than one parameter are available as equally acceptable weighting functions. How do we choose between them?

Cepstral operations on $X_R(Z)$ produce

$$X_R(T) = (2\pi i)^{-1} \int \ln X_R(Z)Z^{T-1} \, dZ. \tag{144}$$

---

[5] Even with exponential weighting there is the question of which is the best weighting to use. Stoffa *et al.* (1974) advocate weighting heavily initially to guarantee an unaliased complex cepstrum and then to try less weighting until aliasing becomes a problem. But this is not a quantitatively precise way of choosing weighting. Also, it does not provide a statement of how to determine out of all such minimum phase functions satisfying the "no aliasing" criteria that particular function which comes as close as it can to modeling reality.

Filtering operations [normally low-pass or high-pass multiplicative filters on $\hat{x}_R(T)$] produce

$$\hat{X}_{R,F}(T) = F(T)\hat{x}_R(T), \tag{145}$$

where $F(T)$ is the filter (e.g., for a high-pass filter one choice is $F(T) = 1$ for $T > T_0$ and $F(T) = 0$ for $T < T_0$, where T_0 is a constant).

The filtered cepstrum is then inverted to yield

$$X_{R,F}(Z) = \exp\left(\sum_{T=-\infty}^{\infty} \hat{X}_{R,F}(T)Z^{-T} \right), \tag{146}$$

and the filtered estimate of $x(t)$ is

$$X_F(t) = [2\pi i R(t)]^{-1} \int X_{R,F}(Z)Z^{t-1}\, dZ. \tag{147}$$

While the unfiltered estimate $x_R(Z)$ is guaranteed to produce a causal estimate of $x(t)$ [since $R(t)$ is chosen so that all poles of $x_R(z)$ are inside the unit circle], no such guarantee is available for the inversion of the filtered cepstrum, $x_{R,F}(Z)$. The only way such a guarantee can be provided is if the filtering function $F(T)$ is chosen so that the poles of the inversion of the filtered cepstrum $x_{R,F}(Z)$, given by Eq. (146), are inside the unit circle in Z space.

Thus, in order to see if a particular chosen form of filter function is causally acceptable, one must check the positions of the poles of $x_{R,F}(Z)$. Any filter function yielding pole positions *outside* the unit circle will produce an acausal estimate of $x(t)$. On physical grounds such filters must be rejected.

What of that class of filtering functions $F(T)$, which provide for all poles of $X_{R,F}(Z)$ lying inside the unit circle? All such filtering operations produce causal estimates of the response $x(t)$. Therefore, we require further criteria to narrow the range of possible filtering functions.

The cepstral filtering function $F(T)$ has been chosen so that, when applied to the cepstrum $\hat{x}_R(T)$, the resulting $\hat{x}_{R,F}(T)$ is regarded as being the estimate of the cepstrum of the intrinsic isotopic signal $w(t)$.

Since the response is regarded as the convolution of the intrinsic signal and the earth's modulation, it follows that the residual cepstrum

$$Y_{R,F}(T) = [1 - F(T)]\hat{x}_R(T) \tag{148}$$

must be regarded as the weighted estimate of the cepstrum of the earth's modulation.

The Z transform of the weighted estimate of the earth's modulation is then given by

$$Y_{R,F}(Z) = \exp\left\{ \sum_{T=-\infty}^{\infty} [1 - F(T)]\hat{X}_R(T)Z^{-T} \right\}, \tag{149}$$

and the filtered estimate of the modulation $r(t)$ is

$$r(t) = [2\pi iR(t)]^{-1} \int Y_{R,F}(Z)Z^{t-1}\, dZ. \tag{150}$$

But the initial premise was that the response at time t was the time convolution of the intrinsic signal and the earth's modulation from all earlier times, hence, the requirement $r(t<0)=0$. This requirement then translates into this statement: *all poles of $Y_{R,F}(Z)$ must also lie within the unit circle in order no to violate the original convolutional premise.* This puts further constraints on the parameters associated with the weighting and filtering functions $R(t)$ and $F(t)$, respectively. Further investigations of these potential problems with the homomorphic deconvolution technique are urgently needed. In summary, the procedure of deconvolution is crucial to the separation of the intrinsic isotopic signal from the observed isotopic record as modified by the earth's response. Every deconvolution technique eventually runs squarely into problems similar to those observed with the homomorphic deconvolution method. We must be clearly aware of the assumptions and approximations made in the deconvolution procedures in order to attach credence to the reconstructed pristine signals. Such awareness has not yet become common in the isotope chronostratigraphy arena.

V. Phase Sensitive Detection of Isotopic Signals

A. Basic Arguments

One of the problems arising in isotope chronostratigraphy is the detection of small signals in the presence of relatively large noise because of postdepositional geologic processes. For this reason it is necessary to analyze the data with algorithms that make the signal-to-noise ratio (SNR) as large as possible. In this section we address this problem. We do not deal with the detection of the continuous trend background of isotopic signals, where the algorithms are limited by the observation time or sedimentological response time, whichever is the smaller. Instead, we consider short bursts of isotopic events that have durations much smaller than any other postdepositional time constant in the geological system, so that the burst events, even if of small magnitude, are recorded in compressed intervals of geologic time (sedimentary depth). Such events mights be caused by glacio-eustatic events, climatic change, orbital effects, etc. The problem is to devise a filtering algorithm that is responsive to such short-term events in an efficient manner in terms of maximizing the SNR.

Over the last decade, methods of handling such problems have been devised in order to measure very rapid pulses of low intensity radiation in the presence of a high thermal background. The algorithmic techniques, referred to as phase sensitive detection (PSD), can be readily transferred to the short isotopic burst problem.

In this section we assume that the level of all signals is referred to some common normalization. The general sense of this section closely follows the work of Amaldi (1979), Blair (1979), and Pallotino (1979). The total signal at the input of the phase sensitive detectors (PSDs) can be written in general in the form

$$v(t) = [V_x(t) + n_x(t)] \cos \omega_R t + [v_y(t) + n_y(t)] \sin \omega_R t, \quad (151)$$

since the signal V_x is already taken to be in quasi-harmonic form with $v_x(t) = V_S \exp[-t/\tau_v]$ and $v_y(t) = 0$, and the noise signal can be expressed according to the Rice decomposition

$$n(t) = n_x(t) \cos \omega_R t + n_y(t) \sin \omega_R t. \quad (152)$$

Here, the exponential decay with time over the fixed time scale τ_v represents the envelope of the short-duration event, and ω_R represents the periodic oscillation frequency of the event within its decay envelope. The terms $v_x(t)$, $v_y(t)$, $n_x(t)$, and $n_y(t)$ of Eq. (151) represents the envelope of the pertinent signals; they contain all the interesting information, and in general their variations are taken to be small over a time scale ω_R^{-1}.

The two PSDs perform these operations on the signals:

$$x(t) = (A/t_0) \int_{-\infty}^{t} dt' \, v(t') \exp\left[\frac{-(t - t')}{t_0}\right] \text{sign}[\cos \omega_R t'], \quad (153a)$$

$$y(t) = (A/t_0) \int_{-\infty}^{t} dt' \, v(t') \exp\left[\frac{-(t - t')}{t_0}\right] \text{sign}[\sin \omega_R t'], \quad (153b)$$

where $\text{sign}(x) = \pm 1$ according as $x \geq 1$, respectively. It can be shown that in order to obtain an output $v(t)$ for an input $v(t) \cos \omega t$, it is necessary that $A = 8/\pi$. The sign functions can be expanded in series, and, if the higher order harmonics of the input signals are negligible, we can replace $A \, \text{sign}(\cos \omega_R t)$ with $2 \cos \omega_R t$ in Eq. (153a) and similarly for $A \, \text{sign}(\sin \omega_R t)$ in Eq. (153b).

The linearity of the operation allows the PSD to be decomposed into two linear subsystems: the first, memoryless and time varying, performs the product

$$x'(t) = 2v(t) \cos \omega_R t, \quad (154)$$

while the second, dynamic and time invariant, performs the integration

$$x(t) = t_0^{-1} \int_{-\infty}^{t} dt' \, v(t') \exp\left[\frac{-(t-t')}{t_0}\right], \tag{155}$$

with transfer function

$$W_p(\omega) = (1 + i\omega t_0)^{-1} = \beta_2/(\beta_2 + i\omega), \tag{156}$$

where $\beta_2 = 1/t_0$. The bandwidth of the dynamic subsystem is very small and in any case much smaller than ω_R.

The total signal immediately prior to this integration can be found by substituting Eq. (151) into Eq. (154):

$$x(t) = [v_x(t) + n_x(t)](1 + \cos 2\omega_R t) + [v_y(t) + n_y(t)] \sin 2\omega_R t, \tag{157a}$$

$$y(t) = [v_y(t) + n_y(t)](1 - \cos 2\omega_r t) + [v_x(t) + n_x(t)] \sin 2\omega_R t. \tag{157b}$$

The second harmonic components in Eq. (157) are eliminated by the integrator [with transfer function (Eq. (156))], and therefore the two outputs $x(t)$ and $y(t)$ depend only on the pertinent envelope functions of the representation in Eq. (151) filtered according to Eq. (156).

The response of the PSD to the noise can be obtained by evaluating the autocorrelation function of the signal $x(t)$ when the input is $n(t)$ with power spectrum $S_{nn}(\omega)$ and autocorrelation $R_{nn}(\omega)$:

$$R_{xx}(\omega) = \langle\{4n(t+\tau)\cos \omega_R(t+\tau)n(t)\cos \omega_R(t)]\rangle. \tag{158}$$

Because of the independence of $n(t)$ and $\cos \omega_R t$, we have from their autocorrelation functions

$$R_{xx}(\omega) = 2R_{nn}(\tau)\cos \omega_R \tau, \tag{159}$$

with a similar result for the y channel. The spectrum can be obtained from the autocorrelation in the form

$$S_{xx}(\omega) = \int_{-\infty}^{\infty} d\tau \, R_{xx}(\omega) \, e^{-i\omega\tau} = S_{nn}(\omega + \omega_R) + S_{nn}(\omega - \omega_R), \tag{160}$$

which shows the effect of transposition of the spectral content of the input from $\omega_R + \omega$, $-\omega_R + \omega$ to the angular frequency ω.

Finally, at the output of the PSD, one has

$$S_{xx}(\omega) = S_{yy}(\omega) = [S_{nn}(\omega + \omega_R) + S_{nn}(\omega - \omega_R)]/(1 + \omega^2 t_0^2). \tag{161}$$

This follows from Eq. (156) and the general theorem that gives the output spectrum of a linear system as a product of the input spectrum and of the square modulus of the transfer function of the system.

Since all further processing is done on the demodulated signals $x(t)$ and $y(t)$, it is useful to characterize the dynamics in terms of equivalent low-pass transfer functions. Equation (153) shows that as long as $t_0 \ll \omega_R^{-1}$, the envelope of the applied input can be well approximated by a Dirac delta function. Since the output envelope is given by Eq. (155), the equivalent low-pass transfer function takes the form

$$W_A(\omega) = e^{-i\pi/2}(1 + i\omega\tau_v)^{-1}. \tag{162}$$

The same result can also be obtained by substituting $\omega \to (\omega - \omega_R)$ in Eq. (153).

We note that the factor $\exp(-i\pi/2)$ has the effect of exchanging the two components (cosine and sine) of the signal, due to the fact that at the resonance frequency $W(\omega)$ is pure imaginary. We neglect this in our model since it can be taken into account by a simple relabeling of the two output channels. We also omit the gain factor in order to refer the level of all the signals to the input. Thus, the equivalent low-pass transfer function becomes

$$W_A(\omega) = (1 + i\omega\tau_v)^{-1} = \beta_1(\beta_1 + i\omega)^{-1}, \tag{163}$$

where

$$\beta_1 = \tau_v^{-1} = \omega_R/2Q, \tag{164}$$

which defines the quality factor Q for the system.

In this representation one can lump all of the resonant noise contributions into an equivalent generator $u_n(t)$ with white spectrum

$$S_{uu} = 2V_{nb}^2/\beta_1, \tag{165}$$

applied at the input, where V_{nb} denotes the rms noise power level.

The wide band noise contributions are lumped into a signal $e(t)$ with white spectrum

$$S_{ee}(\omega) = S_0(\omega + \omega_R) + S_0(\omega - \omega_R) = 2S_0(\omega_R). \tag{166}$$

We recall here that the two output signals are sampled with time interval Δt before being converted from analog to digital form. Normally one might select Δt on the basis of the Nyquist criterion

$$\Delta t < (2f^*)^{-1}, \tag{167}$$

where f^* is the frequency above which the spectrum of the total signal is negligible. Since the cutoff frequency of the wide band spectrum is $1/2\pi t_0$, f^* can be taken to be any value that is large compared to the cutoff value, for example, $10/2\pi t_0$, to fix ideas. However, we are interested here in the detection of a possible isotopic signal addition to the noise energy of the system rather than reconstruction of the original signal. The cutoff frequency of the interesting part of the spectrum, i.e., that which may contain isotopic signals, is not $1/2\pi t_0$ but a much lower value, $1/2\pi\tau_v$. We make the customary choice $\Delta t = t_0$, which, as we will show, has some interesting advantages.

We mention here that it is also possible to model the system and analyze the data with reference to a configuration that does not use the PSD, which has the task, as explained, to translate the frequency band of interest to zero frequency. In that case, instead of using a first-order model with two output variables, we would have a second-order model with a single output variable. By proper application of the sampling theorem to this model, it can be shown that the sampling rate of the data acquisition system becomes only two times larger than the sampling rate for each of the two channels of the standard system; the total number of samples generated per unit time is therefore the same in both cases. Thus, the PSD scheme provides an elegant and attractive method to search a record for short-lived, weak signals superimposed on a noisy, longtime variable system, such as might occur from meltwater situations.

B. Data Analysis Algorithms

In this section we call the total signal $u(t)$ the *innovation*. One of the components of $u(t)$, the noise component, contributes continuously small amounts of energy that maintain an average noise energy. The other, the signal component, provides small amounts of energy that are expected to be concentrated in time so as to allow their detection in spite of having a much smaller average power level.

1. The Direct (D) Algorithm.   The simplest way to try to detect the innovation signal $u(t)$ is to consider the quantity

$$r^2(t) = x^2(t) + y^2(t), \tag{168}$$

where $x(t)$ and $y(t)$ are the quadrature components arising from a PSD operation. In order to compute the signal r_s^2, in the absence of noise, we consider as a standard template innovation the signal

$$x_s(t) = V_s[\exp(-\beta_1 t) - \exp(-\beta_2 t)][1 - \beta_1/\beta_2]^{-1} \tag{169}$$

at the PSD output. For simplicity we take this signal to be in phase with the reference, in which case $y_s(t) = 0$ by definition. Then we obtain

$$r_s^2(t) = V_s^2[\exp(-\beta_1 t) - \exp(-\beta_2 t)]^2[1 - \beta_1/\beta_2]^{-2}. \tag{170}$$

If the input has an arbitrary phase φ, the response of the two channels will be $x(t) = x_s(t)\cos\varphi$ and $y(t) = y_s(t)\sin\varphi$, but r_s^2 would have the same value.

In order to compute the noise, we consider that both $x(t)$ and $y(t)$ have zero mean and normal distribution with variance

$$\sigma_0^2 = R_{xx}(0) = R_{yy}(0) = R(0). \tag{171}$$

Let us now find $R(\tau)$. We have

$$R(\tau) = \left(\frac{1}{2\pi}\right) \int_{-\infty}^{\infty} d\omega \, S(\omega) \, e^{i\omega\tau}, \tag{172}$$

where $S(\omega) = S_{xx}(\omega) = S_{yy}(\omega)$ is the power spectrum of $x(t)$ and $y(t)$. Then,

$$S(\omega) = S_{uu}|W_A|^2|W_P|^2 + S_{ee}|W_P|^2. \tag{173}$$

Thus,

$$S(\omega) = S_{uu}[\beta_1^2/(\beta_1^2+\omega^2)][\beta_2^2/(\beta_2^2+\omega^2)]$$
$$+ S_{ee}[\beta_2^2/(\beta_2^2+\omega^2)]. \tag{174}$$

Substituting in Eq. (172) and integrating we get

$$R(\tau) = S_{uu}[\beta_2 \exp(-\beta_1|\tau|) - \beta_1 \exp(-\beta_2|\tau|)]\beta_1\beta_2/[2(\beta_2^2-\beta_1^2)]$$
$$+ S_{ee}(\beta_2/2) \exp(-\beta_2|\tau|). \tag{175}$$

In particular

$$R(0) = S_{uu}\beta_1\beta_2/[2(\beta_1+\beta_2)] + S_{ee}\beta_2/2. \tag{176}$$

It is convenient to use the quantities v_{nb}^2 and S_0 and to define the new parameters

$$\gamma = \beta_1/\beta_2, \qquad \Gamma = S_{ee}/S_{uu} = \beta_1 S_0/V_{nb}^2. \tag{177}$$

Then, $R(\tau)$ becomes

$$R(\tau) = V_{nb}^2 \exp(-\beta_2|\tau|)$$
$$\times \{[1-\gamma^2]^{-1}[\exp[\beta_2|\tau|(1-\gamma)] - \gamma] + \Gamma/\gamma\}, \tag{178}$$

and

$$R(0) = V_{nb}^2[(\gamma+1)^{-1} + \Gamma/\gamma]. \tag{179}$$

We notice that, neglecting the noise, the variance at the PSD output $R(0)$ is nearly equal to the variances at the PSD input V_{nb}^2. This is because we have considered the ratio of the output to the envelope of the input as equal to one. If the signal is absent or, in general, if its average contribution is negligible with respect to the noise, the variable $r^2(t)$ will have the exponential probability density function

$$F(r^2) = (1/2\sigma_0^2) \exp(-r^2/2\sigma_0^2) \tag{180}$$

and the SNR for this direct algorithm will be

$$(\text{SNR})_d = r_s^2(t)/2\sigma_0^2. \tag{181}$$

The largest SNR value occurs at a time

$$t^* = [\ln(1/\gamma)]/[\beta_2(1-\gamma)]. \tag{182}$$

Thus, we can write

$$(\text{SNR})_d = V_s^2 K_d/\{V_{nb}^2 2[1/(1+\gamma)+\gamma]\}, \tag{183}$$

with

$$K_d = (1-\gamma)^{-2}\{\exp[\gamma(1-\gamma)^{-1}\ln\gamma] - \exp[(1-\gamma)^{-1}\ln\gamma]\}. \tag{184}$$

We consider this enhancement of signal to noise for PSD in comparison to other prediction algorithms for PSD of a short-lived signal later.

2. The Zero-Order Prediction (ZOP) Algorithm.   A better way to detect an innovation signal than that of the direct algorithm is to consider the quantities

$$x_z(t) = x(t) - x(t-\Delta t), \tag{185a}$$

$$y_z(t) = y(t) - y(t-\Delta t). \tag{185b}$$

In fact, on a short time scale, the variations of the output because of the noise are relatively small, while the signal, if present, provides a rather sharp variation according to Eq. (170). At time t, in the absence of a signal, the values of $x(t)$ and $y(t)$ are very close to the values at a previous time $t - \Delta t$ [this is the zero-order prediction for $x(t)$ and $y(t)$, and by taking the difference we try to estimate the possible innovation. This algorithm was suggested by Gibbons and Hawking (1971)].

As far as the noise is concerned, both $x_z(t)$ and $y_z(t)$ have zero mean and normal distribution with variance

$$\sigma_z^2 = \langle x_z^2 \rangle = \tilde{R}(0), \tag{186}$$

and

$$\tilde{R}(0) = 2R(0) - 2R(\Delta t). \tag{187}$$

We introduce the variable

$$\rho_z^2 = x_z^2 + y_z^2, \tag{188}$$

which has normal distribution

$$F(\rho_z^2) = [2\sigma_z^2]^{-1} \exp(-\tfrac{1}{2}\rho_z^2\sigma_z^{-2}). \tag{189}$$

In order to get the SNR for this algorithm, we consider the signal due to the standard innovation

$$\rho_{sz}^2 = V_s^2 K_z, \tag{190}$$

with

$$K_z = (1-\gamma)^{-2}[\exp(-\beta_1 \Delta t) - \exp(-\beta_2 \Delta t)]^2, \tag{191}$$

and the noise $2\sigma_z^2$. Thus,

$$(\text{SNR})_z = \rho_z^2/(2\sigma_z^2) = \tfrac{1}{4}K_z V_s^2 \tilde{R}(0). \tag{192}$$

Clearly, $(\text{SNR})_z$ is function of Δt, which can be chosen conveniently. Also β_2 can be chosen at will, while β_1 is fixed. It seems reasonable to take

$$\beta_2 = (\Delta t)^{-1}, \tag{193}$$

and sample $x(t)$ and $y(t)$ at Δt time intervals. In this way one has nearly complete information made up of nearly independent data. We have

$$R(\Delta t) = e^{-1} V_{nb}^2 [[\exp(1-\gamma) - \gamma](1-\gamma^2)^{-1} + \Gamma/\gamma], \tag{194}$$

and

$$K_z = (1-\gamma)^{-1}[\exp(-\gamma) - e^{-1}]^2, \tag{195}$$

which shows that $(\text{SNR})_z$ depends only on γ for each value of the parameter Γ.

It is possible to find an approximate value of γ that makes $(\text{SNR})_z$ maximum by noticing that, for $\phi \ll 1$, $\sigma_z(\gamma)^2$ varies with γ much more than K_z. Neglecting, with respect to unity, terms of the order of γ^2, we find

$$\gamma_m = [\Gamma(e-1)]^{1/2}, \tag{196}$$

corresponding to

$$\Delta t_m = [(e-1)\tau_v S_0 V_{nb}^{-2}]^{1/2}. \tag{197}$$

We can derive the spectral energy density of the isotopic signal which produces a value of ρ_{sz}^2 by replacing r_s^2/K_d with ρ_{sz}^2/K_z.

Again, we defer further discussion of the signal-to-noise enhancement until later when we perform numerical comparisons of the various PSD prediction algorithms.

3. The First-Order Prediction (FOP) Algorithm.   An improvement over the ZOP algorithm can be achieved by making a better prediction of $x(t)$ and $y(t)$ (first-order prediction). Indicating the predicted values by $x_p(t)$ and $y_p(t)$, we put

$$x_p(t) = ax(t-\Delta t), \qquad y_p(t) = ay(t-\Delta t), \tag{198}$$

and find that value of a that minimizes the square deviation

$$\sigma_p^2 = \langle [x(t) - x_p(t)]^2 \rangle. \tag{199}$$

It can be easily demonstrated that

$$a = R(\Delta t)/R(0),$$

and thus

$$\sigma_p(\Delta t)^2 = R(0)[1 - R(0)[1 - R(\Delta t)^2/R(0)^2]. \tag{200}$$

Let us now consider

$$X(t) = x(t) - x_p(t), \tag{201a}$$

$$Y(t) = y(t) - y_p(t). \tag{201b}$$

These variables have zero mean and normal distribution with variance $\sigma_p(\Delta t)^2$. The quantity

$$\rho^2(t) = X^2(t) + Y^2(t), \tag{202}$$

had distribution

$$F(\rho^2) = (2\sigma_p^2)^{-1} \exp\left(\frac{-\frac{1}{2}\rho^2}{\sigma_p^2}\right). \tag{203}$$

The signal due to the standard innovation is given again by

$$\rho_s^2 = V_s^2 K_p, \tag{204}$$

with $K_p = K_z$.

The noise will be $\langle \rho^2 \rangle = 2\sigma_p^2$ and thus

$$(\text{SNR})_p = K_p V_s^2 / 2\sigma_p^2. \tag{205}$$

Also in this case it is convenient to put $\Delta t = \beta_2^{-1}$ and then $(\text{SNR})_p$ is only a function of Γ and γ.

The approximate value γ_m, which maximizes $(\text{SNR})_p$ is obtained by neglecting terms of the order of $(\Gamma\gamma)$ and γ^2.

We derive the spectral energy density of the isotopic signal, which produces a value of ρ^2, by replacing r_s^2/K_d with σ_s^2/K_p.

Once more, further discussion of this PSD algorithm for signal-to-noise enhancement is deferred to the Chapter 16.

4. The Wiener–Kolmogorov (WK) Algorithm. The prediction can be further improved by using more data samples, and, in general, the best linear prediction can be obtained by using all past and future data, weighted according to the Wiener–Kolmogorov theory. However, we are not interested in the prediction of the variables $x(t)$ and $y(t)$ but rather in the estimation of the innovation acting at the input of the system.

The best linear estimate $\tilde{u}_x(t)$ for the x channel (in-phase) component of the innovation is

$$\tilde{u}_x(t) = \int_{-\infty}^{\infty} w(\tau)x(t-\tau)\,d\tau, \tag{206}$$

where $w(\tau)$ is the weighting function (the impulse response of the optimum filter), which is determined by minimizing the square deviation

$$\sigma_w^2 = \langle[u_x(t) - \tilde{u}_x(t)]^2\rangle. \tag{207}$$

We remark here that for normal processes, as in our case, the linear mean square estimation gives the same results as a more general nonlinear estimation algorithm.

By applying the orthogonality principle of linear mean square estimation

$$\langle[u(t) - \tilde{u}(t)]x(t)\rangle = 0 \tag{208}$$

between the deviation and the observation, it can be shown that the connection between the autocorrelation functions $R_{ux}(\tau)$ and $R_{xx}(\tau)$ is provided by

$$R_{ux}(\tau) = \int_{-\infty}^{\infty} R_{xx}(t-\alpha)w(\alpha)\,d\alpha, \tag{209}$$

where $\tau = t - t'$.

Then, by applying the Fourier transform to Eq. (209), we obtain the transfer function of the optimum filter $W(\omega)$ as

$$W(\omega) = S_{ux}(\omega)/S_{xx}(\omega), \tag{210}$$

where $S_{ux}(\omega)$ is the cross-spectrum of the signals $u(t)$ and $x(t)$. i.e., the Fourier transform of their cross-correlation function.

The solution of Eq. (209) is very simple because of our choice of the estimator [Eq. (206)], which uses all past and future data. A more complex procedure is required if one wants to work in real time, i.e., to use only past data. However, this is not a problem because in practice the analysis is performed on data stored on magnetic tapes so that one has access to both the past and the future; a quasi-real time analysis is also possible without modifying the algorithm by using a moderate amount of future data with a time scale determined by the correlation times of the data.

We also have that

$$S_{ux} = S_{uu}W_A^*(\omega)W_P^*(\omega), \tag{211}$$

since the cross-spectrum between the input and the output of a system can be obtained by the product of the input spectrum with the complex conjugate of the transfer function of the system. Here, $W_A(\omega)$ is the low-pass transfer function acting on the raw innovation to produce a filtered input signal to present to the PSD algorithm, and $W_P(\omega)$ is the transfer function acting on the PSD output signals to convert them to filtered time signals $x(t)$ and $y(t)$ in quadrature.

We obtain

$$W(\omega) = (\gamma/\Gamma)(\beta_2 - i\omega)(\beta_1 + i\omega)/(\omega^2 + \beta_3^2), \qquad (212)$$

with

$$\beta_3 = \beta_1(1 + \Gamma^{-1})^{1/2}. \qquad (213)$$

The transfer function can also be put in the form

$$W(\omega) = (W_A W_P)^{-1}(1 + \Gamma/|W_A|^2)^{-1} \qquad (214)$$

to show that it operates as an inverse filter.

We have from Eq. (212)

$$W(t) = (2\pi)^{-1} \int_{-\infty}^{\infty} W(\omega) \exp(i\omega t) \, d\omega$$

$$= (\gamma/\Gamma)[\delta(t) + (2\beta_3)]^{-1}(\beta_2 \pm \beta_3)(\beta_1 \pm \beta_3) \exp(\pm\beta_3 t)], \qquad (215)$$

where the negative sign is for $t > 0$ and the positive sign is for $t < 0$, and where $\delta(t)$ is the Dirac δ function.

This algorithm operates just as a filter that can be added to the filtering chain. In order to find the mean square deviation of the estimation $\tilde{u}(t)$, we must compute the autocorrelation $R_{uu}(\tau)$, which, in turn, is obtained from the power spectrum $S_{uu}(\omega)$ of the quantity $\tilde{u}(t)$. We have

$$S_{uu} = S_{xx}|W(\omega)|^2 = S_{uu}[1 + \Gamma\beta_1^{-2}(\omega^2 + \beta_1^2)]^{-1}, \qquad (216)$$

which gives

$$R_{uu} = (2\Gamma\beta_3)^{-1}\beta_1^2 S_{uu} \exp(-\beta_3|\tau|). \qquad (217)$$

Thus, the mean square deviation is

$$\sigma_w^2 = R_{uu}(0) = S_{uu}\beta_1^2/(2\Gamma\beta_3^2) \equiv V_{nb}^2\beta_1/(\Gamma\beta_2). \qquad (218)$$

In order to find the SNR, we now compute the response to the standard innovation at the PSD output. We have, considering only the variable $x(t)$,

$$S(t) = \int_{-\infty}^{t} d\alpha \, w(\alpha) x_s(t - \alpha). \qquad (219)$$

Solving the integral with $w(\alpha)$, given by Eq. (215), we get

$$S(t) = V_s\beta_1(2\Gamma\beta_3)^{-1} \exp(-\beta_3 t). \qquad (220)$$

Finally, considering the two variables $x(t)$ and $y(t)$, we have

$$(\text{SNR})_w = S(t)^2/(2\sigma_w^2)$$

$$= (V_s^2/V_{nb}^2)\beta_1(8\Gamma\beta_3)^{-1}\exp(-2\beta_3 t), \tag{221}$$

which is a maximum for $t = 0$. Using Eq. (213) with $t = 0$, we have

$$(\text{SNR})_w = (V_s^2/V_{nb}^2)8(1+\Gamma)\Gamma^{1/2}. \tag{222}$$

The interesting feature of this result is that $(\text{SNR})_w$ does not depend on the integration time of the PSD, $t_0(\equiv \beta_2^{-1})$, which can therefore be determined by other considerations. For instance, we can take t_0 very small in order to analyze in greater detail a possible isotopic signal, or it can be taken very large if a small number of samples are wanted.

This result is, however, only an approximation because of the finite sampling Δt. Thus, if the data are sampled at Δt time intervals, the variables $x(t)$ and $y(t)$ become discrete and the result will then only be approximate. Clearly, it will maintain its accuracy for

$$\Delta t \ll \beta_3^{-1}, \tag{223}$$

since β_3^{-1} is the characteristic time of this filter.

The largest value of the response of the Wiener–Kolmogorov filter is obtained at $t = 0$. Considering the two variables $x(t)$ and $y(t)$, we have

$$\rho_{sw}^2 = S_x^2 + S_y^2 = V_s^2 K_w, \tag{224}$$

with

$$K_w = \gamma(1-\gamma)^{-1}(\Gamma\beta_3^2)^{-1}(\beta_2+\beta_3)(\beta_1+\beta_3)\sinh(\beta_3/2\beta_2)$$

$$\times \exp(-\beta_3/\beta_2)\{\exp(-\beta_1/\beta_2)[1-\exp[-(\beta_1-\beta_3)/\beta_2]]^{-1}$$

$$-e^{-1}[1-\exp[-(1+\beta_3/\beta_2)]]^{-1}\}. \tag{225}$$

Again, we defer further discussion on the signal-to-noise enhancement of the Wiener–Kolmogorov filter applied to PSD of innovation until Chapter 16.

C. Discrete Time Model and Analysis

The methods of system modeling and of data analysis considered in previous sections, with particular reference to the Wiener–Kolmogorov filter, are based on a continuous time approach, while the actual data samples are available only at discrete values of time. This is an intrinsic weakness of these methods, which usually becomes apparent only when the sampling time Δt becomes comparable with some natural time scale for geologic change.

More general results can be obtained by directly considering the data as functions of a discrete time abscissa. The discrete time sequences $x(n\,\Delta t)(\equiv x_n)$ are defined by their statistical properties, such as the autocorrelation function, and by the response to a standard reference signal. We introduce the envelope notation

$$W = \exp(-\Delta t/\tau_v), \qquad V = \exp(-\Delta t/t_0), \qquad \eta = t_0/\Delta t, \qquad (226)$$

where the parameter η provides an additional degree of freedom since it allows the choice of a sampling time interval Δt different from the integration time t_0 of the PSDs. The autocorrelation can be rewritten as

$$R_{xx}(n) = K(w^{|n|} + Av^{|n|}), \qquad (227)$$

with K and A constants as previously given, and the response to an isotopic signal, which is chosen to be a Dirac delta function synchronized with the sampling time, is

$$x_s(n) = V_s c(w^n - v^n). \qquad (228)$$

A discrete model for the generation of the data in the form of an autoregressive, moving average (ARMA) system is

$$x_i = a_0 u_i - a_1 u_{i-1} + (v + w)x_{i-1} - vwx_{i-2}, \qquad (229)$$

where u_i is an uncorrelated (white) zero mean normal sequence, and the coefficients a_0 and a_1 are chosen to obtain an autocorrelation with the parameters of Eq. (227).

To this model we can now apply the various data analysis algorithms considered previously. However, in this case, a remarkable simplification is obtained by following the matched filter approach. (We remark that this approach can also be applied when dealing with models based on continuous time data.) The matched filter aims at recovering signals of known shape, $S(t)$, but of unknown time of arrival. As we have shown previously, this is done by using a filter with an inpulse response equal to a time reversed version, $S(-t)$, of the signal. We assume a background of white noise. Therefore the proper matched filter section must be preceded in general by a filter section that whitens the noise.

As the signal at the input of the system is assumed to be a Dirac delta function, then its spectrum is white, as is the spectrum of the noise, and the matched filter provides the same results as the Wiener-Kolmogorov filter. However, the matched filter approach is more general and can be used to detect events of finite duration as well as to search for signals of given shape.

In our case the whitening filter is the inverse of Eq. (229), that is,

$$\lambda_i = (1/a_0)[x_i - (v + w)x_{i-1} + vwx_{i-2} + a_1\lambda_{i-1}], \qquad (230)$$

and the response due to the standard signal is

$$\lambda_{sn} = (2cV_s/a_0)(w-v)(a_1/a_0)^n. \tag{231}$$

In the absence of noise, the response would be a delta function and no further filtering would be required.

The filter matched to the response [Eq. (231)] is

$$z_i = (c/a_0)(w-v)\lambda_{i+1} + (a_1/a_0)z_{i-1}, \tag{232}$$

where the scaling factor $2V_s$ has been neglected. We note that this is an anticausal filter; since the whitening is based in principle on all past samples and the matching is based on all future samples, the total matched filter depends on both the past and future data, as is the case for the Wiener–Kolmogorov filter. However, the structure of the discrete time matched filter, as well as the computations required, is remarkably simpler than the continuous time, Wiener–Kolmogorov filter.

When considering *both* channels, the SNR at the output of the matched filter is given by the expression

$$(\text{SNR})_M = V_s^2(2V_{nb}^2)^{-1}(1+\gamma)(1-\gamma)^{-1}(w-v)^2$$
$$\times[(1-w^2)^2(1-v^2)^2(1+A)^2 + 4A(w-v)^2(1-v^2)(1-w^2)]^{-1}, \tag{233}$$

where

$$A = (\Gamma/\gamma)(1-\gamma^2), \tag{234}$$

of the Wiener–Kolmogorov filter.

Many tests of the matched filter algorithm have been made on data. Results are in close agreement with the theory and provide a considerable improvement over those obtained with a continuous time Wiener–Kolmogorov filter, thereby providing significant improvements in the capability of detecting weak, short-lived signals in the presence of a noisy, time-varying record—a problem of considerable concern in the desire to detect unequivocally the signature of rapidly varying geological events in the isotopic record.

Chapter 15 | Maximum Entropy and Q-Model Methods

While time series analyses and frequency domain methods have proven inordinately useful in general, geological analyses of data records have a set of problems that suggest other methods of analyses are more easily interpretable. In the seismic case, one often has an extremely good idea of the source (dynamite, Vibroseis, etc.) and an equally good idea that it is the reflection coefficients of stratigraphic layers that are responsible for the observed seismic signals. In addition the record sampling can often be done at equal time intervals and with a good density of coverage and high resolution across the whole of the seismic record. By way of contrast, the geologic record of isotopic chronostratigraphy is not uniformly dense in time and is not at the same resolution even when dense time coverage exists, and we really do not know the intrinsic source of the isotopic signal or the geological–biological filtering done to the isotopic signal by the organisms and sediments in which we record the isotopic signature.

This should not be construed as a statement that the standard time series–frequency domain analyses are not to be attempted or trusted. Indeed quite the contrary is true. Such methods have the virtue of supplying quantitative error information for a detailed interpretation of results, as well as providing superb quantitative tools for problems of quality assessment of data, the processing of data, and the interpretation of robustness of the signals thought to be in the records. Yet the contrasting behavior of geologic data relative to seismic data suggests that techniques not so beholden to equal time sampling, source knowledge, etc., should also be of value in assessing signals and potential patterns of signals in isotope chronostratigraphic records.

Such techniques were introduced in the field of astrophysics by Gull and Daniel (1978). Since then, the methods, called maximum entropy methods

(MEM) [not to be confused with the maximum entropy algorithms for predicting autocorrelation coefficients introduced by Burg (1967)], have advanced significantly. In essence the sense of the argument, and its later developments, owes much to Shannon and Weaver's (1963) information theory concept, as we shall see. To explain the development in as simple a manner as possible, we first consider the unbiased MEM, and then the effect of biasing the MEM to account for known, or suspected, signals in the data set. We then look at the resulting uses of factor analysis and vector analysis to account for the commonality of response of a group of records.

I. Unbiased MEM

Consider a set of M records $x(k, t)$ $(k = 1, 2, \ldots M)$ sampled at times t_i $(i = 1, \ldots, P)$, where the sampling times for each record may be different, and the number of sampling times of each record may also be different, so that, as said, we deal with a very different situation than is usually available in a uniformly sampled set of records over the same total time interval.

The records need to be searched in some way for a common signal. To provide measures indicative of a common response, we first pool the data[6] obtaining

$$x(t) = M^{-1} \sum_{k=1}^{M} x(k, t). \tag{235}$$

If we now normalize $x(t)$, such that $\int_0^T x(t)\, dt = 1$, where T is the total length of the record, then $x(t)$ measures directly the probability of finding the value $x = X$ at time t.

The MEM now defines a pooled variable over a time interval $T_i < t < T_{i+1}$

$$\varphi_{i+1} = \int_{T_i}^{T_{i+1}} x(t)\, dt, \tag{236}$$

with

$$\varphi_1 = \int_0^{T_1} x(t)\, dt, \tag{237}$$

where we take the measurements to start at $t = 0$.

Then we construct the entropy of the pooled data set where N is the number of intervals (we will soon discuss the question of how to choose

---

[6] We take the data to be of like type (say $\delta^{18}O$). If we have to deal with different data types, we first convert them to a common scale, as was done previously.

the number of intervals):

$$E_{\text{pool}} = -\sum_{j=1}^{N} \varphi_j \ln \varphi_j. \tag{238a}$$

The entropy of the pooled data set is maximized when all the φ_j are identical,[7] i.e.,

$$\int_{T_i}^{T_{i+1}} x(t) \, dt = \varphi = \text{constant.} \tag{238b}$$

Then,

$$E_{\text{pool}}(\max) = -N\varphi \ln \varphi. \tag{239}$$

It is clear that if we have prechosen the number N of intervals, then it is a relatively simple matter, starting at $t = 0$, to construct from the relevant φ the sequence $T_1, T_2, \ldots$. (A pragmatic way of doing this is to pick the desired value of φ and then figure how many N values will cover the pooled data set.)

That the pooled data set has a maximum entropy of a fixed N does not mean that the entropy of the individual records is maximized. Indeed, the entropy of the kth record taken over the intervals appropriate to the pooled data is

$$\mathscr{E}(k) = -\sum_{j=1}^{N} \varphi_j(k) \ln \varphi_j(k), \tag{240}$$

where

$$\varphi_j(k) = \int_{T_j}^{T_{j+1}} x(k, t) \, dt. \tag{241}$$

The relative entropy RE of record k is given by

$$RE(k) = \mathscr{E}(k) / \mathscr{E}_{\text{pool}}(\max)$$

$$= -(\ln N)^{-1} \sum_{j=1}^{N} \varphi_j(k) \ln \varphi_j(k). \tag{242}$$

The average relative entropy of *all* the M records is

$$RE = M^{-1} \sum_{k=1}^{M} RE(k). \tag{243}$$

A measure of statistical significance of a given relative entropy is determined by how far RE departs from unity. The greater the departure from unity, the greater is the statistical significance.

---

[7] If we deal with a normalized pooled record, then $\varphi = N^{-1}$ and $E_{\text{pool}}(\max) = +\ln N$.

Many tests are available for deciding on the significance of information in given bins. For instance, one can define the normalized matrix of relative entropy

$$RE(k; j) = -(\ln N)^{-1}\phi_j(k) \ln \phi_j(k);$$

$$k = 2, \ldots, M; \qquad j = 1, 2, \ldots, N; \qquad (244)$$

and regard this matrix as conveying all of the statistically meaningful information on both the pooled data set and the individual records. Standard techniques, as used in the section on spectral analysis, can then be gainfully employed with respect to the data $RE(k; j)$ in order to determine a common signal.

However, the problem of how to determine the correct number of intervals, N, into which to subdivide the pooled data set is clearly of importance. If, arbitrarily, N is chosen to be unity, then the pooled maximum entropy is zero;[8] while if N is arbitrarily increased to an extremely large value, then most of the intervals contain no measure at all or contain only one measurement at a time from only one record at a time.

In either event the pooled entropy of the set of records is then not a very relevant quantity. For $N = 1$, we have too coarse a resolution to measure anything of significance, while for $N \to \infty$, we have over resolved the data and are looking at the "graininess" of the system. Somewhere between these two extremes is the answer sought for the best N value.

Following standard residual error analysis techniques, we choose that N which minimizes the variance in the average relative entropy per record. Thus, for a choice of N, the relative entropy of a record is

$$RE(N; k) = -(\ln N)^{-1} \sum_{j=1}^{N} \varphi(k) \ln \varphi(k). \qquad (245)$$

The average relative entropy of the M records is

$$R_{\text{mean}}(N) = M^{-1} \sum_{k=1}^{M} RE(N; k). \qquad (246)$$

The variance in the average relative entropy per record is

$$\delta R(N)^2 = (M-1)^{-1} \sum_{k=1}^{M} [RE(N, k) - R_{\text{mean}}(N)]^2. \qquad (247)$$

As N varies monotonically in $\infty \geqslant N \geqslant 1$, $\delta R(N)^2$ will be large at small N (too few bins for any resolution) and large again at large N (too much "graininess" in the attempted data resolution). Hence, somewhere in

---

[8] Without loss of generality, for the remainder of this chapter we deal with the normalized system.

between is the best value to choose for the number of bins, N. The nonlinear dependence of entropy on the functional φ_j, and its nonlinear dependence on the interval positions $T_1, \ldots, T_j, \ldots$, (which are data structure dependent), precludes the availability of a simple functional form for expressing N directly in terms of operations on the data, as we did in the time series analysis and spectral series analysis methods where *linear, additive* noise was taken to be present.

Once having determined N, the best number of intervals into which to split the pooled data, from minimal variance in the average relative entropy per record, we know the size of the relative entropy data matrix, $RE(k, j)$. Questions regarding the "visibility" of signals in the data can then be answered. For instance, using standard vector analysis we can take the relative entropy matrix of data and diagonalize it, thereby obtaining the eigenvectors Λ and associated eigenvalues λ_n, which characterize common signal information in the data set of M records. The corresponding uncertainties in the eigenvalues and eigenvectors are then determined from conventional residual least squares techniques.

Alternatively, one can choose to Fourier transform the relative entropy data matrix (in the space of the N intervals) and treat it with the powerful spectral methods of analysis of the previous sections, or one can regard the relative entropy matrix (which is already positive-definite) as the matrix to which to apply spectral methods directly.

The point is that once the entropy of the pooled data records is defined and maximized, and the optimal number of intervals determined for the pooled entropy, then the rest of the data can be organized into a relative entropy matrix on which standard analysis techniques can be used.

In the applications of the MEM to petrographic image analysis, the usual mode of operation is to perform a linear unmixing analysis (see later).

II. Biased MEM

A set M of records $x(k, t)$ gives the probability $f^{(k)}(t)$ for each record that some quantity lies in the range x to $x + dx$ at time t (take $x > 0$ for simplicity and without loss of generality). The pooled data give

$$f(t) = M^{-1} \sum_{k=1}^{M} f^{(k)}(t). \qquad (248)$$

We have already seen in the previous sections how to "bin" the pooled data to maximize the entropy of the system. But in answering this question we assumed that no information was available about the data, i.e., we wanted to determine the maximally unbiased answer.

However, in reality we often know a tremendous amount about the data probability (e.g., whether it is flat, curved up, curved down, or peaked, or if it wiggles). Such information is normally seen by eyeball inspection, i.e., we know quite a bit about the low-frequency behavior of the data. Now, how do we use the information available to better determine the bins for the pooled data that will maximize the entropy of the system, subject to the constraint that we know some of the low-frequency "shaping" of the data?

Let the bin coordinates be at $b_1, b_2, \ldots, b_N$ with $f(t) = 0$ in $t > b_N$ and $b_1 = 0$, and let

$$\varphi_i = \int_{b_i}^{b_{i+1}} f(t) \, dt; \qquad \sum_{i=1}^{N} \varphi_i = 1 = \int_0^{\infty} f(t) \, dt. \qquad (249)$$

With no constraint, the binned, pooled data has the entropy

$$E(N; \mathbf{b}) = - \sum_{i=1}^{N} \varphi_i \ln \varphi. \qquad (250)$$

Let the original samples of data be discretely sampled on a regular grid of interval width Δt ("padding" with zeros in the usual manner if necessary to fill the grid uniformly). Then a discrete Fourier transform algorithm will yield the Fourier transform of the pooled data

$$F(\omega) = (2\pi)^{-1} \Delta t \sum_{l=1}^{P} f(l \, \Delta \tau) \exp(-i\omega l \, \Delta \tau), \qquad (251)$$

with $P \, \Delta \tau = b_N$, and with a similar expression for each record:

$$F^{(k)}(\omega) = (2\pi)^{-1} \Delta \tau \sum_{l=1}^{P} f^{(k)}(l \, \Delta \tau) \exp(-i\omega l \, \Delta \tau). \qquad (252)$$

The standard error $\sigma(\omega)$ in each Fourier transform component $F(\omega)$ can then be estimated by[9]

$$\sigma(\omega)^2 = (M-1)^{-1} \sum_{k=1}^{M} [|F^{(k)}(\omega) - F(\omega)|^2]. \qquad (253)$$

We would normally regard only those values of $F(\omega)$ as significant for which $|F(\omega)| \geq \sigma(\omega)$. Let this requirement produce the significant values $F(\omega)$, with $\omega \in \Omega$ describing the set of values contained in the domain Ω that satisfy the requirement.

---

[9] A similar analysis can be performed in the entropy space of φ. Since the pooled data are uniformly spread over intervals of width φ, Fourier transforms of individual records in φ space can be taken, etc. We have not followed this procedure here, since it is more difficult to visualize the physical significance of the operations.

Now if we have binned the pooled data at coordinates $b_1, b_2, \ldots, b_N$, we have traded resolution on $f(t)$ relative to the original gridding of interval width $\Delta\tau$ in favor of precision. Basically, we have replaced $f(t)$ in the interval b_i to b_{i+1} by the single value $(b_{i+1} - b_i)^{-1}\varphi_i$.

In particular the binned Fourier transform of the pooled data at angular frequency ω is

$$F_b(\omega) = (2\pi)^{-1}i\omega^{-1}\sum_{j=1}^{N}\varphi_j[\exp(-i\omega b_{j+1}) - \exp(-i\omega b_j)]. \qquad (254)$$

We now want to extremize the entropy subject to the constraint that $F_b(\omega)$ come as close as it can to being the more correct $F(\omega)$ value in $\omega \in \Omega$, i.e., we want to extremize

$$E(\text{constrained}) = -\sum_{j=1}^{N}\varphi_j\ln\varphi_j + \tfrac{1}{2}\lambda\sum_{\omega\in\Omega}|F(\omega) - F_b(\omega)|^2\sigma(\omega)^{-2}, \qquad (255)$$

where λ is a Lagrange undetermined multiplier and Ω is such that $|F(\omega)| \geq \sigma(\omega)$.

The unknown variables in Eq. (235) are the bin positions b_i. The extremization of Eq. (255) for fixed λ with respect to bin positions b_i is

$$\frac{\partial E(\text{constrained})}{\partial b_i} = 0 = f(b_i)\ln\left(\frac{\varphi_i}{\varphi_{i-1}}\right)$$

$$+ \lambda(2\pi)^{-1}\sum_{\omega\in\Omega}\Bigg\{ -(\varphi_i - \varphi_{i-1})\cos(\omega b_i)/\sigma(\omega)^2$$

$$+ f(b_i)(\omega\sigma(\omega)^2)^{-1}[\sin(\omega b_{i+1}) + \sin(\omega b_{i-1})$$

$$- 2\sin(\omega b_i)]\Bigg\}. \qquad (256)$$

Correspondence with standard difference formulations in radio astronomy (Gull and Daniel, 1978) and with crystallographic difference maps (Ramachandran and Srinivasan, 1970) implies for λ, the value

$$\lambda = 2[VF(0)]^{-1}\left[n(\Omega)^{-1}\sum_{\omega\in\Omega}\sigma(\omega)^{-2}\right]^{-1}, \qquad (257)$$

where $n(\Omega)$ is the number of terms in the sum, $\sum_{\omega\in\Omega}$ and where V is the volume factor $M\Delta\tau$.

Now, in the *absence* of the constraint (i.e., $\lambda = 0$), we know the minimization occurs when $\varphi_i^{(0)} = \varphi_{i-1}^{(0)} = $ constant, where N is the number of bins. Let these bin positions be at $b_i^{(0)}$.

If, as desired, the agreement between $F(\omega)$ and $F_b(\omega)$ is good, then the factor multiplied by λ, which measures the disparity, will be small. We can then write Eq. (256) in the form

$$\ln \varphi_i - \ln \varphi_{i-1}$$

$$= \lambda (2\pi)^{-1} \left\{ [\varphi_i - \varphi_{i-1}] f(b_i)^{-1} \sum_{\omega \in \Omega} \sigma(\omega)^{-2} \cos(\omega b_i) \right.$$

$$\left. - \sum_{\omega \in \Omega} [\omega \sigma(\omega)^2]^{-1} [\sin(\omega b_{i+1}) + \sin(\omega b_{i-1}) - 2\sin(\omega b_i)] \right\}. \quad (258)$$

The lowest order approximate solution to Eq. (258) is given by writing $b_i = b_i^{(0)} + \Delta b_i$ so that

$$\varphi_i = \varphi_i^{(0)} - \Delta b_i f(b_i) + \Delta b_{i+1} f(b_{i+1}), \quad (259a)$$

$$\varphi_{i-1} = \varphi_{i-1}^{(0)} - \Delta b_{i-1} f(b_{i-1}) + \Delta b_i f(b_i). \quad (259b)$$

Then Eq. (258) gives the approximate equation

$$[N - \lambda (2\pi f(b_i^{(0)}))^{-1} \sum_{\omega \in \Omega} \sigma(\omega)^{-2} \cos \omega b_i^{(0)}]$$

$$\times [\Delta b_{i+1} f(b_{i+1}^{(0)} - \Delta b_{i-1} f(b_{i-1}^{(0)})]$$

$$= -\lambda (2\pi)^{-1} \sum_{\omega \in \Omega} [\omega \sigma(\omega)^2]^{-1}$$

$$\times [\sin(\omega b_{i+1}^{(0)}) + \sin(\omega b_{i-1}^{(0)}) - 2\sin(\omega b_i^{(0)})], \quad (260)$$

which expresses the shift in the $(i+1)$th bin position in terms of the $(i-1)$th bin position shift. Note that for the lowest bin position $(b_1 = 0)$ so that for $i = 2$ we have to replace Eq. (260) by

$$\left[N - \lambda (2\pi f(0))^{-1} \sum_{\omega \in \Omega} \sigma(\omega)^{-2} \right] \Delta b_3 f(b_3^{(0)})$$

$$= -\lambda (2\pi)^{-1} \sum_{\omega \in \Omega} [\omega \sigma(\omega)^2]^{-1} [\sin(\omega b_3^{(0)}) - 2\sin(\omega b_2^{(0)})]. \quad (261)$$

Also note that since

$$\sum_{i=1}^{N} \varphi_i = 1, \quad (262)$$

it follows that

$$\sum_{i=1}^{N} \Delta b_i f(b_i^{(0)}) - \sum_{i=1}^{N} \Delta b_{i+1} f(b_{i+1}^{(0)}) = 0. \quad (263)$$

From Eqs. (261) and (262) we see that the odd values $\Delta b_3, \Delta b_5, \ldots, \Delta b_{2n+1}$ are all explicitly given in terms of the lower odd values. The even values

Δb_4, Δb_6, ..., Δb_{2n} are all given implicitly in term of linear functions of Δb_2, so that we can eventually eliminate all even b values between b_{2n} and b_2 to obtain a linear relation of the form

$$\Delta b_{2n} = G(n) \, \Delta b_2, \tag{264}$$

where $G(n)$ is a complicated function of n, but obtainable analytically. Hence, by using Eq. (264) and the explicit forms for the Δb_{2n+1} in Eq. (263), we can determine explicitly Δb_2 and hence obtain explicit values for the remaining even Δb_{2n}.

This completes the solution, because with first-order approximations for the b_i, one can now return to Eq. (258) and iterate again if necessary. This procedure provides an iterative, constrained entropy maximization.

Once the bin positions b_i are known (and fixing the number of bins N by the requirement of minimal residual error in the relative entropy), we again have available the arsenal of methods based on time series analysis and power spectral analysis from which to assess the records for commonality of signals. The point here is that MEM has the advantage of maximally using knowledge of the low-frequency signal components in order to concentrate its resolving power on the more subtle information extractable from the high-frequency components. Repeated iterative applications then determine the capability of the system at all frequencies for signal retention, identification, and extraction in the presence of noise.

III. End Members and Linear Unmixing

From the relative entropy matrix, or some version of it which may have been filtered or otherwise manipulated, we seek a set of vectors that carry the correlated information of similar signals. The eigenvectors and associated eigenvalues of the matrix carry this information, and a conventional method of assessing the eigenvectors of significance from the relative entropy matrix is to include more and more eigenvectors until, for instance, 100% of the variance, or as much as is possible, has been accounted. EXTENDED CABFAC (Klovan and Imbrie, 1971; Klovan and Miesch, 1976) is a computer algorithm specifically designed for such a purpose.

Linear unmixing algorithms (in the sense of Full *et al.*, 1981, 1982) take the eigenvectors and data matrices resulting from an EXTENDED CABFAC application and seek to determine end-member compositions and their relative proportions in each record. Choosing criteria for the selection of the "right" number of variables to feed to the linear unmixing algorithms is still somewhat of an art. Conventionally, those eigenvectors accounting for almost 99% of the variance, but determined to a high degree of precision

(85% accuracy or more is a usual rule), and a commonality throughout the set of records of greater than 95%. The number of potential end members that arises from such linear unmixing algorithms with these criteria is typically in the two to five range.

IV. Q-Model Types of Analysis

Most of the previously discussed techniques are derived from the theory of time series analysis and power spectral analysis. Such methods are a natural choice for most isotopic data analysis, as usually one or a relatively few types of isotopes are measured. The questions asked are usually of optional display, identification of events and correlation of these events to other cores or water temperature changes. Recently, however, more isotopic variables measured at the same sample location have been routinely calculated in addition to other types of supplementary information. Commonly, the goal is to use all the available information, establish patterns of variability from sample location to sample location, and then infer some geologic reality from these patterns. Some problems with this approach are that the correlations between such data tend to be variable (noisy) and it is often difficult to know how to interpret the derived patterns of variability.

The most successful techniques to date that attempt to solve these problems would appear to be those that make use, to a greater or lesser extent, of vector analysis (a form of linear algebra). With such methods, each sample location is considered a vector that consists of a set of measurements taken at that location. From these data, or some version of the data that may have been filtered, manipulated, or optimized via MEM, we seek a new set of references vectors that carry the correlated information of a similar sample location (i.e., similar signals). The relationship of the individual sample locations to this new set of reference vectors are used to define the patterns of variability from which interpretations are made.

However, before we go into the detail of these methods, we discuss some shortcomings of this approach. First, in order to be effective, many variables should be measured at each sample location and many sample locations should be used in the analysis. A large number of variables and samples ensure that any underlying geologic realities can be fully expressed in the data. Second, sampling becomes important. It is imperative with these techniques that the diversity of variability be equally sampled, or else either the obvious will be "discovered," or the pattern of variability most sampled will dominate the final outcome. Third, for most of the techniques, it is implicitly assumed that the changes in ratios between measurements are important and not the absolute quantities of any one measurement. This latter assumption will be of dubious validity in many isotopic studies.

Despite these shortcomings, and the fact that the results can rarely be used as a "final product" when the underlying assumptions do not hold, meaningful insights can be gleaned, suggesting appropriate application of many of the previous time series or power spectral techniques. Therefore it is instructive to review some of these techniques in order to illustrate their power for providing different insights. In this light, we now discuss one such popular method, QMODEL, and then discuss alternative approaches.

A. QMODEL Family of Algorithms

The QMODEL family of algorithms has been used to define end members and sample proportions for data sets considered to be mixtures. That is, individual samples of a data set can be viewed as mixtures of a relatively few end members. For example, oxygen isotopes have been interpreted traditionally to be related functionally to ice mass. In the QMODEL sense, most oxygen isotope data can be modeled as a mixture of two extreme cases: one where maximum ice mass is present and the other where minimum ice mass is present.

Normally the QMODEL family of algorithms is applied to data that sum to a constant value, such as whole rock compositional data or samples of a continuous spectra, which is made to sum to 100%. Relative to isotopic data, which normally do not sum consistently from sample to sample to any constant value, the data must be transformed to a constant value in such a way as to preserve the ratio information between all the variables. Studies using these algorithms include those by Imbrie and Van Andel (1964), who unmixed heavy mineral assemblages; Klovan (1966) and Fillon and Full (1984), who unmixed grain-size distribution data; Miesch (1976a) who used geochemical and petrologic data: Mazzullo and Ehrlich (1980, 1983), Brown et al. (1980), Hudson and Ehrlich (1980), Riester et al. (1982), and Mazzullo et al. (1984) who studied shape analysis; Healy-Williams et al. (1985), Healy-Williams (1983), Healy-Williams and Williams (1981) who studied micropaleontology; and recently, Ehrlich et al. (1984) who used these algorithms in the field of petrographic image analysis.

The roots of these techniques lie in classical factor analysis, a group of methods wrought with statistical pitfalls for the unwary. We will review the history of these algorithms to show that they are not subject to the same objections as "classical" factor analysis, and we will further show, by example, how proper mathematical models can be chosen via analysis of the results from these algorithms. Furthermore, in the course of our discussion, it will be shown that the QMODEL algorithms cannot force the data into any particular model, but rather give the data opportunity to express an appropriate model. However, before this can be done, a brief discussion of the type of data analysed by these algorithms will be presented, followed by a history of the algorithms.

1. Constant Sum Data. A data matrix consisting of M rows (M the number of individual samples) and N variables (N the number of measurements defining each sample) constitutes the raw input for the QMODEL family of algorithms. If the N measurements of each row are summed and the sum defines the same number for each row, the data matrix is said to consist of constant-sum data. Such data commonly arise when proportions or percentages are used to define each sample. Data matrices of constant-sum data have also been termed closed arrays. The problems inherent with such data arrays are well described by Chayes (1971) and are not discussed here.

Geometrically closed arrays can be visualized to lie in a space of dimensionality of at least one less than the number of measurements. This can be easily seen in Fig. 89 in two and three dimensions. In two dimensions (Fig. 89a), the data must lie on a line; in three dimensions (Fig. 89b), all possible data must lie on a plane; and in three or more dimensions, all possible data are constrained to lie on a space of one less than the number of measurements. This relatively simple observation is fundamental to the QMODEL family of algorithms and provides for an elegant escape from the problem of closure in the data matrix. This observation will also form

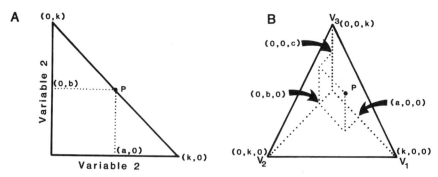

Fig. 89. Conceptual example of data in a closed array. (A) In two dimensions, any point P must lie on a line through $(k, 0)$ and $(0, k)$ as with $P = (a, b)$ and $a + b = k$, the summation constant of the array. (B) In three dimensions, any point P must lie on the plane defined by $(k, 0, 0)$, $(0, k, 0)$ and $(0, 0, k)$ as $P = (a, b, c)$ and $a + b + c = k$.

the distinction between factor analysis and vector analysis, which will be elaborated in Section IV, A, 2. In the meantime, it is assumed that the data input into the QMODEL family of algorithms consist of constant-sum data. With this in mind, a brief "algorithmic" history-of the QMODEL algorithms will be presented.

2. Factor versus Vector Analysis. Traditional factor analysis was first developed by psychologists in order to interpret intelligence tests in terms of a relatively small number of factors (e.g., intelligence and environment). The usual procedure was to summarize the raw data matrix (not necessarily confined to constant-sum data) via a correlation matrix, calculate the eigenvalues and eigenvectors of this correlation matrix, and then rotate the axes defined by the correlation matrix to coincide with the eigenvectors. Usually the first one or two eigenvectors were then used as an estimate of the variability expressed in the data matrix and subsequently given an interpretation. A more complete treatment of traditional factor analysis can be found in Joreskog *et al.* (1976). With this type of data modeling, serious considerations must be given to the underlying statistical assumptions in order to make the factor approach scientifically acceptable (see Matalas and Reiher, 1967; Temple, 1978).

Q-mode factor analysis differs from R-mode factor analysis (traditional factor analysis) in that the relationships between samples are analyzed as opposed to the correlation between variables. The similarity between each sample is usually measured by the cosine of the angle measured from the origin to the pair of samples. If the data consisted of 50 samples, a matrix of 50×50 cosines would comprise the similarity matrix. The analysis proceeds exactly like traditional factor analysis except that the cosine similarity matrix replaces the correlation matrix.

In both R-mode and Q-mode factor analysis, the data may be expressed in terms of oblique vectors representing actual variables or samples. Procedures to do this are numerous and have been called oblique rotations. The rotation are usually performed using a subset of the variability expressed by the set of eigenvalues. Imbrie (1963) makes the distinction between factor analysis and vector analysis on whether such oblique rotations are used (i.e., the use of the oblique rotation in the analysis implies vector analysis). In the Imbire (1963) algorithms, factor analysis is used to create a reduced space of dimensionality equal to the number of eigenvalues preserved in the factor model wherein vector analysis can be applied.

Relative to Q-mode analysis, this interpretation defines the point where the confusion between vector analysis and factor analysis lies. There is little question that an oblique rotation is vector analysis; however, if the eigenvectors ("factors") are not directly interpreted, then the factor analysis is purely a data reduction technique—a technique common in vector analysis. Therefore, the distinction between factor analysis and vector analysis lies not only in the application of an oblique rotation, but also in the interpretation of the reduced space. In the QMODEL algorithms, the reduced space is used only as a reference space for various oblique rotations and is not used for eigenvector interpretation. That is, the eigenvectors retained in this first step of the analysis serve only as a set of basis vectors defining the space used in subsequent analysis. In this light, the real problem becomes one of determining the correct number of dimensions that adequately define this reduced space. It is this question that the QMODEL family of algorithms addresses first. In the terminology of time series and spectral analyses, the reduced space is assumed to have captured the strongest signal, while the basis vectors not used in the space reduction are assumed to define the noise spectra.

3. CABFAC Family of Algorithms.   Before the question of how many eigenvalues to preserve was adequately answered (i.e., the "true" dimensionality of the reduced space), the work of Imbrie (1963) was extended to large data sets by Klovan and Imbrie (1971). This program laid the basis from which the other vector analysis technqiues could be applied practically to data sets consisting of many measurements over a large number of samples. This extension of Imbrie's work was made possible by the realization that the number of positive eigenvalues of the cosine similarity matrix and the correlation matrix was the same. Hence, the eigenvalues used in the Imbrie Q-mode analysis could be effectively extracted from the correlation matrix. This feature was incorporated into the program CABFAC (Klovan and Imbrie, 1971).

The question of the proper number of eigenvalues to preserve was directly addressed by Miesch (1976b). Before Miesch (1976b), the proper number of eigenvalues was determined by how much of the total variability was preserved in the reduced space (usually 95%) and by how much of the individual sample variance was preserved (measured by sample communality). Miesch, recognizing that the eigenvalues and sample communalities can be misleading, used the coefficient of determination to help decide the proper number of eigenvectors to retain. That is, each vector representing a sample in the reduced space can be projected to the space defined by the original metric and compared, via the coefficient of determination on a variable to variable basis, to the original sample measurements. Examples of the superiority of this approach over the usual arguments were given in Miesch (1976a). More simply, the set of coefficients of determination is a measure of how well the reduced space reflects the raw data. In another sense, the true dimensionality of the data has been determined. That is, even though the raw data set consisted of many measurements, the data actually lies on a plane or hyperplane (a figure in a space of more than three dimensions corresponding to a plane in ordinary space) of less dimensionality. A simple example depicting a space of lower dimensionality is depicted in Fig. 90. The data consist of three variables but, in reality, lie on a line segment at an oblique angle to the original axis. In this example, the resulting eigenvalues are $\#1 = 6.63$, $\#2 = 1.37$, and $\#3 = 0.0$. The coefficients of determination (Table III) verify that the data truly lie on a line segment.

The definition of this reduced space makes further analysis computationally easier. Large data sets can be reduced to a more manageable space, where all or most of the relationships between samples are preserved, further analysis performed, and the results brought back into the original data space with minimal loss of information. The algorithm EXTENDED CABFAC defines the transformations needed to functionally transfer the sample data from one space to another. EXTENDED CABFAC, when applied in this sense, cannot be construed as factor analysis but rather can only be considered vector analysis.

One final point to discuss is the use of the varimax rotation. The varimax rotation is an orthogonal rotation that has as its objective the definition of a new set of orthogonal axes such that projections from each vector onto these axes are either near the extremities or near the origin. The relative relationships of the data to each other are preserved and are commonly

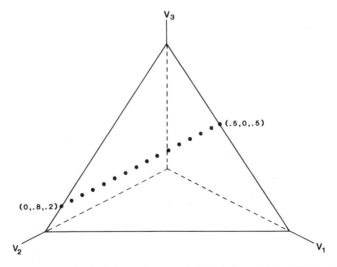

Fig. 90. Simple example depicting a data set defined in 3 variables (V_1, V_2, V_3), which actually lie in a space of lesser dimensionality.

called varimax loadings relative to this new set of axes. If each varimax loading is squared (preserving signs) and divided by the sum of squares for a particular vector (i.e., communality), the matrix of normalized varimax loadings is defined. The varimax loadings (defining varimax space) and normalized varimax loadings (defining normalized varimax space) are convenient spaces to visualize the relationships of the data vectors with each other in a realistic manner. Vectors in these spaces can be easily plotted and trends, or lack of trends, noted. For the most part these spaces are not critical to the QMODEL analysis. The loadings, however, have been used to identify very extreme samples in a data set, although extremes in these spaces do not always define extremes in the original data space. Hence, the use of the varimax or normalized varimax spaces for identification of extreme samples must be judged in the context of the entire analysis (i.e., additional criteria for extremeness must be used).

4. *QMODEL Algorithms.*   Once the true dimensionality of the data could be determined, Miesch (1976b) applied EXTENDED CABFAC to Imbrie's (1963) oblique rotation in order to define samples that would represent potential end members. Once located, the composition of these end members in the original space could be defined, and the relationships of all the remaining data to these end members could be defined via linear algebra. If the data happen to be linear mixtures of the defined end members, the relative location coordinates (sometimes called oblique loadings) of the data to these end members, could be interpreted as proportions or percentages. It is important to note that the mixing model, as applied here, is an interpretation and not an inherent attribute of those procedures. When the data are not inherently mixtures of end members, the relative coordinates (loadings) of the data about these end members represent a linear scaling of the variability displayed in the reduced space as opposed to proportions. The precise interpretation of such scaling lies with the interpreter and not with the algorithms.

The computer programs EXTENDED CABFAC and QMODEL (Klovan and Miesch, 1976) incorporate the procedures for the definition of the proper reduced space and for various oblique rotations. Further extensions of the QMODEL procedures were made by Full *et al.* (1981, 1982), wherein end members not contained within the original data set could be defined. The exact procedures used and the appropriate choice of technique can be found in the references. The computer programs that incorporate the extensions of Full *et al.* (1981, 1982) are EXTENDED QMODEL and FUZZY QMODEL. It is important to point out that FUZZY QMODEL incorporates the fuzzy clustering technique described by Bezdek (1981). The results of this clustering technique will be shown to be quite useful to detect cases where the data may be inherently clustered. Furthermore, when the data cluster, the proper number of clusters (i.e., cluster validity) can be ascertained easily using entropy considerations applied to the set of resulting membership functions (for a more complete treatment see Bezdek, 1981). Because of the importance of the fuzzy clustering technique in this analysis, a brief overview will be presented.

5. Fuzzy c Means (FCM). Basically a fuzzy c partition (c clusters) of a data set X is one that characterizes the membership of each sample point in all the clusters by a membership function that ranges between zero and unity, and the sum of the memberships for each sample point must itself be unity. One such algorithm available to resolve a data set into fuzzy c partitions, and which defines an appropriate membership, is the fuzzy c Means (FCM) algorithm of Bezdek (1981). The clustering criterion upon which FCM is based is the weighted sum of squared errors objective function J_m, defined as

$$J_m(U,\mathbf{V}) = \sum_{k=1}^{N} \sum_{i=1}^{c} (u_{ik})^m \|\mathbf{X}_k - \mathbf{v}_i\|^2, \tag{265}$$

where

1. $U = (u_{ik})$ is a fuzzy c partition of X (i.e., U is a matrix whose element u_{ik} represents the membership of vector $\mathbf{X}_k$ in the ith fuzzy cluster). The function that defines the u_{ik} is a nonlinear inverse function and is the key element in determining cluster validity.
2. $\mathbf{V} = (\mathbf{v}_1, \ldots, \mathbf{v}_c)$ is the set of cluster centers (i.e., $\mathbf{v}_i$ is the center or core of fuzzy cluster i). The location of the cluster centers is determined by a function that weights the contribution of each vector to a particular cluster by the memberships U.
3. $\|\cdot\| =$ any inner product norm. The choice of inner product norm essentially weights the contribution of any one variable in the clustering. In the examples given, the Euclidean norm is used.
4. $m \in [1, \infty) =$ a weighting exponent called the fuzziness exponent. Usually 2.0 is commonly used. A value of 1.0 will produce conventional or "hard" clusters.

The formula for the calculations of U and $\mathbf{V}$ can be found in Bezdek (1981). Some observations, however, can be gleaned from Eq. (265). Outlying data have virtually no effect on the locations of the centers v_i, because $(u_{ik})^m$ (the weighted membership of X_k in cluster i, raised to the mth power) approaches zero much faster than the $(u_{ij})^m$ of the other points X_j. In other words, the effects of noise can be essentially "filtered out" by adjusting m, the fuzziness weighting exponent.

Interest in FCM centers around three facts, which are well established in both theory and practice:

1. The centers $\{v_i\}$ generated by FCM lie well inside the convex hull of data set X. This is important because the edge adjustment procedure used in EXTENDED QMODEL always expands initial convex polytopes that have vertices that are putative end members.

2. The convex hull of the $\{v_i\}$ is, loosely speaking, a skeleton of the main structure of X itself. In particular, iterative expansion of the faces of this polytope will proceed toward the type of unmixing solution described in Full *et al.* (1981).

3. Noisy points (outliers in X) will exert little effect on the spatial location of any center v_i. This is one of the primary advantages of FCM—it is relatively insensitive to the presence of outliers.

The implications of 1–3 for the linear unmixing problem are simple: the centers $\{v_i\}$ generated by FCM always occupy spatial locations in end-member space, which are favorable for the initialization of the maximum edge translation procedure. The FUZZY QMODEL procedure extends the usefulness of the overall methodology of this set of techniques. Furthermore, the success or failure of the fuzzy solution will determine the appropriateness of the linear unmixing algorithms. That is, if the DCM algorithm shows that the data are indeed clustered, then the unmixing results are suspect.

6. Summary of Underlying Assumptions. Before a discussion of the appropriate model can ensue, a brief summary of the underlying assumptions of the QMODEL family of algorithms is in order. The violation of any assumption does not invalidate the mathematics, but rather confounds the interpretation of the results in terms of simple underlying models. It is interesting to note that when such violations are detected the QMODEL algorithms can still produce meaningful insights into the underlying realities measured in the data.

The major assumption is that each variable is completely and linearly determined by the subset of end members (Imbrie, 1963). Violation of this assumption causes a "discrepancy in proportion to the deviation from linearity" (Imbrie, 1963, p. 7). However, "if the relationships are not linear but are monotonic, then some simple transformation of raw data can usually be found to solve the problem. In the case of nonmonotonic relationships, however, where a parabolic or other peaked function is involved, the problem demands a more elaborate transformation procedure" (Imbrie, 1963, p. 7). With this assumption comes the responsibility of demonstrating the appropriateness of the techniques.

The demonstration of the appropriateness of the QMODEL family of techniques can be aided by the assumption that the geometry (position) of the data in *all* the spaces (raw data space, varimax space, oblique space) contains information. That is, if the data defines some geometric object in space (dimensionality can be greater than three but less than or equal to the number of variables), then the fact that a geometric figure is defined has some meaning. Given this assumption, the appropriateness of the QMODEL family of techniques can be approached from a geometric viewpoint, and techniques that define the geometric structure of vectors in *n* space can be used as a type of "front-end" processor to define collections of vectors with apparently common geometries. One such promising algorithm is the FUZZY *n* varieties algorithm of Bezdek (1981). It should be pointed out that if the geometry of the structure of the data has no meaning, the QMODEL family of algorithms (and any other similar multivariate approach for that manner) is not appropriate.

Two more assumptions are also made. The first is that the subset of end members must be equal to, or less than, the number of variables. This assumption must be carefully watched with isotope data as it may be commonplace that there are more end members than there are variables measured. Under such circumstances, there is no unique modeling solution. The second assumption is that the data sum to a constant value (constant-sum data). It is instructive to note that it is *not* assumed that the data are inherently mixing nor is it assumed that the underlying sample distributions are simple. The data samples can be complex and only have to reflect some interpretable relationship with respect to each other.

Given that these assumptions are satisfied, what remains is to choose an appropriate model for the interpretation of the output. If the underlying characteristics of the data are known *a priori*, then the choice of an appropriate model is stratightforward. However, if the inherent properties of the data are unknown, or if these properties are not known with absolute certainty, then the problem becomes one of clear recognition of these properties and the definition of the appropriate form of analysis. That is, can any incorrectly chosen model or a violation of some underlying assumption be detected? The simple way to detect such cases is to understand the geologic system samples in the data set. An alternative is to use some of the mathematical criteria described in the following sections.

7. Choice of an Appropriate Model. Three general classes of models can be defined that are relevant to the QMODEL family of algorithms: mixing, clustering, and scaling. Each of the models is, of course, chosen *a priori* before the QMODEL analysis is performed and is based on an understanding of the system under analysis. However, as can be gleaned from the previous discussion, the data are not forced to express any particular model. Therefore, this discussion will focus on the attributes of the QMODEL algorithms that direct the interpreter to question or revise the choice of model. Under such circumstances, the power of the QMODEL algorithms is displayed.

Mixing Model. The ideal mixing situation is one where all the data are linear combinations of a limited number of end members and that have structures in data space that exhibit uniform distributions of sample points. Additionally, the ideal situation is one where the number of end members is less than or equal to the number of measurements recorded for each sample. Figure 91 displays such an ideal case where three end members (Fig. 91a) are mixed in the proportions depicted in the triangular diagram (Fig. 91b). Note that the ideal end members are polymodal (number of modes greater than the number of end members). This data set was analyzed by EXTENDED CABFAC and produced the coefficients of determination shown in Fig. 92. It is clear from Fig. 92 and Table III that the data can be viewed in a space defined by three end members (i.e., the data lie in a plane) despite the fact that the data were originally defined in 10-dimensional variable space. The QMODEL algorithm (Miesch, 1976a) was applied to the reduced data set defined by the EXTENDED CABFAC algorithms and exactly reproduced the end members and proportions, as would be expected in such an ideal situation.

With less than perfect data, commonly encountered in geology, there is usually not as dramatic a convergence of the coefficients of determination to the ideal value of 1.0. Normally in geology, the coefficients of determina-

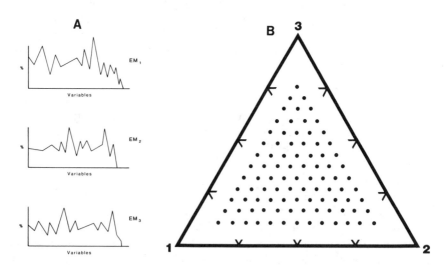

Fig. 91. Ideal mixing example wherein the three end members (EM) depicted in (A) are mixed in the proportions shown in (B).

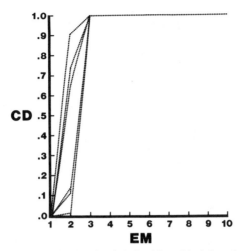

Fig. 92. Coefficient of determination (CD) derived from ideal data shown in Fig. 91 versus the number of end members (EM) in the reduced space. This figure strongly suggests that the data can be modeled using a three end-member solution.

tion converge to 1.0 at a slower rate, with increasing end members reflecting the random noise in most natural systems. Figure 93 displays an example of this taken from geochemical data analyzed in a rhyolite–basalt complex on the Gardiner River, Yellowstone National Park, Wyoming (Miesch, 1976b).

In each of these cases, it was known *a priori* that the data were inherently mixtures of a relatively few end members. What happens if a data set is analyzed where this assumption is violated? An example of a data set violating this assumption is one where the data consist of unimodal distributions (Gaussian normal in this example), where each sample differs from each other sample only in the mean for each distribution. Such a situation

Table III

Coefficient of Determination versus Variable Number[a]

Variable	Coefficient of determination	
	2 end members	3 end members
V_1	1.0	1.0
V_2	1.0	1.0
V_3	1.0	1.0

[a] Use this table with Fig. 92. The coefficients of determination were calculated by EXTENDED CABFAC.

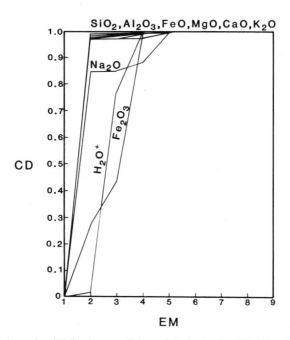

Fig. 93. End-member (EM) versus coefficient-of-determination (CD) diagram for a rhyolite-basalt complex on the Gardiner River, Yellowstone National Park, Wyoming (after Miesch, 1976a).

can arise in biologic morphometrics if it is assumed that the shape variability exhibited by a particular species is gradual through time or in changing environments. Several data sets consisting of 10 normal distributions with different amounts of overlap were created (Fig. 94). These data were recast as constant sum in the maximum entropy format with 10 intervals (Full *et al.*, 1984) to maximize differences between the samples (10 intervals used) and then analyzed via the EXTENDED CABFAC-QMODEL family of algorithms.

A typical set of coefficients of determination are depicted in Fig. 95. Note the slow convergence and the difficulties in determining the proper number of end members to choose. If the three end-member solutions were chosen and the normalized varimax loadings plotted on a triangular plot, the data would form arcs (Fig. 96). Each of the arcs is defined by a different overlap value that was held constant for each individual data set. When the overlap constant exceeded 9.2%, the data could not be realistically plotted on a triangular diagram.

Choosing the maximum number of end members equal to the number of variables (10), the QMODEL family of algorithms could not define suitable

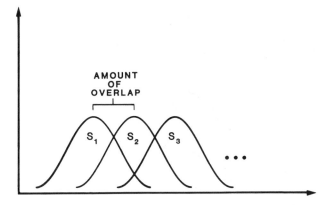

Fig. 94. Creation of data sets used in the example and depicted in Figs. 95 and 96. A unimodal curve was offset at multiples of a fixed distance to create subsequent samples.

end members to regenerate the data even in this ideal example (i.e., the algorithms could not converge). The conclusion derived from this example is that although the data were contained on a 10-dimensional hyperplane (as per the constant-sum constraint), a greater number of end members are needed to accurately and completely reproduce the original data (actually the number needed is equal to the number of samples), and this is a violation of the basic assumptions. Therefore, at least in this case, violations of basic assumptions can be detected, and from the failure of the QMODEL algorithms, information can be gleaned from which some of the inherent properties of the data can be ascertained.

Clustering Model.   Data that are modeled as clusters are usually not analyzed via the QMODEL family of algorithms. Techniques such as the FUZZY *c* Means algorithm (Bezdek *et al.*, 1984) are more appropriate for such data. The situation may arise, however, where the structure of the data was previously unrecognized to form discrete groupings in data space, which can happen when many variables are analyzed.

Within limits, cases of clustering can be detected using the QMODEL family of algorithms. The limits are (1) that the number of clusters be less than or equal to the number of variables and (2) that the clusters display discrete groupings in the variable space. Detection of clustering in data sets that fall outside of these limits must be made with either *a priori* understanding of the inherent properties of the data or via pure clustering techniques.

An ideally clustered data set was created and consisted of 10 variables and 4 clusters. EXTENDED CABFAC indicated that a 3 end-member solution was appropriate (Fig. 96). EXTENDED QMODEL and FUZZY QMODEL were used to model the data. Both techniques produced similar

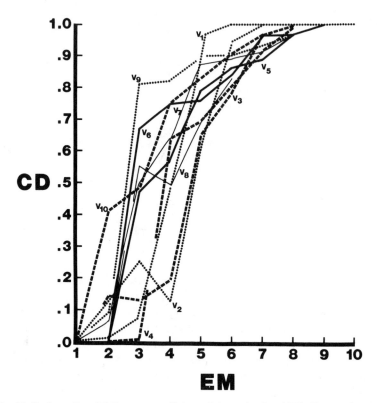

Fig. 95. End-member (EM) versus coefficient-of-determination (CD) diagram for the data described in Fig. 94. Note the slow convergence of the variables ($V_1, V_2, \ldots, V_{10}$); overlapping data, 6.0%, 10 intervals.

solutions. FUZZY QMODEL, however, produced the clues needed to suggest that the data might be better modeled as a clustering situation. The clues were found in the fuzzy memberships. The fuzzy memberships measure the similarity between a sample and every cluster. When the data are truly clustered, the memberships are generally 0.9 or greater, or less than 0.1, for all samples, whereas in mixing situations no such trends can be observed. Therefore, within limits, clustering can be detected.

Situations can arise where the data may appear clustered because of sampling problems or may appear to be mixing because of overlapping clusters. The use of FUZZY QMODEL and other members of the QMODEL family of algorithms does not replace solid scientific thinking, but rather these are tools to produce new insights into the underlying realties measured in a data set.

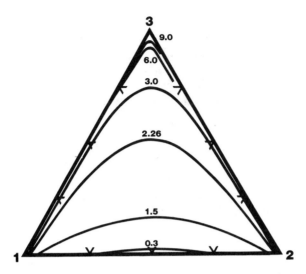

Fig. 96. Plot depicting curves defined by the data of Fig. 94 with varying percent overlap (above each curve). The curves are defined in three end-member varimax space (each apex of the triangle represents a different varimax reference axis).

Scaling. The scaling model, when appropriate, can only be detected via insight into the true nature of the data. Geometrically, the scaling model can appear to be similar to the mixing model or the cluster model. An example of linear scaling would be the measuring of the degree of abrasion of a sand sample as it is traveling down a river. An example of nonlinear scaling would be the chemical dissolution of grains of feldspar, which is a function of surface area. In the latter example, the loadings could no longer be considered proportions but rather some other measure related to the rate of change.

Another example of scaling can be found when the compositions of liquid hydrocarbons are analyzed. With migration, the proportions of the various components can change with distance, while at the same time bacteria can also affect the compositions of the oil samples. Only with knowledge of the geochemical nature of the resulting end members can a meaningful model be applied to the results of the QMODEL analysis in such a situation.

B. Nonconstant Sum

One active area of research is to model data in a fashion similar to the above procedures using vectors that are inherently not constant sum. While progress appears to be imminent, no sets of algorithms to date have been able to duplicate the same results as previously described. Such algorithms would preserve the absolute quantity for each sample location while still being able to accurately relate each sample to a smaller subset of reference vectors. However, in such a nonconstant analysis the underlying assumptions will be similar to the constant-sum case. As with the constant-sum algorithms, the appropriateness of any model will have to be proved and not assumed.

Relative to clustering models, however, the FCM algorithms offers much hope, in that this technique does not have the constant-sum constraint. If clustering is suspected, then the optimal number of clusters can be easily ascertained via entropy-based statistics. If clustering is not appropriate, then these statistics will warn the researcher that another form of analysis should be used.

V. Summary

The application of sophisticated methods of time series analysis, spectral analysis, or maximum entropy analysis, followed by some form of signal identification and determination, are the workhorses of modern signal processing techniques. Two basic types of technique have been examined here: linear and nonlinear. The linear techniques seek to design a linear filter to apply to the data in some way to maximize the signal-to-noise ratio on the assumption of additive noise. The nonlinear techniques seek to apply a prescribed nonlinear filtering operation to the data, which will boost the signal relative to the noise, and to then examine the consequent signal-to-noise behavior of the modified data as a consequence of the applied nonlinear operation. Both types of technique have their strengths and weaknesses, and both types yield different ways of examining records for signal signature.

Lest the reader get the idea that we have examined exhaustively methods of signal extraction from data records, we point out here that we have done no more than present a cursory overview of some of the more common methods of analysis in use today. Other techniques designed for more specialist data collection devices, or for more particular signal formats, are available. Among such techniques not looked at here are: (1) *coherent* noise filtering methods; (2) pulse shape and minimum phase deconvolution methods; (3) tomographic inverse schemes; and (4) generalized inverse methods (see, e.g., Menke, 1984).

We have included those quantitative methods of analysis that, in our opinion, are likely to be most applicable in the immediate future to data records of the isotope chronostratigraphic variety. As data collection techniques improve, and as more sophisticated methods of analysis are needed to examine the records for increasingly subtle suggestions of a commonality of signal, we have the comfort of knowing that such more highly sophisticated methods of quantitative data analysis have already been developed in other fields in order to accommodate precisely the same problems in sophistication of data collection and signal identification.

Chapter 16 | Numerical Examples and Case Histories

In the previous chapter, we saw developed the mathematical tools for an array of processes that can be used in attempts to extract a signal from a record, or set of records, in the presence of noise. In this chapter we provide a sampling of data from several fields to which the techniques have been applied. In order to keep the length of this chapter within reasonable bounds, we do not present here examples illustrating each and every technique, but rather we use illustrative examples from a fraction of the techniques. The "flavor" of the developments can then be more readily recognized, and some idea seen of the sort of graphs one ought to expect to obtain from individual processes applied to data in particular situations.

Since the introduction of the semblance technique into seismic geophysics, by Neidell and Taner (1971), it has had an excellent track record as a procedure for determining seismic velocities with travel time and so of converting seismic travel time to true subsurface depth. We report here on an illustrative example taken from Al-Chalabi (1979). Velocity analysis techniques are based on the assumption of straight-line (refraction-free) raypaths so that the relation between travel time T and offset x is supplied through the classical hyperbolic form

$$T^2 = t_0^2 + x^2/V^2 \tag{266}$$

where t_0 is the zero offset time, and V is the velocity to be determined.

The basic scheme in velocity analysis consists of performing a stack across the common depth point (CDP) gather along hyperbolic trajectories defined by Eq. (266). To illustrate the principle of these techniques, let us consider a hypothetical CDP gather (Fig. 97). Suppose that the noise-free reflection in Fig. 97 forms an exact hyperbola H_{op}, and that the zero offset time

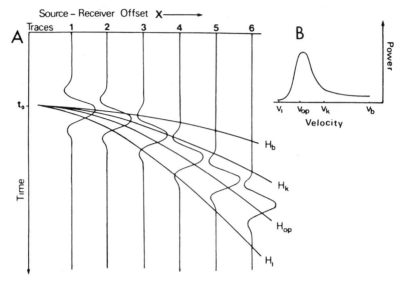

Fig. 97. A sketch (A) showing a series of hyperbolic trajectories. Hyperbola H_{op} produces the highest semblance at zero offset time t_0. The power spectrum (B) for t_0 as the velocity is varied is also depicted (after Al-Chalabi, 1979).

corresponding to the peak of the reflection is t_0. Suppose also that the velocity analysis is to be carried out with reference to t_0 and that the range of velocity to be covered by the analysis is V_a and V_b. The analysis is carried out as follows:

1. An initial stacking velocity $V_1 = V_a$ is assumed. This velocity corresponds to hyperbola H_1, i.e., align the traces according to hyperbola H_1.
2. The degree of semblance between the traces at this alignment is measured, for example, by summing the amplitudes at t_0 and determining the output power, i.e., the absolute value of the summation.
3. The velocity is then incremented by an appropriate step, then semblance is again measured.
4. Step 3 is repeated until V_b is reached.
5. The zero offset time is then incremented from t_0. Steps 1-4 are repeated.
6. These steps are repeated until the appropriate range of time down the CDP record has been covered.

In practice, the amplitude summation is carried out within a time gate, usually centered about the reference zero offset time. Thus, in a gate of width $\Delta\tau$ centered about t_0, the coherency is based on samples between

$t_0 - 0.5 \, \Delta\tau$ and $t_0 + 0.5 \, \Delta\tau$. The gate width is generally of the order of 0.75–1.5 times the predominant wavelet period.

The results of analysis in steps 1–4 are displayed as a plot of semblance measure versus velocity. Such a plot is known as a velocity spectrum. A new spectrum is produced at every time increment, so that the analysis of the whole CDP record is displayed as a series of these spectra. The spectrum corresponding to t_0 is shown in Fig. 97. This spectrum illustrates the increase in power buildup as the stacking velocity increases from V_1 and the traces are gradually brought in phase. Maximum semblance is attained when all the traces are maximally in phase, i.e., when the stacking velocity reaches a value that corresponds to the hyperbola H_{op}. As higher velocities are applied, the traces begin to get out of phase again and the power diminishes.

Figure 98 shows a variable density display. The density varies in steps; each step encompasses a range of semblance values. A number of distinct, high-coherency buildups can be seen. Each of these buildups corresponds with a strong primary reflection on the seismic section. At time 1.456 s, the maximum coherency occurs at a velocity of 2060 m s^{-1}.

The interpretation of a velocity spectra display is basically a process of identifying coherency buildups corresponding to primary reflections and isolating those due to multiplies and other spurious coherent events. The interpretation of the display of Fig. 98 shows that the primary semblance buildups are readily recognizable down to about 2 s. It can be seen that the contours become increasingly stretched along the velocity axis with increasing travel time. This stretching represents a deterioration in the sensitivity of the semblance measure with increasing time and causes a loss in resolution. It results from the fact that, with increasing time, large stacking velocity changes corresponding to low frequencies become increasingly predominant.

To illustrate the results of the semblance technique applied to isotope data, in Fig. 99(a) we show contours of constant semblance in a $(\Delta\tau, t)$ plot for isotopic data from sites V28239 and 502B for times t, ranging from the present to 800 KY ago with window width $\Delta\tau$ up to 170 KY. To be noted on the plot is a broad, high semblance (>1.9) plateau ranging from approximately 200 to 500 KY, nearly independent of the window interval, indicating broad, long-term trends in both isotopic records over the 200–500-KY period with the same sign of trend direction. A gradual roll-off of this common semblance trend is seen at younger and older times. Superimposed on this long-term trend behavior are more rapid fluctuations in semblance peaks and valleys, with a loss of resolution at window intervals larger than 20–40 KY, indicative of short-lived events on both records, which span a 20–40 KY width. A very rough quasi-periodicity in time t of these short-lived semblance events can be seen over a scale (estimated by eye) to be around

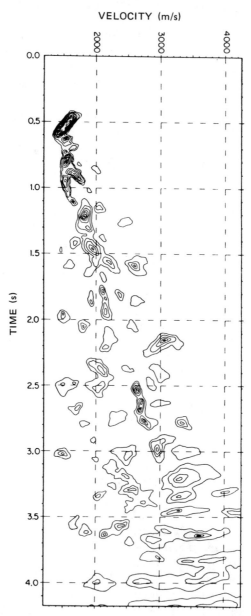

Fig. 98. A contoured velocity display spectrum with travel time. Below 2 s there is a deterioration in quality caused by a decrease in the signal-to-noise ratio and interference from spurious events (after Al-Chalabi, 1979).

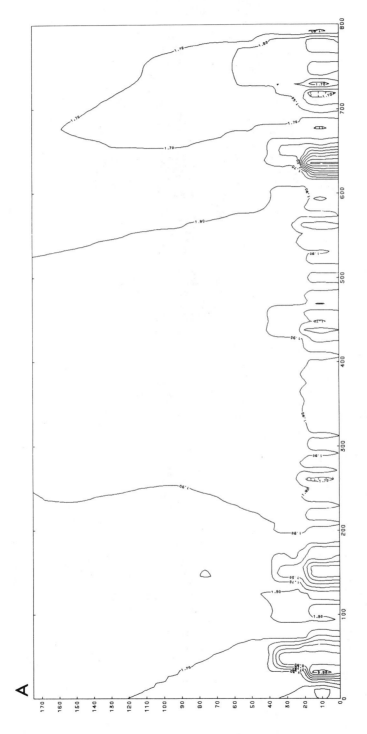

A

Fig. 99. (A) Contoured semblance plot of isotope records from sites 502B and V28239 to 800 KY with a window width to 170 KY. (B) The same as (A) but the window width is reduced to 50 KY and the scale expanded so that the localized fluctuations in semblance are better displayed.

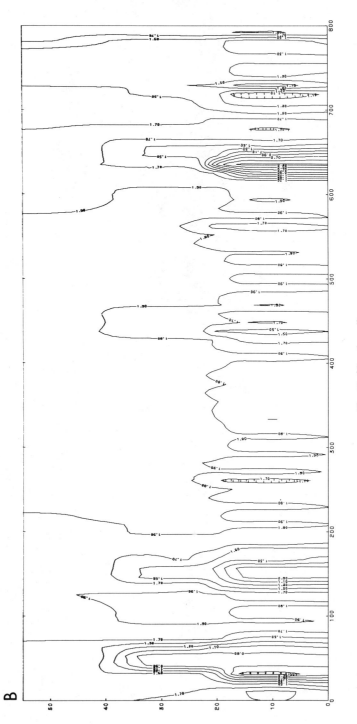

B

Fig. 99 (Continued)

20-50 KY—indicative, perhaps, of some recurring sequence of geological events occurring on this time scale, lasting about 20-40 KY and being registered in the isotopic record.

A similar semblance plot for isotopic data from both sites is shown in Fig. 99(b), but now the the window interval scale only goes to 50 KY to enhance the visibility of the small-scale, rapidly fluctuating semblance contours. The same broad, large-scale plateau of high semblance is seen between approximately 200-500 KY roughly independent of the window interval. Once more we see the rapidly fluctuating semblance events over time windows on the order of 20-50 KY, with a rough periodicity in time of approximately 20-50 KY, again indicative of some recurring sequence of geological events.

Chapter 17 | Filter and Deconvolution Methods

Following the original work of Wiener (1949), a vast body of literature has evolved dealing with the effects of sampling, filtering, and deconvolution procedures on data. Since our main interest is in providing only a few illustrative examples, and not in subverting the main thrust of this book to one dealing with all of the fundamental aspects of filter and deconvolution methods, we content ourselves here with quoting just a few examples. The interested readers can extend the applications almost indefinitely by consulting the references or by constructing their own examples.

I. A Filter Response

Figure 100 shows a sonic log of velocity with depth sampled at every 5 ft, while Fig. 101 shows the same log but smoothed over a 20 ft interval using a boxcar filter in *frequency* space of constant unit weight up to 60 Hz and zero at higher frequencies. The time delay of a seismic signal is influenced by such sampling and filtering. Figure 102 shows the normalized time delay for the logs of Figs. 100 and 101. We see that the effect of the smoothing (filtering) operation is to reduce the time delay by over a factor of two at low frequencies (<60 Hz), while retaining the correct shape of the time delay with frequency curve. However, above 60 Hz the low-frequency smoothing operation has seriously misrepresented the high-frequency information. These results were taken from Banik *et al.* (1985b). We have already seen in Chapter 16 an example (taken from Lerche and Komesaroff, 1979) of a prediction filter applied to data, and there seems no need to repeat that example here.

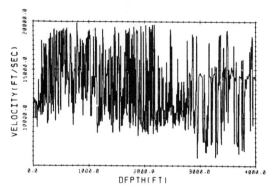

Fig. 100. Sonic log from the Persian Gulf. The recorded velocity was smoothed before resampling to 5 ft (1.5 m) (after Banik et al., 1985b).

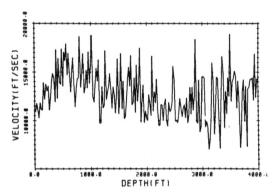

Fig. 101. Same as Fig. 100 except the log was smoothed over 20 ft (6.1 m) before resampling (after Banik et al., 1985b).

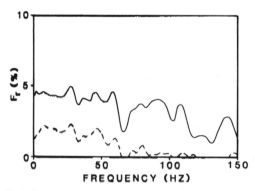

Fig. 102. Normalized time delay for the logs of Figs. 100 (heavy line 5-ft internal) and 101 (dashed line 20-ft interval). Note the decrease (by a factor of about 2) for the larger sample spacing log below 50 Hz and the complete loss of information and shape above 60 Hz (after Banik et al., 1985b).

II. Mapping Function Deconvolution, Noise, and Nonlinearity

Of relevance to the case of isotope chronostratigraphy is the question of the degree of matching that can be extracted from an intercomparison of two signals so that the stretching and compression of one signal in relation to another (the template) can be estimated and so relative rates of sedimentation determined. Of particular importance here is the degree to which a match can be obtained in the presence of noise and the degree to which the noise can be suppressed. We shall follow the nonlinear technique presented by Martinson, Menke, and Stoffa (1982, hereinafter MMS), so that techniques of *additive* noise suppression are not as helpful here as they are in the conventional prediction filter, smoothing, and deconvolution methods.

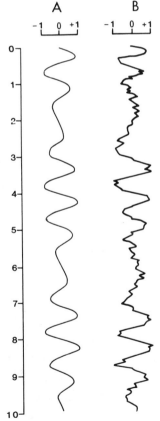

Fig. 103. The effect of noise on mapping function recovery. The distorted signal (B) is constructed by mapping the reference signal (A) with a known function (see Fig. 104) and adding 20% noise (after Martinson *et al.*, 1982).

Figure 103 illustrates synthetically the effect of noise on a comparison between two records where the noisy record has been produced by nonlinearly mapping the template and then adding 20% noise. Recovery of the mapping function using these two records then follows the nonlinear prescription given in Chapter 16 and expounded upon in detail in MMS. Figure 104 shows the results of the inverse nonlinear mapping procedure using 30

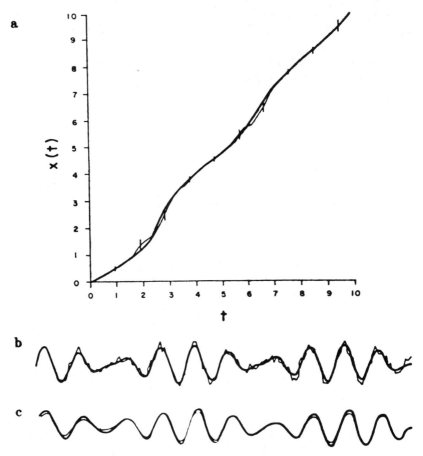

Fig. 104. (a) Mapping function recovered (light line) by correlating signals of Fig. 103 with and without suppressing the noise by filtering. The recovered mapping function differs from the true mapping function (heavy line) by small perturbations. The recovery was made using 30 coefficients and the error bars are magnified ($\times 100$) to illustrate relative size differences in errors as a function of t. (b) Match between reference signal (heavy line) and unfiltered recovered signal (light line). (c) Match between reference signal (heavy line) and filtered recovered signal (light line). The filtered signal was used to determine the mapping function (after Martison et al., 1982).

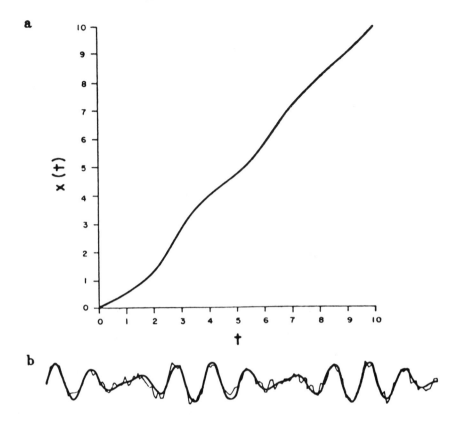

Fig. 105. (a) Mapping function for signal shown in Fig. 103 recovered by using only 21 coefficients and without noise suppression by filtering. The recovered and reference mapping functions overlay. (b) The match between the recovered signal (light line) and the reference signal (heavy line) is not as good as in Fig. 104(b) because the noise was not correlated (after Martinson *et al.*, 1982).

coefficients in the map routine and with a filter applied to suppress the noise and so determine the mapping function. By way of contrast, when fewer coefficients (21) are used in the mapping procedure, a significantly poorer fit to the template signal is found (Fig. 105), particularly when the noise is not suppressed by filtering. The strong dependence of the mapping function on both the number of coefficients used in attempts to recover the map and on the degree of noise present in the records is clearly seen and dominantly exemplified in applications of the inverse mapping technique to oxygen isotopic records RC11-120 and V19-30, as shown in Fig. 106.

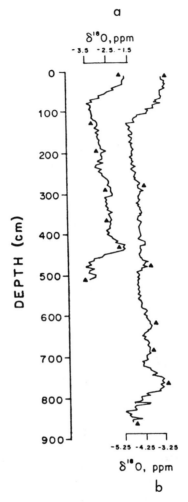

Fig. 106. Example of applying the mapping function method to stratigraphic correlation using oxygen isotope data from two marine cores. RC11-120 (a) is from the subpolar Indian Ocean and V19-30 (b) is from the equatorial Pacific. Triangles indicate points correlated by eye (after Martinson *et al.*, 1982).

Figure 107 shows the relative depth correlation of record V19-30 using record RC11-120 as a template when the triangle points (correlated by eye) in Fig. 106 are used as the underpinning basis for the mapping reconstruction. The spectral amplitude (not power) of each of the two records is shown in Fig. 108, using the eyeball correlation.

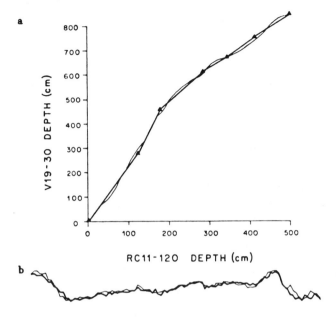

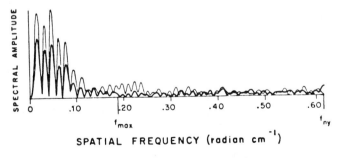

Fig. 107. (a) Mapping function recovered (light line) for the oxygen isotope data of Fig. 106 based upon fine tuning of the initial guess (bold line) chosen by eye. The triangles represent the points correlated by eye in Fig. 106. (b) Match between the reference signal RC11-120 (heavy line) and the recovered signal V19-30 (light line) has a coherence of 0.94 (after Martison *et al.*, 1982).

Fig. 108. Spectra of the reference signal RC11-120 (heavy line) and the recovered signal V19-30 (light line) using the mapping function of Fig. 107. The maximum frequency used in the mapping function is f_{max}, and the Nyquist (alias) frequency for the data is f_{ny} (after Martinson *et al.*, 1982).

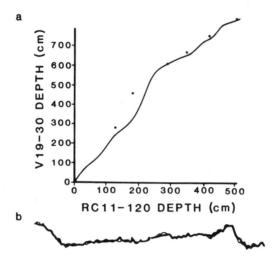

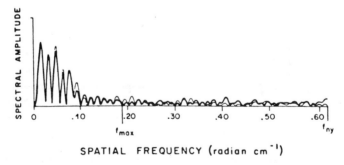

Fig. 109. (a) Mapping function for the oxygen isotope data of Fig. 106 using the objective mapping function technique (solid line). The triangles show the comparison of the points correlated by eye. (b) The match between the reference signal RC11-120 (heavy line) and the recovered signal V19-30 (light line) has a coherence of 0.97 (after Martinson *et al.*, 1982).

For comparison, a systematic and objective use of the MMS inverse mapping technique (rather than by subjective eyeball correlation of events) produces the mapping function shown in Fig. 109, which is a noticeable improvement (even by eye!) over that shown in Fig. 107. In addition the relative spectral amplitudes of the two records shown in Fig. 110 are now markedly improved in degree of fit over those obtained in Fig. 108, illustrat-

Fig. 110. Same as Fig. 108 but here we have used the objective mapping function method and the mapping function of Fig. 109(a). Note the significant improvement relative to Fig. 108 (after Martinson *et al.*, 1982).

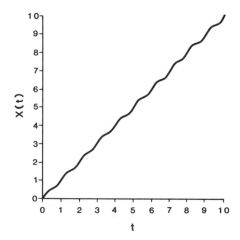

Fig. 111. Effect of periodicities in the mapping function on the properties of the distorted signal. The mapping function is a linear trend overlain by a sinusoid of unit period (after Martinson *et al.*, 1982).

ing both the high caliber of objective methodologies and the subjective bias inherent in visual attempts to produce the correct results.

In addition, a second factor is disturbing and can have a profound impact on attempts to tie the isotopic variations with depth to global and orbital periodic events (see Pisias and Shackleton, 1984; Pisias *et al.*, 1984; Shackleton and Pisias, 1985). This factor, while present to some degree in the spectral amplitude data from sites RC11-120 and V19-30, is perhaps best exemplified by a synthetic illustration. Consider the mapping function in Fig. 111, which shows the effect of periodicities in the mapping function on the properties of the distorted signal. Now apply this mapping function to the pure sinusoidal signal given in Fig. 112. The true power spectrum of the sinusoid contains just one prominent peak. But the distorted signal contains many peaks in its power spectrum, most of which have nothing to do with the intrinsic periodicity of the pristine signal. Hence, in order to tie together correctly the events recorded on a template record and those recorded on a distorted record (relative to the template) *it is absolutely crucial that the nonlinear distortion effects of the mapping function be removed with the greatest degree of precision*, subject only to noise in the record. If this is not done, then spurious periodicities could easily be mistaken for reality. In addition, even if this mapping removal has been properly done, we have one more major problem to face: we must now take the template isotope record and convert it from a record with respect to depth to one with respect to time using yet another mapping function. It should come

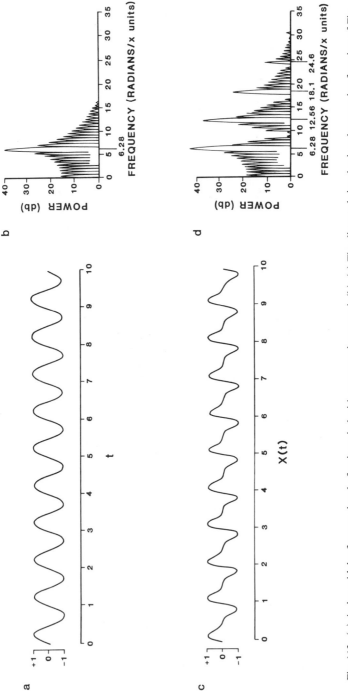

Fig. 112. (a) A sinusoidal reference signal of unit period with one prominent peak (b). (c) The distorted signal using the mapping function of Fig. 111 has a spectrum (d) containing several spurious peaks at harmonics of the unit period. This example makes apparent the paramount necessity of correctly removing the effects of the mapping function from the distorted signal (after Martinson et al., 1982).

as no surprise that a finite number of depth points, with measured ages, leave us in the unenviable state of reconstructing the mapping function at all intervening points *without* impressing spurious periodicities onto the record from incorrect decompaction, assumed depositional rates, bioturbation, and allied interpolations of sedimentation rate between the age-dated depths, that is, the problems of sample interpolation and filtering of data are still very much with us. Thus, if we were to obtain correlations on the scale of a sample spacing, as has been known to happen, we would clearly recognize that we are beyond the pale of any resolution except that which we have impressed upon the data from our own devices (but see, e.g., Pisias and Moore, 1981).

Chapter 18   Frequency Domain Methods

As we have remarked previously, frequency domain methods are probably one of the most detailed and developed modes of analysis available to us.

I. Linear Interpolation and Power Spectra

A simple example of the application of autocorrelation and power spectral methods is provided by the data presented in Fig. 113, which shows the δO^{18} isotopic data with depth for site V28239. *For illustrative purposes only,* the connectivity of the data is supplied by assuming that the datum points can be connected by straight-line segments (alternate assumptions such as triangular filtering, for example, based on the filtering and deconvolution procedures outlined in the previous section can, and should, also be invoked in order to determine the degree of robustness and reproducibility of subsequent analyses).

In any event with the record as interpolated, it is then possible to construct the power spectrum as shown in Fig. 114 (normalized so that the power at zero frequency is unity and with 170 harmonics of spacing $1/T$, where T is the record length, 1.36 MY). To be noted in the power spectrum is the general decay of power as the frequency increases, but also to be noted is that superimposed upon the overall decay are spikes of statistical significance to at least harmonic 33. A rough periodicity can be seen in the power spectrum with dominant peaking roughly every sixth harmonic on either side of harmonic 15 (the most dominant harmonic) but of width on the order of two to three harmonics. Thus, some sequence of events in the

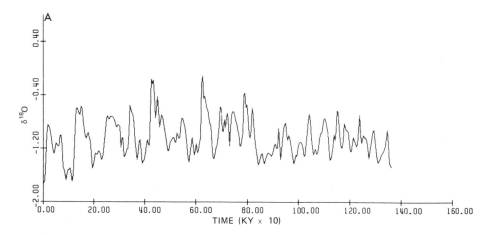

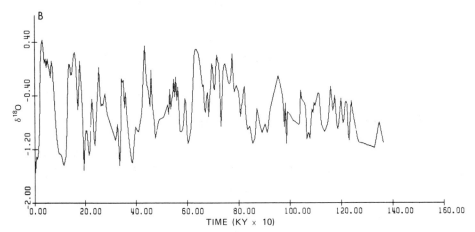

Fig. 113. Oxygen isotope data from a marine core at (A) site V28239 and (B) site 502B. The line through the data is made up of linear segments between each pair of datum points.

original record is occurring in a roughly cyclic manner, but neither the cycle time nor the duration of the events is holding precisely constant.

For comparison, in Fig. 115 we show the power spectrum of the δO^{18} record taken from site 502B over the same record length, so that the power levels at each harmonic are comparable on a one-to-one basis. Again, the power spectrum is normalized to unity at zero frequency.

The similarities and differences between the two power spectra are illuminating. Figure 115 shows basically the same pattern of peaks and troughs at about the same harmonics (to within the resolution limits) as Fig. 114. However, the strength of the dominant harmonic 15 is reduced in

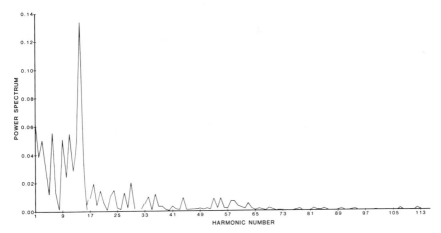

Fig. 114. The power spectrum of the linearly interpolated data of Fig. 113(a) normalized to unit power at zero frequency.

Fig. 115, while the harmonics lower than 15 are all more uniform in size than in Fig. 114, as is also the case for the harmonics higher than 15, but there is an abrupt change in the uniformity scale size around harmonic 15.

Thus, the events, of near cyclicity but variable duration, being recorded by the isotopic data at V28239 were also being recorded simultaneously (to

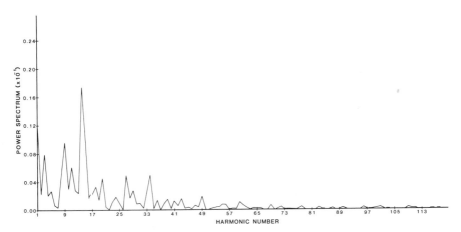

Fig. 115. Same as Fig. 114 but for isotope data from site 502B [Fig. 113(b)] taken over the same record length. The conversion of depth to time in Figs. 113, 114, and 115 is done by radioactive age dating and the assumption of constant deposition rate between age-dated horizons.

within about two or three harmonic widths) by the isotopic data at well 502B. But the low-frequency data at site 502B are less "ragged" than those shown in Fig. 114 for site V28239, and the high-frequency difference [since the high-frequency (above harmonic 15) power spectrum shows an abrupt drop in magnitude] is indicative of some process erasing the high-frequency information. Alternatively, there never was any significant level of high-frequency information deposited in the isotopic record and, above harmonic 15, what we are seeing in both power spectra is the result of the sampling interpolation of the data.

II. Autocorrelation and Cross-Correlation in Time

A different way to illustrate the information content of the data is to produce autocorrelation and cross-correlation plots in time. Figure 116 shows the autocorrelation function for site V28239 as a function of lag time, while Fig. 117 does the same for well 502B. Inspection of the two figures shows striking differences as well as a few similarities. Both autocorrelations record a general decrease as lag time increases, but whereas the record for

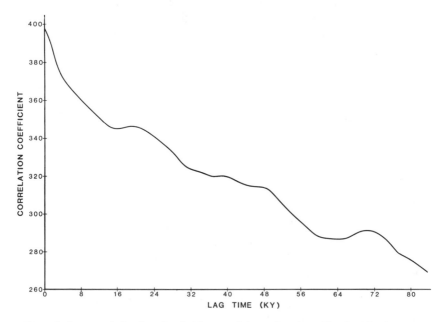

Fig. 116. Autocorrelation function (arbitrary units) as a function of lag time for the oxygen isotope data from site V28239.

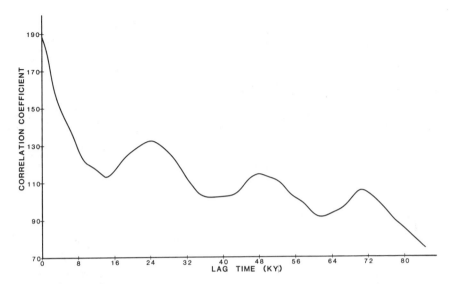

Fig. 117. Same as Fig. 116 but for oxygen isotope data from site 502B.

site V28239 shows minor perturbations on the general decreasing background, Fig. 117 for well 502B shows three extremely noticeable peaks at approximately 24, 48, and 72 KY, each with a width of approximately 4–8 KY. This is the low-frequency information so sharply recorded in the power spectrum. The same information of cyclicity is present in the autocorrelation record for site V28239 (as is shown by the corresponding power spectrum) but it is much more difficult to discern by eye. Crudely speaking, slow variations in time show up as rapid variations in frequency and vice versa, which is why it is important to investigate both the time and frequency domains for information content of the records. What can be seen easily in one domain is often obscured in the other.

To examine the extent to which the two records are representing the same events (within the resolution of simultaneity of some 4–8 KY) Fig. 118 shows the cross-correlation function with respect to lag time in thousands of years. We see the same broad peaks at about 24, 48, and 72 KY, of width about 4–8 KY, but they are no longer so sharply defined as they were for the autocorrelation record for site 502B, representing the blurring of the commonality and simultaneity of the events by the less pronounced record of the events shown by site V28239. In particular note that the recorded event at about 24 KY had its peak shifted to a shorter lag time and has broadened somewhat. Clearly, as long as the records are sampled on a finer time scale than the 4–8 KY width of the significant peaks, no harm is done by any filtering operation or resampling to a coarse scale, as long as that scale is

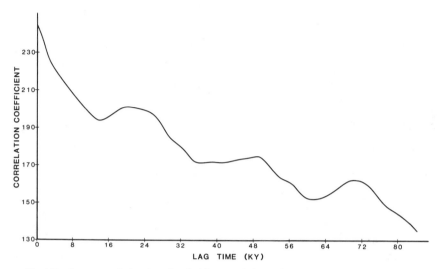

Fig. 118. Cross-correlation function (arbitrary units) as a function of lag time for the oxygen isotope data from sites 502B and V28239.

much less than the 4–8-KY width of the observed peaks and troughs. Filtering, smoothing, and resampling at a larger interval will seriously distort the information content of a given isotopic record in particular and of the cross-correlation information betwen isotopic records in general.

III. Window Effects on Power Spectra

The degradation of an isotopic record with increasing time after deposition can also be investigated by a variant on the power spectrum technique. There is, of course, always the degradation or coarsening of the record because of larger time intervals between isotopic sampling points as our capability for age dating lessens the further back in time we go. But even if this effect is ignored [and it can always be accommodated to some extent by (1) resampling of the coarsest record or (2) using a predictive filter technique of the Burg (1967) algorithmic type, or some other variant thereof, to interpolate between the coarse time samples and so "fine up" the record], there will, nevertheless, be some systematic effect in the isotopic record with time because of either a long-term depositional trend or modulation by the earth in response to the pristine isotopic signal with time.

To illustrate this point we borrow an example from reflection seismology. Figure 119 shows the power spectra (the zero frequency power value has been suppressed) for several seismograms taken in the same region with

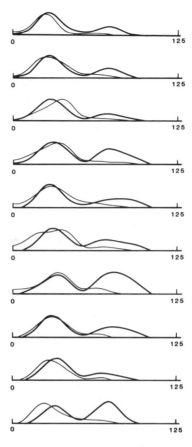

Fig. 119. Power spectra for reflection seismograms at increasing offset. Heavy line gives the spectra windowed in 750–1750 ms, light line gives the spectra windowed in 2750–3750 ms. Note the systematic attention at all offsets of the secondary peak around 80 Hz for the deeper windowed power spectra.

the same shot but recorded at increasing offset from the shot. The solid line shows the power spectrum to 125 Hz for a portion of each record between 750 ms and 1750 ms, while the dashed line shows the power spectrum for each record between 2750–3750 ms. It is clear by inspection that the deeper record has lost (or never had) high-frequency information content relative to the shallower record. The secondary peak around 80 Hz is severely attenuated in the deeper record. The perception is that the earth is modulating the behavior of the record. In order to extract the unmodulated response, we must extract the pristine signal from the observed record by deconvolving the earth's response.

IV. Homomorphic Deconvolution

As we have seen in a previous section, deconvolution is a very different type of problem than that of additive noise suppression, since the modulation effect multiplies the original signal in some nonsimple manner (normally assumed to be in the convolutional sense, but that assumption, too, needs to be viewed circumspectly unless, and until, justified by comparison against some other possibility, possibly in a maximum likelihood sense). In any event some deconvolution procedure is needed. Here, we report on an illustrative example for a synthetic seismic record taken from Tribolet (1979), using homomorphic deconvolution techniques and then comment on its drawbacks.

Figure 120(a) shows a synthetic seismogram with time, which corresponds to the convolution of the intrinsic signal, given in Fig. 120(b), with the earth's response shown in Fig. 120(c). In this example it is assumed that the earth's response does not attenuate the incident signal but merely distorts it. The seismogram was exponentially weighted with a 0.99 weighting and the complex cepstrum of the weighted seismogram constructed as depicted in Fig. 120(d). The cepstrum was then split into low-time and high-time parts by cutoff at $T_c = 293$ ms, as shown in Figs. 120(e) and (f), respectively. The weighted seismogram time sequences associated with the low-time and high-time cepstral division were then constructed as shown in Figs. 120(g) and (h), respectively. The results of deweighting exponentially produces the two seismograms shown in Figs. 120(i) and (j), respectively, with Fig. 120(i) taken to correspond to the intrinsic signal and Fig. 120(j) to the earth's response. The obvious conclusion to draw from this illustration is that it is possible to recover the intrinsic signal by appropriate weighting, cepstral time gating, and deweighting. Also, note that a good recovery of the earth's response by high-time gating is also present. There is, however, a noise component present in both estimates, which is observed after the cutoff time T_c and is because of the presence of a high-time cepstral component in the intrinsic signal, which is overlapping the earth's response. Thus, we need to *assume* that the intrinsic signal and the earth's response occur on very disparate time scales in order to effect a clean recovery of both; however, we have no knowledge that this is indeed the case in the isotopic signal records, leading to an obvious possibility of signal confusion. In addition there is no manifest reason for choosing an exponential weighting of 0.99 or a time cutoff of 293 ms as opposed to any other values nor is it clear how sensitive the recovered intrinsic signal and the earth's modulation response are to variations of these parameters.

In addition, the presence of an attenuative component to the earth's response function has also been looked at by Tribolet (1979; see, e.g., his

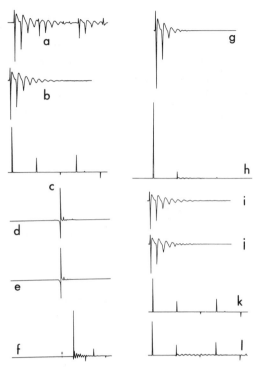

Fig. 120. Time-invariant homomorphic analysis of a synthetic seismogram (noise and attenuation effects excluded): (a) synthetic trace produced by convolving (b) source pulse with (c) earth's response function to an impulse; (d) complex cepstrum of the exponentially weighted trace; (e) low-time cepstral component; (f) high-time cepstral component (×40 magnification); (g) low-time component of weighted trace, (h) high-time component of weighted trace; (i) original source pulse [as in (b)] compared with (j) deweighted low-time component; (k) earth's response function [as in (c)] compared with (l) deweighted, high-time component (after Tribolet, 1979). Reproduced with permission from Prentice-Hall.

Fig. 8.2). The intrinsic signal is corrupted even further since it now changes in time as a result of attenuation in the earth's response. The extra complication of noise in the observed seismogram corrupts even more the separation of the intrinsic signal from the earth's response (Tribolet, 1979, Fig. 8.30).

These problems of deconvolution are not unique to homomorphic methods. Every deconvolution technique makes an assumption about the time (or frequency) separation of the intrinsic signal and the earth's modulation effect. The recovery of the signal is then always tainted by the post-assumption processing technique, parameter dependence of the technique, noise, and the implicit or explicit assumptions made. This is of particular concern when temporal variations of the observed isotopic signatures with depth are associated with climatic and earth orbital effects in time (Pisias and Moore, 1981). Extreme care must be taken to ensure that any temporal variation recovered from the observed isotopic record is indeed as best a representation as it can be (to within noise limits) of the intrinsic isotopic signal and not a variation determined by, and magnified and distorted by, the assumptions intrinsic to the deconvolution procedure itself. Such care has not yet been properly taken in isotopic chronostratigraphy.

V. Multispectral Wiener Filtering

We pointed out previously that multiple cross-power spectral tensor analysis provides an elegant and powerful method of constructing not only self-consistent definitions of signal-to-noise ratios when three or more signals are present, but also of producing the optimal broadband Wiener filter to use on the raw data. In Fig. 121 we show the results of applying such a Wiener filter to the linearly interpolated (i.e., raw) isotope data from sites V28239 and 502B. Figures 121(a) and (b) present the linearly interpolated data, while Figs. 121(c) and (d) show the corresponding Wiener filtered data using the multispectral method in the text. (Our current program is set up to handle as many as 10 isotopic records at a time for cross-power spectral filter estimation.)

Comparison of the raw and filtered records for site V28239 shows that the Wiener filter has added a slight amount of high-frequency structure to

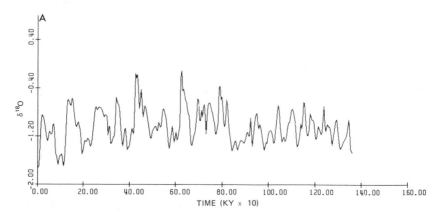

Fig. 121. (A) and (B) linearly interpolated isotope data from sites V28239 and 502B, respectively; (C) and (D) the same data, but after applying a data-determined, multichannel, cross-spectral Wiener filter on the data.

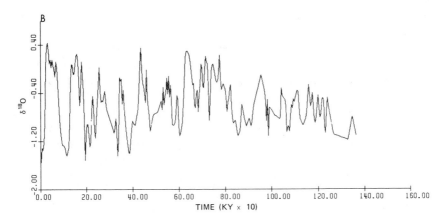

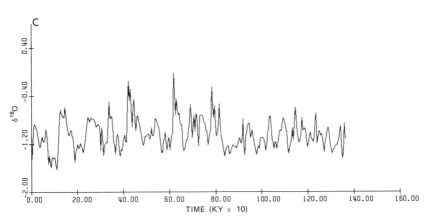

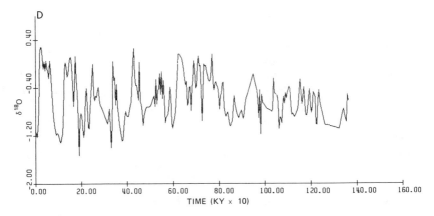

Fig. 121 (Continued)

the raw record, and this is in accord with the mathematics underlying multispectral filtering. In the case of site 502B, comparison of the raw and filtered records shows that the Wiener filter has not produced as large an effect on the high-frequency content of the raw record, in fact it has reduced slightly the high-frequency noise, and allocated that component of the noise to the isotope record from site V28239 through the Wiener filter determination. Thus, the two records, after multispectral Wiener filtering, have a more common high-frequency structure. At the same time, the individual character of each record is maintained through the lower frequency components of the record, which have their identities preserved by the filter at the level at which the multispectral techniques distinguish the signal from the noise, as exemplified by the resulting Wiener filter.

VI. Phase Sensitive Detection—Noise Analysis

In the case of searching for small, short duration, isotopic signals impressed on the record by geologically rapid events, such as ice-water melts, the techniques of phase sensitive detection (PSD) are of importance. Two factors are of dominant concern. First, we need to estimate how the signal-to-noise recovery procedures of the direct, zero-order predictive, first-order predictive, and Wiener–Kolmogorov (WK) algorithms vary as the parameters in the PSD algorithms are varied. Second, we need to estimate how the relative contrast in detectability for each algorithm depends on the number of events and amplitude of each event.

To answer the first question, in Fig. 122 we present a signal-to-noise sketch for each of the algorithms (for $\Gamma = 10^{-3}$) as a function of the data analysis parameter γ (here the parameters have the same meaning as previously). To be noted in the figure is that the Wiener–Kolmogorov algorithm provides the maximum sensitivity of signal-to-noise detection over the broadest range of γ values, while the direct algorithm provides the lowest sensitivity with decreasing capability of signal–to–noise resolution at low and high values. The zero-order predictive and first-order predictive algorithms provide a higher signal-to-noise sensitivity than the direct algorithm, but they are "tuned" to a peak γ value of about 3×10^{-2} with rapid roll-off in sensitivity on either side of this best γ value. The conclusion to be drawn is that unless there is some compelling constraint that absolutely forbids the use of the WK algorithm (and we cannot imagine such a constraint), then signal-to-noise detectability by PSD is best served using the Wiener–Kolmogorov algorithm.

Guided by this sensitivity analysis, in Fig. 123 we show the results of Monte Carlo random noise events handled according to the various algorithms. Inspection of the figure shows that the WK algorithm is at least

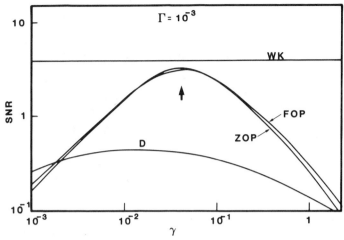

Fig. 122. Comparison of signal-to-noise sensitivity as a function of γ for each of the PSD algorithms reported in the text (after Pallottino, 1974). Reproduced with permission from Springer-Verlag.

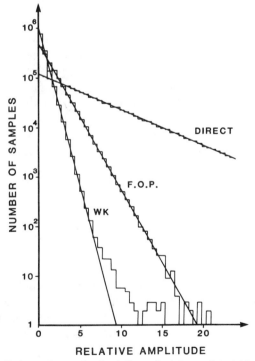

Fig. 123. Monte Carlo random noise simulations of minimum detectable short-time signals using the various PSD algorithms in the text. The ordinate is the relative number of events needed in order to detect by PSD methods a signal of relative amplitude given on the abscissa (after Pallottino, 1974). Reproduced with permission from Springer-Verlag.

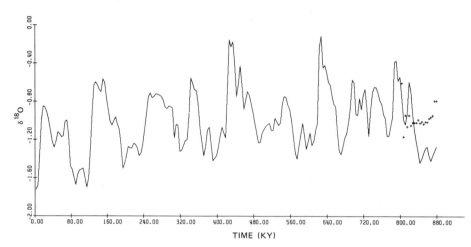

Fig. 124. Comparison of the result of applying a data-determined, Wiener filter in the extrapolative prediction mode to isotope data. Note that beyond about 2% from the end of the record used to generate the filter the predicted values (crosses) are far removed from the observed values.

twice as sensitive to small amplitude disturbances as its nearest competitor, the first-order predictive algorithm, with the direct algorithm an extremely poor device. For instance, at a level of 10^3 samples the WK algorithm will measure amplitudes of about 200–250 units, while the first-order predictive algorithm is only capable of amplitude resolution at about 400 units—twice as poor a device.

In short, in order to extract, in a maximally efficient fashion, small amplitude, short-lived (in a geological sense) events from an isotopic record, it is most expeditious to use the broadband, high-sensitivity, Wiener-Kolmogorov data analysis algorithm, unless some compelling constraint forbids its use.

VII. Predictive Wiener Filtering

In order to illustrate the manner in which filtering can be used in the predictive mode, in Fig. 124 we examine the effect of using 200 datum isotope points at 4-KY spacing, starting at 2.6 KYBP, to construct a Wiener filter in the manner previously described. This filter is then used to predict the expected values of datum points 201 to 220. From Fig. 124 we see that as the prediction distance increases, the predicted values depart further and further from the observed values. For this particular set of data, it would seem difficult to trust the Wiener prediction for more than about 2-3% in the extrapolation mode beyond the end of the 200 point record.

The point here is not the precise prediction distance over which the Wiener filter is accurate, but rather that, in situations in which we do not have a "ground truth" of data against which to compare the prediction, it is impossible to know how well the prediction can be trusted to represent reality. Any attempt to estimate the intrinsic isotopic signal based on using extrapolative filter techniques will most likely include a significant degree of error based on the incorrectly predicted filter estimate. Unfortunately, we have no *a priori*, or *a posteriori*, way of determining the size of this error. The same argument then applies to uses of an extrapolated, isotopic signature to infer geological causes for the putative isotopic signal. There is no way to guard against an incorrectly estimated signal from coloring one's perception of the root causes of the isotopic variations.

Chapter 19 | Maximum Entropy Methods

Finally in this compendium of illustrative examples of some of the quantitative tools available in attempts to "unscramble" signals from observed isotopic variations, we turn our attention to the maximum entropy and linear unmixing (Q-model) methods of analysis.

I. Unbiased MEM

As we have remarked earlier, the fact that geological data are, most often, neither sampled uniformly in time nor in sedimentary depth is a strong reason in and of itself to consider methods of data analysis that are less beholden to techniques dependent upon uniform sampling. Relative entropy measures fill such a role. In Fig. 125 (from Full *et al.*, 1984) we show the

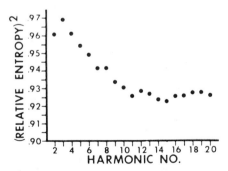

Fig. 125. Plot of the relative entropy (squared to enhance separation) versus harmonic number. Harmonic 15 has the lowest relative entropy followed by harmonics 14, 16, and 17 (after Full *et al.*, 1984).

relative entropy of a set of Fourier harmonics for a distribution of sedimentary grain shapes (200 grains in this instance and 20 Fourier harmonics per grain). The data set represented stream sediment samples from drainage basins on the west flank of the Big Horn Mountains in Wyoming. Clearly, harmonic 15 has the lowest relative entropy followed by harmonics 14, 16, and 17. When analyzed further using the EXTENDED CABFAC–FUZZY QMODEL algorithms, these four harmonics produced similar cluster solutions (end-member sources), which strongly reflected the four important source terrains of the stream sediment samples.

II. Biased MEM

Biased maximum entropy methods have also proven inordinately useful in a variety of fields for reconstructions that yield the best image model derivable from the available data, in the sense that the model so produced is likely to need the least correction when new measurements and additional data are added to the system. For instance, in Fig. 126 (after Collins, 1982) we see the effect of reconstructing the electron density image of the tyrosine-11 residue of the rubredoxin of *Clostridium pasteurianum* under two conditions: first, with a resolution of 1.5 Å using standard least squares phase methods and second, using coarser data taken at 2 Å (and so 33% less resolution) but with the biased maximum entropy method. Inspection of the two components of Fig. 126 shows virtually no difference in the electron

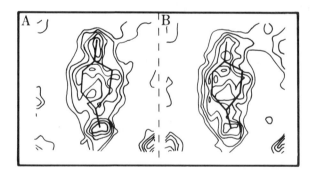

Fig. 126. Reconstructions of the electron density images of Tyrosine-11 (A) based on high resolution 1.5 Å data using least-squares phase determination and (B) based on coarser 2 Å data using biased maximum entropy methods of reconstruction. Note that the molecular bond shape, bond lengths, and bond angles (heavy lines) are almost identical for both reconstructions (after Collins, 1982).

density contours of tyrosine-11; the bond lines, shapes, and angles are identical to within resolution at the 1.5 Å level. The maximum entropy method, or its biased version, shows a strong proclivity for extracting the maximum amount of information in the available data and organizing the information in a maximum entropy sense so that its further manipulation by other techniques is maximally facilitated.

III. QMODEL, EXTENDED QMODEL, and FUZZY QMODEL Unmixing

The use of factor analysis techniques in conjunction with maximum entropy methods is rapidly becoming one of the dominant techniques in geological data management. In Fig. 127 we show the results of iterations designed to extract the best positive loading scores for three end members using both QMODEL and its modified form EXTENDED QMODEL. The original QMODEL could not produce entirely positive mixing proportions for all samples, in contrast to the end members produced by EXTENDED QMODEL. Figure 128 shows the end-member histograms generated by QMODEL and EXTENDED QMODEL. Note that while there is little difference in the end member E_2 and E_2^1, there is considerable difference in the histograms of the first (E_1, E_1^1) and third (E_3, E_3^1) end-member

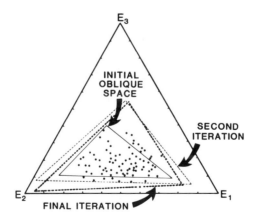

Fig. 127. Results of an iterative solution providing positive composition loadings. Edges of the final polytope are not parallel to the QMODEL polytope. Note that each edge of the EXTENDED QMODEL polytope includes at least one datum point, a common result of the algorithm (after Full *et al.*, 1981).

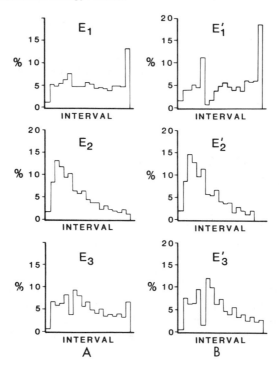

Fig. 128. Composition of end-member histograms generated by QMODEL (a) and EXTENDED QMODEL (b). Note the contrast between the end members E_1, E_1^1 and E_3, E_3^1, respectively (after Full *et al.*, 1981).

histograms. Thus, any interpretations of the end members that rely on the histograms will clearly be influenced by the capability of the unmixing algorithm to produce positive mixing proportions for all samples. Presumably this fundamental difference between the QMODEL and EXTENDED QMODEL algorithms is responsible for the differences in the end-member histograms.

However, as noted earlier, even though the EXTENDED QMODEL algorithm provides a significant improvement over the original QMODEL algorithm, it, too, suffers from a problem (but of lesser severity than the nonpositive mixing proportion problem suffered by the original QMODEL). The EXTENDED QMODEL includes *all* of the data within the triangle of end members, as can be seen in Fig. 127 for the three end-member case. Yet outliers in the data base may be the result of poor measurement or transcription error, or may not even properly belong to the end-member distribution space. Full *et al.* (1982) overcame this problem by using a FUZZY QMODEL algorithm as outlined earlier. Figure 129 (from Full *et*

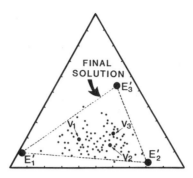

Fig. 129. Location of the fuzzy centers $\{V_i\}$ and final mixing polytope for the same data set as in Fig. 126. E_1^1, E_2^1, and E_3^1 denote the locations of the FUZZY QMODEL end members. Note the presence of three ignored outliers (after Full *et al.*, 1982).

al., 1982) illustrates the effect of applying the FUZZY QMODEL algorithm to the same data set as shown in Fig. 127 and, as can be seen from Fig. 129, we now have several outlier data points that are not included in the final solution for the end members.

The contrast in the end-member histograms derived from EXTENDED QMODEL and from FUZZY QMODEL is shown in Fig. 130 (from Full *et al.*, 1982). In this case we see slight changes in the second and third end members, (E_2, E_2^1) and (E_3, E_3^1), respectively, but we see a great contrast in the first end-member behavior (E_1, E_1^1), implying that the outlier points were having undue weight in influencing the first end member.

In turn this influence could have produced a significant change in interpretation of the cause of the end-member histogram behavior if it had not been recognized and allowance made for it.

IV. End-Member Behaviors for EXTENDED and FUZZY QMODELS

In order to illustrate the end-member variability, we consider the four sets of same site data shown in Fig. 131, representing variations of the calcium carbonate percentage with depth [Fig. 131(a)], the coarse grain fraction percentage with depth [Fig. 131(b)], the $\delta^{18}O$ variation with depth [Fig. 131(c)], and the $\delta^{13}C$ variation with depth [Fig. 131(d)]. The quesiton we ask is: If we look for a three end-member solution, what degree of significance do we obtain, and are we consistent with the requirements of a mixing or a clustering behavior?

In Table IV, part A, we show the three end-member relative proportions and the relative coefficients of determination for the three end-member

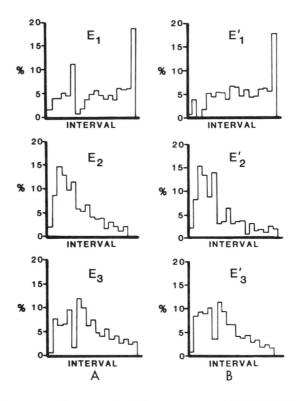

Fig. 130. Comparison of end-member histograms generated by EXTENDED QMODEL (A) and FUZZY QMODEL (B). Note the contrast between end members E_1, E'_1, because of the sensitivity of E_1 to the outlier datum points ignored in the FUZZY QMODEL but included in the EXTENDED QMODEL (after Full *et al.*, 1982).

solution using the FUZZY QMODEL algorithm with a fuzziness exponent $m = 2$. In part B of the table, we show similar results obtained for three end members when using the EXTENDED QMODEL algorithm. Note that the three end members from the FUZZY QMODEL do not have to correspond, even distantly, to those from the EXTENDED QMODEL algorithm. One way to check for degrees of correspondence is, of course, to construct the various end-member behaviors with depth for both algorithms, search for high degrees of correlation between any pair of end members using the power spectral or time series methods of the previous section, and assign degrees of significance of commonality of signal behavior in the end members relative to the noise.

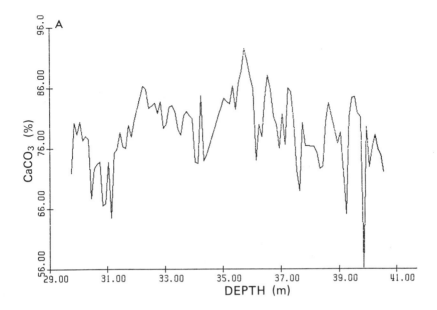

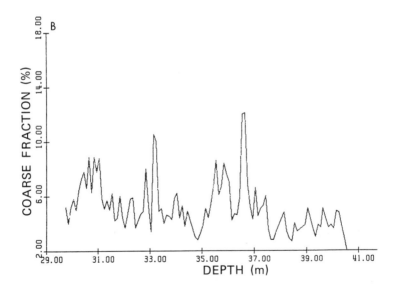

Fig. 131. Raw data used in the EXTENDED QMODEL and FUZZY QMODEL unmixing algorithm. (A) $CaCO_3$ percentage with depth, (B) coarse fraction percentage with depth, (C) $\delta^{18}O$ isotope variation with depth, and (D) $\delta^{13}C$ isotope variation with depth. All data are from site 572C.

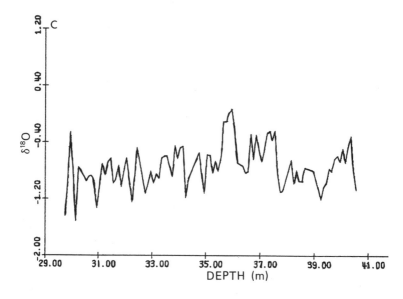

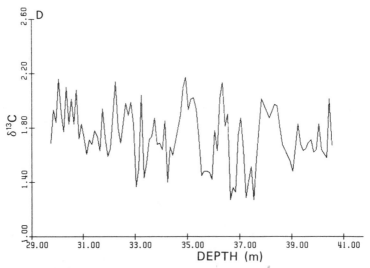

Fig. 131 (Continued)

It is also clear from the FUZZY QMODEL membership table (Table IV) that the results generally seem to be consistent with a cross between a mixing and a clustering situation since the end-member values are either low (0–0.2) and high (0.8–1.0), indicative of a clustering mode, or intermediate (0.2–0.8), more indicative of a mixing mode. Note in particular the presence of small

Table IV

FUZZY QMODEL Membership

$V_1 = $ Coarse fraction (%)
$V_2 = $ Calcium carbonate (%)
$V_3 = \delta^{18}O$ (‰)
$V_4 = \delta^{13}C$ (‰)

A. Three End-Member Best Solution, FUZZY QMODEL, Relative Proportions

E_1	E_2	E_3	Coefficients of determination for three end-member solution
$V_1 = 0.222$	0.075	0.001	0.9758
$V_2 = 0.767$	0.911	0.989	0.9933
$V_3 = -0.018$	0.008	-0.017	0.7559
$V_4 = 0.029$	0.006	0.027	0.8147

B. Three End-Member Best Solution, EXTENDED QMODEL, Relative Proportions

E_1	E_2	E_3	Coefficients of determination for three end-member solutions
$V_1 = -0.030$	0.103	0.138	0.9758
$V_2 = 1.019$	0.887	0.849	0.9933
$V_3 = -0.008$	-0.023	0.003	0.7559
$V_4 = 0.019$	0.033	0.010	0.8147

Note: 99.04% total variance accounted for by three end members.

but negative loadings for E_3 in the $\delta^{18}O$ proportions using FUZZY QMODEL and of negative contributions to V_1 and V_3 using EXTENDED QMODEL. These negative loadings are clear indications that an intrinsic underlying requirement has been violated by a small amount.

Curves of the three end-member behaviors with depth for both the FUZZY QMODEL [Fig. 132(a)] and the EXTENDED QMODEL [Fig. 132(b)] are shown so that an initial inspection can be made (by eye) to see if any coarse grade similarities or trends are present in the end members of the two cases. Table V illustrates the percentages of each of the three end-members used in the FUZZY QMODEL that are to be associated with each depth—the first column is the sample number, the second is the depth of sampling (in m), and the last three columns represent the membership assignment (fractional) of each sample in each of the three end members.

These membership assignments will, of course, change as the fuzziness exponent m is varied and as more and different types of data are added.

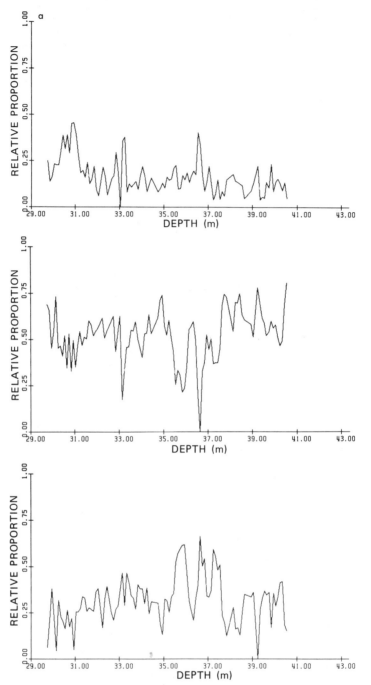

Fig. 132. Plots of the three end members with depth for the data of Fig. 131 using (a) FUZZY QMODEL and (b) EXTENDED QMODEL. Note that the end members show a strong difference, one with respect to the other, both within the framework of a particular QMODEL ALGORITHM and between the FUZZY and EXTENDED QMODEL algorithms.

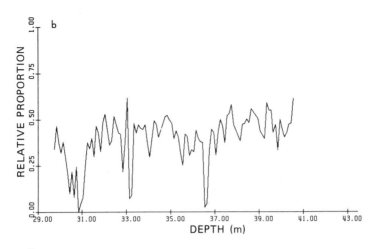

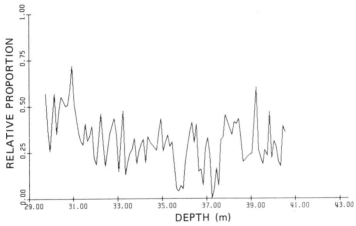

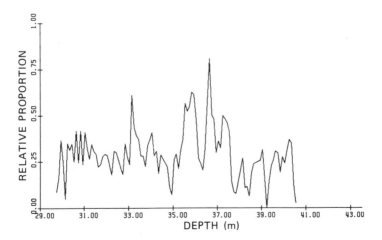

Fig. 132 (Continued)

283

Table V

FUZZY QMODEL and Member Percentages

| Sample | Depth (m) | Memberships | | |
		Group 1	Group 2	Group 3
1	29.74	0.1209	0.0965	0.7826
2	29.84	0.0319	0.0434	0.9247
3	29.94	0.1065	0.6254	0.2681
4	30.04	0.2136	0.1437	0.6427
5	30.14	0.0982	0.0915	0.8104
6	30.24	0.3967	0.2927	0.3106
7	30.34	0.8773	0.0543	0.0684
8	30.44	0.9490	0.0256	0.0254
9	30.54	0.7221	0.1019	0.1760
10	30.64	0.9216	0.0419	0.0365
11	30.74	0.5800	0.1354	0.2846
12	30.84	0.8512	0.0783	0.0705
13	30.94	0.9178	0.0388	0.0434
14	31.04	0.9216	0.0416	0.0368
15	31.14	0.7587	0.1051	0.1363
16	31.24	0.0555	0.0747	0.8698
17	31.34	0.2036	0.3765	0.4199
18	31.44	0.0743	0.1886	0.7371
19	31.54	0.3670	0.1925	0.4405
20	31.64	0.0026	0.0047	0.9926
21	31.74	0.0018	0.0029	0.9952
22	31.84	0.2086	0.1677	0.6237
23	31.94	0.0504	0.1910	0.7587
24	32.04	0.0559	0.2270	0.7170
25	32.24	0.0813	0.0727	0.8460
26	32.34	0.0850	0.1974	0.7176
27	32.44	0.0651	0.3243	0.6106
28	32.64	0.0047	0.0069	0.9883
29	32.74	0.0248	0.0307	0.9444
30	32.84	0.9068	0.0454	0.0478
31	32.94	0.0628	0.0935	0.8437
32	33.04	0l0447	0.1379	0.8174
33	33.14	0.7850	0.1378	0.0772
34	33.24	0.9298	0.0383	0.0319
35	33.34	0.0322	0.8790	0.0887
36	33.44	0.0516	0.7819	0.1665
37	33.54	0.0347	0.1069	0.8584
38	33.64	0.0296	0.0820	0.8884
39	33.74	0.0007	0.0012	0.9981
40	33.84	0.0698	0.5294	0.4008
41	33.94	0.1013	0.6232	0.2665
42	34.04	0.3141	0.4945	0.1914
43	34.14	0.0507	0.0918	0.8575

Table V (Continued).

Sample	Depth (m)	Memberships		
		Group 1	Group 2	Group 3
44	34.24	0.0644	0.3171	0.6186
45	34.34	0.0157	0.0244	0.9599
46	34.44	0.0438	0.0925	0.8637
47	34.74	0.0152	0.0320	0.9528
48	34.84	0.0588	0.0811	0.8602
49	34.94	0.0784	0.0963	0.8252
50	35.04	0.0191	0.0490	0.9319
51	35.14	0.0573	0.1219	0.8208
52	35.24	0.0041	0.0063	0.9896
53	35.34	0.0747	0.2167	0.7086
54	35.44	0.2787	0.4422	0.2790
55	35.54	0.2013	0.6875	0.1111
56	35.64	0.0601	0.8613	0.0786
57	35.74	0.0772	0.8287	0.0941
58	35.84	0.1139	0.7855	0.1006
59	35.94	0.1035	0.7946	0.1018
60	36.04	0.0637	0.8798	0.0565
61	36.14	0.0172	0.0404	0.9424
62	36.24	0.0163	0.0228	0.9608
63	36.34	0.0450	0.0478	0.9072
64	36.44	0.1153	0.2720	0.6126
65	36.54	0.8534	0.0857	0.0609
66	36.64	0.4431	0.4243	0.1325
67	36.74	0.0479	0.9061	0.0460
68	36.84	0.0377	0.9045	0.0577
69	36.94	0.0516	0.1584	0.7900
70	37.04	0.3499	0.3518	0.2983
71	37.14	0.0747	0.3599	0.5654
72	37.24	0.0669	0.8317	0.1014
73	37.34	0.0459	0.8831	0.0710
74	37.44	0.0066	0.9848	0.0086
75	37.54	0.0362	0.8793	0.0845
76	37.64	0.0399	0.0619	0.8982
77	37.74	0.0680	0.0972	0.8348
78	37.84	0.0771	0.0911	0.8318
79	38.14	0.0462	0.0687	0.8851
80	38.24	0.0605	0.0755	0.8640
81	38.34	0.0575	0.0729	0.8696
82	38.44	0.0809	0.0995	0.8196
83	38.54	0.0154	0.0246	0.9599
84	38.64	0.0339	0.0947	0.8714
85	38.94	0.0243	0.0657	0.9100
86	39.04	0.0675	0.2799	0.6526
87	39.24	0.0997	0.0983	0.8021

Table V (Continued).

Sample	Depth (m)	Memberships		
		Group 1	Group 2	Group 3
88	39.34	0.0414	0.0710	0.8876
89	39.44	0.0260	0.0635	0.9105
90	39.54	0.0446	0.1483	0.8071
91	39.61	0.0595	0.1978	0.7427
92	39.74	0.0511	0.1918	0.7571
93	39.84	0.1321	0.0996	0.7684
94	39.94	0.0410	0.1384	0.8216
95	40.04	0.0011	0.0021	0.9967
96	40.14	0.0734	0.2113	0.7153
97	40.24	0.0556	0.7558	0.1885
98	40.34	0.0636	0.6430	0.2934
99	40.44	0.0505	0.0660	0.8835
100	40.54	0.0896	0.1195	0.7908

The assignment of memberships clearly shows the degree to which significance can be attached to end-member behavior. A low membership fraction of most samples at most depths shows the presence of a rare breed of sample, one that can be culled for a more detailed investigation. In this sense the membership assignment is a useful "quick look" tool.

Chapter 20 | An Integrated Application of Multiple Techniques

Unconventional and innovative approaches are needed for more accurately locating hydrocarbon source rocks and their related oil and gas reservoirs in deep-water tracts in both traditional and frontier exploration areas, such as those of the northwestern Gulf of Mexico (Williams and Trainor, 1986). Isotope chronostratigraphy is one such approach to increase stratigraphic resolution across a basin, allowing for increased accuracy in hydrocarbon prediction. While empirical correlations, either visually or graphically, have proven successful (Williams and Trainor, 1986), this Chapter introduces how some of the quantitative approaches, discussed in the previous Chapters, can be used to correlate isotopic records both across a basin and on a global scale. In these examples four Pleistocene oxygen isotopic records are treated as time series and some signal analysis techniques are used simultaneously to provide criteria concerning (1) the identification of common events in isotope records and (2) the level of accuracy with which such events can be recognized and interpreted. Use of these techniques allows us to address *quantitatively* these questions:

1. Do isotope records at different locations show common events, or events that have shifted spatially in time?
2. What degree of accuracy can we ascribe to the common isotope events?
3. How can the best isotope signal be extracted in comparing one record against other records?
4. What is the best way to allow for variable time sampling in different records when attempting to determine common events or spatially migrating events?

5. How do we extract an underlying lower frequency signal in an isotope record that is highly variable and quasi-periodic in the amplitude and frequency domains?

Another impetus for the quantitative analysis of the Pleistocene $\delta^{18}O$ signal is its applicability as a new form of high-resolution chemical stratigraphy in petroleum exploration wells (Williams and Trainor, 1986).

As we have detailed already, numerous factors play a role in these important problems. Limits to our knowledge on the evolutionary behavior of foraminifera, the evolution of the oceans, isotopic exchange processes with time, nonlinearity in the ocean–atmosphere climate system, disturbance during the recovery process, and diagenetic effects during and after sedimentary burial mean that there is a certain amount of noise in every isotope record. In addition, the sampling of a record is done at a set of discrete times over a finite space of geologic time.

For this initial analysis of the Pleistocene $\delta^{18}O$ record, we shall discuss several basic methods of data reduction, each of which performs a different type of operation, in order to extract a signal from the data. We then apply these operations to a suite of four Pleistocene $\delta^{18}O$ records.

I. Signal Processing Strategy

Our initial quantitative analysis of Pleistocene $\delta^{18}O$ records involves these techniques to process the raw data: (1) interpolative filters, (2) semblance and coherence analysis, (3) autocorrelation and cross-correlation methods, (4) power spectral and cross-spectral methods, and (5) noise minimization filters. These basic techniques in data analysis serve as good starting points for quantitative isotope chronostratigraphy.

Application of these methods is addressed in two sections: (1) formulation of a $\delta^{18}O$ type record for the Pleistocene and (2) comparison of the type record with other $\delta^{18}O$ records in the same time range. In (1) we transform a high-quality $\delta^{18}O$ record from piston core V28-239 into what we define as TYPE record. The TYPE record has much of the noise in the original record removed (we must, of course, *define* the noise level). In (2) we compare our TYPE record to $\delta^{18}O$ records from DSDP sites 502B, 504, and 552A. The location of the core material and the foraminifera analyzed to generate the isotope records are given in Table VI. As discussed previously, it is widely accepted that much of the isotopic signal represents global isotopic changes in ocean chemistry driven by the volume of Pleistocene ice sheets of the Northern Hemisphere (Shackleton, 1967; Shackleton and Opdyke, 1973; Williams, 1984). We use this idea to search for a common global signal in the aforementioned Pleistocene $\delta^{18}O$ records. The results

Table VI

Summary of Input Records

$\delta^{18}O$ Record	Latitude	Longitude	Specimen	Average sedimentation	Reference
V28-239	3°15'N	157°11'E	*Globigerinoides sacculifer* (planktonic)	1 cm/KY	Shackleton and Opdyke (1976)
DSDP 502B	11°30'N	79°23'W	*Globigerionoides sacculifer*	2.2 cm/KY	Prell (1982)
DSDP 552A	56°02.56'N	23°13.39'W	*Cibicidoides wuellerstorfi* (benthic)	1.9 cm/KY	Shackleton and Hall (1984a)
DSDP 504	1°13.6'N	83°43.9'W	*Globigerinoides ruber* (planktonic)	4.2 cm/KY	Shackleton and Hall (1984c)

of subtracting a global signal from an individual record must therefore represent other components in the $\delta^{18}O$ signal, such as temperature, meltwater impulses, the accuracy of datum horizons in individual records, dissolution, and sediment disturbance. Having defined our noise component, we address the recognition of global isotopic signals and the degree of uniqueness, resolution, and precision with which the $\delta^{18}O$ records can be correlated in both time and spectral domains (frequency).

II. Brief Explanation of Data Processing Methods

In this section we describe briefly the theoretical basis for the methods used to treat isotope data as time series information. Each method can contribute to locating common isotopic events, correlating isotope records of varying quality and completeness, and determining the spectral character of isotope records in a proper manner.

Depth–Time Conversions and Interpolative Filters

The first problem that arises when using isotopic and other geologic records as time series lies with the conversion of the record from depth to equally incremented time records. For the most part, depth–time conversions of isotope data have been based on time horizons determined by radiometric dating, biostratigraphy, and paleomagnetic stratigraphy, with constant sedimentation rates assumed between these known time horizons (Shackleton and Opdyke, 1973, 1976; Kominz et al., 1979). This method is also used for this initial analysis, although, in the future, we will investigate alternate methods, such as those described earlier, where we use semblance and power spectra techniques to determine the relative distortion of a record when compared to a $\delta^{18}O$ template.

The second problem is to convert the time record to a time series with equally incremented datum points in time. We assume that the points can be connected by straight-line segments, and that this continuous record can be sampled at equally spaced increments. This assumption is obviously more or less robust depending on errors in the original time conversion of records with different and internally varying accumulation rates. Alternate methods using other interpolative filters, such as triangular filters, for example, can and should be invoked to determine the degree of robustness and reproducibility of subsequent analyses.

For our analysis of the Pleistocene $\delta^{18}O$ record, we employ several of the more common types of filters. We assume that each isotope record contains both a signal and a *linear* additive component, i.e., noise. The task is to design a filtering operation on the records that will extract the best signal while minimizing the noise content of the record without altering the primary signal in a fundamental way.

A common thread in all such methods dates back to the work of Wiener (1949), who designed a least squares noise minimization filter that is data determined. Once the Wiener filter has been applied to a record, one obtains a component of the power spectrum represented by a constant power level (after a Fourier transform has transformed the data from the time domain to the frequency domain). We take this constant level to be the level of the noise power, which has been extracted from the individual isotope record by the Wiener filter.

After the noise level has been defined using the Wiener filter, we next apply an "on–off" (low-pass) filter to the Fourier coefficients in frequency space of the original record (pre-Wiener filter). This simple filter only has values of one or zero (on or off). The power in the frequencies above the determined signal–noise frequency boundary is multiplied by zero, while the power in the frequencies below is multiplied by one (low-pass filtering). This treatment destroys high-frequency defined noise. Alternately, one can also filter the power spectrum by power level, subtracting the power of the noise at all frequencies (white noise filter). The former filtering is used in this analysis of Pleistocene $\delta^{18}O$ records, while application of the latter will be explored in future research.

The last type of filtering used in this study is a triangular smoothing filter. This filter slides over a record, smoothing the data by weighting of neighboring data points and producing a moving weighted average. A particular triangular filter (Hamming and Tukey, 1949) is used to smooth power spectra over three harmonic widths. Because of the discrete nature of the applied Fourier transform, the power level of each harmonic is only an estimate of the true frequency (i.e., the true frequency may lie between harmonics). This filter serves to compensate for the nature of the discrete Fourier transform. With these basic tools in time series analysis and with a knowledge of the resolution and precision available with each tool, it is now possible to apply these methods to an isotopic scenario for the Pleistocene (0–1.8 MYBP).

III. Formulation of a TYPE Pleistocene $\delta^{18}O$ Record with Time

In order to better correlate oxygen isotope records in time and frequency, the need arises for a type record (TYPE) that we assume, initially, to be the template against which other Pleistocene oxygen isotope records are to be compared. The initial choice of a TYPE record is based on (1) a relatively fine, continuously sampled record with depth; (2) a sediment accumulation rate that preserves isotopic shifts at *specific* times (rather than small thickness of sediment covering large time ranges); (3) a relative abundance of biostratigraphic and paleomagnetic time horizons; and (4) a simple structure and stratigraphy in the core sample. Piston core V28–239 fulfills these basic requirements relative to most other Pleistocene records and is chosen as our initial TYPE section. Figure 133 shows the location of V28-239 along with DSDP sites 502B, 504, and 552A used in this analysis. All four records are from open ocean areas. The $\delta^{18}O$ record of site 552A is derived from a benthic foraminifera, while those of the other sections are based on planktonic foraminifera (Table VI). No attempt has been made to normalize the benthic data to the planktonic data because of $\delta^{18}O$ temperature fractionation. The transformation of V28-239 into our TYPE record is shown schematically in Fig. 134(a), in accordance with some of the methods discussed previously.

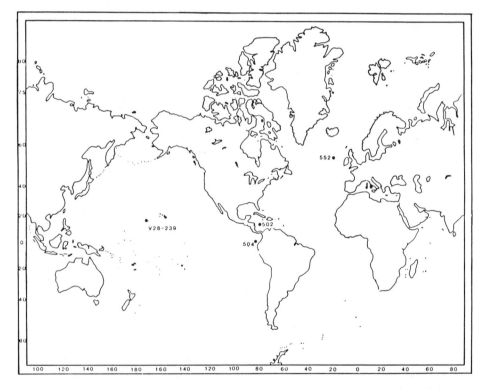

Fig. 133. Location map showing the approximate location of the DSDP sites used in this analysis.

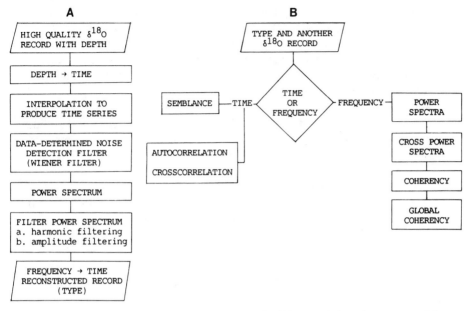

Fig. 134. Flowcharts showing sequence of analysis. (A) Formulation of a type record for $\delta^{18}O$; (B) comparison of the TYPE record with other $\delta^{18}O$ records.

Table VII
Well Sites and Forams Used for Age Dating

Datum	Age (MY)	Depth (m)			
		V28-239	502B	552A	504
Base of Brunhes	0.73	7.40	16.25	14.60	
Top of Jaramillo	0.91	8.80	18.60	18.60	
Top of *Gephyrocapsa*	0.92	9.10	20.25		41.10
Base of Jaramillo	0.99	9.40	21.40	19.90	
Base of *Gephyrocapsa*	1.13	10.90	24.60	22.30	50.80
Large *Gephyrocapsa* LAD	1.30	12.45	30.70		57.40
C. macintyrei LAD	1.45	13.60	33.30		64.30
G. oceanica FAD	1.60	14.90	35.80	30.00	66.50
Top of Olduvai	1.66	15.50	37.50	32.10	
D. brouweri LAD	1.88	17.85	42.00	34.30	84.20

First, the $\delta^{18}O$ depth record of V28-239 is converted to time using the time horizons listed in Table VII and assuming constant sedimentation rates between these known time horizons. We now sample the constructed time record at 5000-yr intervals, thereby producing a $\delta^{18}O$ time series, with a time range of 5-1870 KY (Fig. 135). This $\delta^{18}O$ record contains the basic input data for the ensuing analyses.

Visually, the oxgen record of V28-239 in Fig. 135 shows a distinct break in character at approximately 930 KYR (Shackleton Opdyke, 1976; Kominz et al., 1979; Pisias and Moore, 1981). The top (younger half) is composed of relatively broad peaks and troughs while the bottom (older half) has narrow peaks and troughs. As have others (Kominz et al., 1979; Pisias and Moore, 1981), we subsequently treat the two sections as separate time series of equal length, V28-239T being the top of the record (5-935 KY) and V28-239B being the bottom of the record (940-1870 KY).

The first step in determining the signal-to-noise ratio is to use the Wiener least squares noise minimization filter. Figure 136(a) illustrates V28-239T before application of the Wiener filter. After application of the Wiener filter [Fig. 136(b)], notice the increased high-frequency structure. By transforming the two pre- and post-Wiener filter data sets from the time domain to the frequency domain via a Fourier transform, we can determine the power at *discrete* frequencies in the records as shown by the power spectra in Fig. 137. There are 94 harmonics in these spectra with harmonic spacing equal to

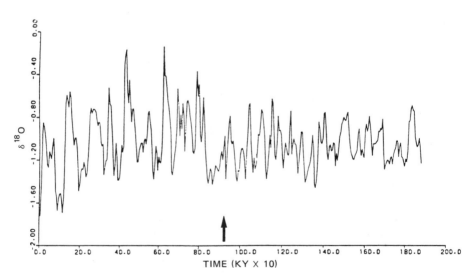

Fig. 135. Linearly interpolated ($\Delta t = 5000$ yr) $\delta^{18}O$ time series for V28-239. Arrow indicates split time for analysis.

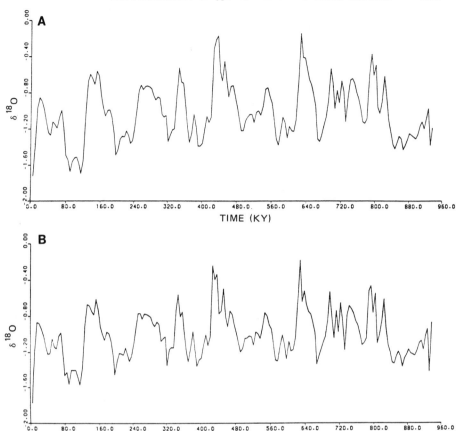

Fig. 136. Top half (5–935 KY) of V28-239 (V28-239T) δ^{18}O record: (A) original interpolated record and (B) post-Wiener filtering record.

$1/T$, where T is the total length of the record (930 KY). The zeroth harmonic is suppressed for scaling clarity. Some of the dominant peaks are labeled with their corresponding periods (in thousands of years). Note the roughly constant low power above harmonic 44 in Fig. 137(b) (arrow). This is the noise power that has been extracted by the Wiener filter. We can define noise as being all frequency components above harmonic 44 (period of 21 KY). We realize that this destroys a component of the precession signal interpreted by Hays *et al.* (1976), Kominz *et al.* (1979), and Imbrie and Imbrie (1980), but we want to attack this problem with objective data analysis. By zeroing these harmonic Fourier coefficients with an on-off harmonic filter (low-pass filter), we have revised Fourier coefficients in harmonics 0–44 with a spacing $1/T$. An alternate method of filtering by

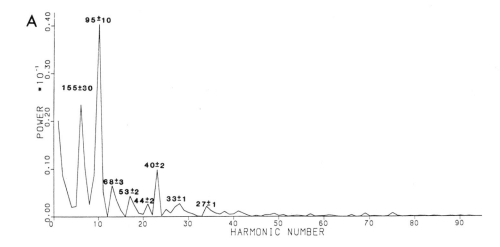

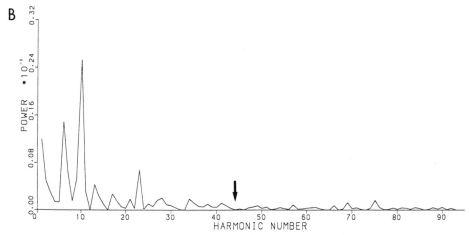

Fig. 137. Power spectra for V28-239T (5–935 KY) for (A) original interpolated record and (B) post-Wiener filtering record. Arrow indicates signal–noise break point. Some of the corresponding periods for peaks are also shown.

power level (a white noise filter) rather than harmonic number will be explored in future research.

The data are next transformed from frequency space back into time via an inverse Fourier transform in order to reconstruct the time record from the revised Fourier components (Fig. 138). A comparison of this reconstructed record with the input "raw" interpolated record shows that most of the basic features in the input record have been preserved while removing high-frequency components defined as noise by the Wiener filter.

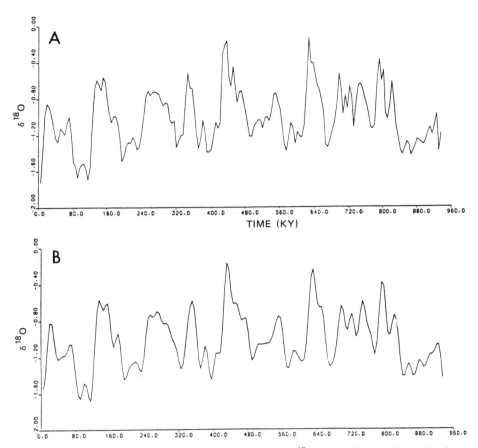

Fig. 138. Top half (5–935 KY) of V28-239 (V28-239T) $\delta^{18}O$ record: (A) raw, interpolated record and (B) post-on–off harmonic filtering record.

The same process is applied to the older part of the Pleistocene record (V28-239B) as used for V28-239T (Fig. 139). A comparison of the V28-239B $\delta^{18}O$ record with time before and after application of the Wiener filter illustrates the addition of a high-frequency component after applying the Wiener filter [Fig. 139(b)]. A power spectum for V28-239B is generated before and after the Wiener filter (Fig. 140). Again, harmonic spacing is $1/T$ where T is the total length of the record (930 KY), with the zeroth harmonic suppressed for clarity. Note the very different power spectra for V28-239T [Fig. 137(a)] and V28-239B [Fig. 140(a)]. The power spectra for V28-239B show nearly an order of magnitude decrease in power for the dominant harmonics, and also the power is more evenly distributed than in V28-239T. This result strengthens the *a priori* decision to treat the top

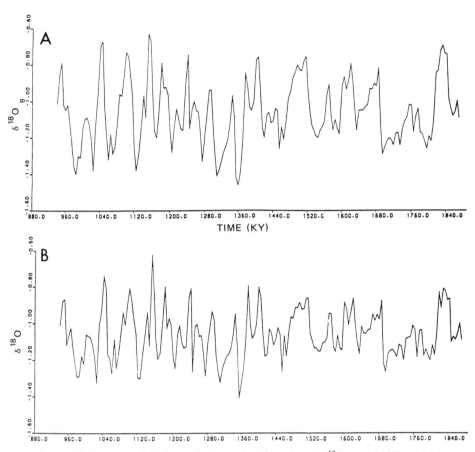

Fig. 139. Bottom half (940-870 KY) of V28-239 (V28-239B) $\delta^{18}O$ record: (A) original interpolated record and (B) post-Wiener filtering record.

and bottom portions of V28-239 as separate time series. Figure 140(b) shows the increase in the noise level by the Wiener filter. We have chosen harmonic 40 (period = 23 KY) as our noise cutoff level. Harmonics above 40 show a roughly constant power level, and their corresponding Fourier coefficients are zeroed by the on–off filter. The data are transformed from frequency back to the time domain using the revised Fourier coefficients. The results of this transformation are shown in Fig. 141 with the raw interpolated record for comparison.

Now that we have independently removed noise from both the top and bottom halves of V28-239; we can link them together again to form a noiseless $\delta^{18}O$ record with time from 5-1870 KY. This reconstructed record

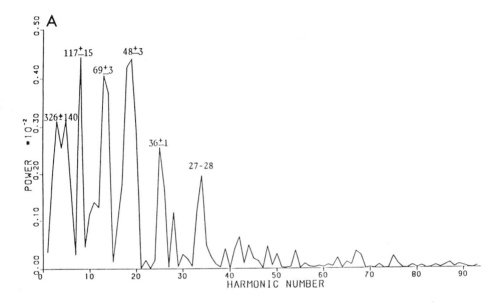

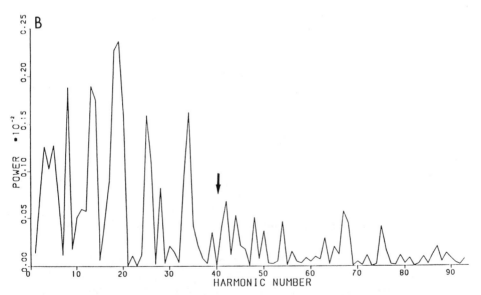

Fig. 140. Power spectra for V28-239B for (A) original interpolated record with some of the corresponding periods for dominant peaks and (B) post-Wiener filtering record. Arrow indicates signal–noise break point.

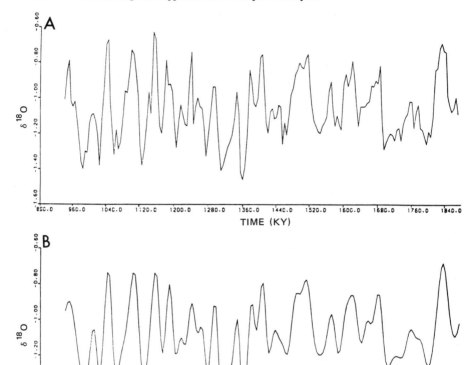

Fig. 141. V28-239B $\delta^{18}O$ record (940–1870 KY): (A) raw, interpolated record and (B) post-on-off harmonic filtering record.

is designated as TYPE for the purposes of this illustration (see Fig. 142) and is used as our reference record for the remainder of this analysis.

IV. Comparison of the TYPE Record with Other Pleistocene $\delta^{18}O$ Records

The TYPE reference record established previously is used with some of the time series methods to determine with what degree of resolution and precision we can correlate this TYPE record with other $\delta^{18}O$ records from DSDP cores 502B, 504, and 552A (Fig. 133). Each record is converted from depth to time assuming constant sedimentation rates between the time horizons shown in Table VII, and converted to a time series by sampling

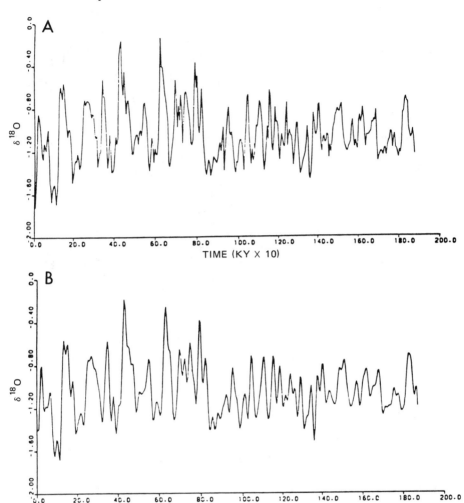

Fig. 142. (A) Raw, interpolated $\delta^{18}O$ record for DSDP V28-239 (5–1870 KY) and (B) $\delta^{18}O$ TYPE record constructed from DSDP V28-239.

each new time record at a 5-KY interval (Fig. 143). As in constructing TYPE, we assume that data points of each time record are connectable with straight-line segments. TYPE, 502B, and 504 have $\delta^{18}O$ values obtained from planktonic foraminifera, while 552A $\delta^{18}O$ values are from benthic foraminifera (note scale change). TYPE, 502B, and 552A are sampled in the time range of 5–1360 KY, while 504 is in the time range of 185–1540 KY. All records therefore have an equal time range of 1355 KY, allowing for easy comparison, since they will have equal harmonic spacing in power

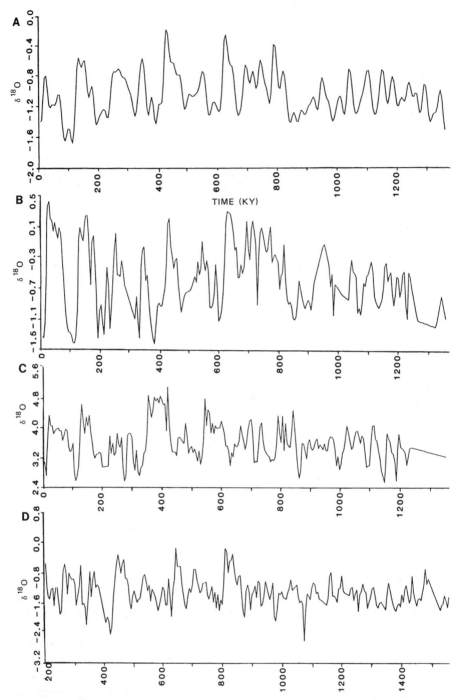

Fig. 143. $\delta^{18}O$ records for TYPE (A) and DSDP cores 502B (B), 552A (C), and 504 (D). Note time shift for DSDP 504 and $\delta^{18}O$ scale difference for DSDP 552A.

spectra. We assume that the same frequency signal(s) in TYPE, 502B, and 552A (5–1360 KY) are included in 504 in the time range 185–1540 KY.

The ensuing analysis follows the basic format of the flowchart shown in Fig. 134(b). We compare the records in the time domain using autocorrelation, cross-correlation, and semblance methods, and then use frequency methods such as power spectra, cross-spectra, and spectral coherency at discrete frequencies to provide different comparisons of the records.

A. Autocorrelation and Cross-Correlation

The autocorrelation versus lag time of each record is shown in Fig. 144, along with the autocorrelation for the original V28-239 (before reconstruction). Common to all records is (1) the general decrease in the autocorrelation coefficients as lag time increases and (with the possible exception of 552A) (2) the occurrence of peaks in correlation at lag times 95 ± 30, 180 ± 40, and 280 ± 30 KY. These lag times represent preferred lag times for each record, common to most records, and so indicate that a common signal is indeed present in the records. Also, there is a common periodic signal in these autocorrelograms of 95 ± 35 KY (peak to peak).

Cross-correlation versus lag time plots are shown in Fig. 145 for TYPE versus each record. Recall that we are now sliding the TYPE record against the other records at increasing lags and looking for common preferred lag times. Preferred lag times are seen at 95 ± 30, 195 ± 30, and 280 ± 40 KY, similar to those for the autocorrelations for each record (Fig. 144). Note that for TYPE versus 552A, preferred lag times are troughs rather than peaks because TYPE has mostly negative (planktonic) $\delta^{18}O$ values, while 552A has positive (benthic) $\delta^{18}O$ values. Hence, when multiplied within the cross-correlation function, preferred peaks would be the most negative peaks or troughs. The peaks in this particular cross-correlogram are not significant; rather, they result from the signal in the TYPE record. Again, the periodic signal of 95 ± 30 KY is common to TYPE, 502B, and 504 as in the autocorrelation treatment. Thus, these analyses enable us to identify a global $\delta^{18}O$ signal, along with its uncertainty range; that is, a periodicity of 95 ± 30 KY. This is in accordance with the periodicity for V28-239 and other Pleistocene $\delta^{18}O$ records desribed by Kominz et al. (1979) and Pisias et al. (1984).

B. Semblance Methods

Semblance methods provide an alternative technique to locate common signals and determine the duration and strength of these signals between record pairs, in *specific* time ranges. The results of the semblance technique

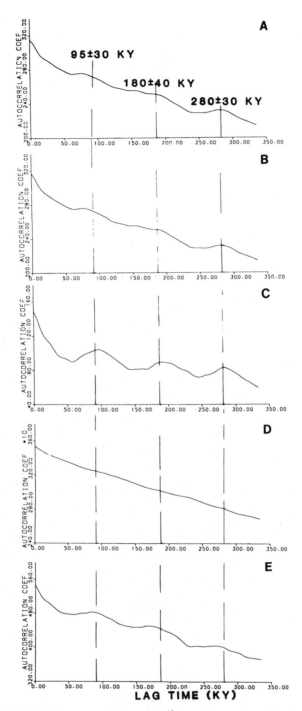

Fig. 144. Auto-correlation versus lag time for $\delta^{18}O$ records for TYPE (A) and cores V28-239 (B), 502B (C), 552A (D), and 504 (E).

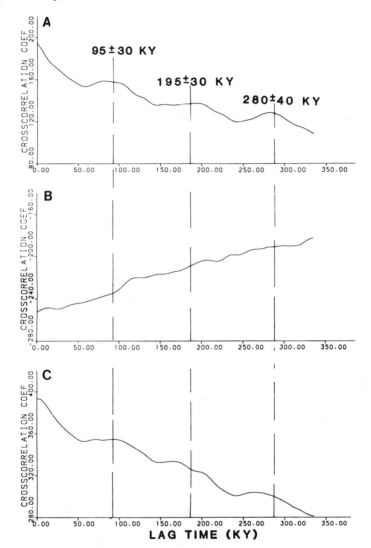

Fig. 145. Cross-correlation versus lag time for δ^{18}O records for TYPE versus DSDP cores 502B (A), 552A (B), and 504 (C).

(Fig. 146) show contours of constant semblance for δ^{18}O data from TYPE versus sites 502B, 504, and 552 for times ranging from 200 to 850 KY with window widths 5–100 KY (502B and 504) and 5–50 KY (552A).

The semblance plot for TYPE and site 502B (Fig. 146a) shows relatively high coherence (semblance >1.7) centered on times 435, 630, and 840 KY,

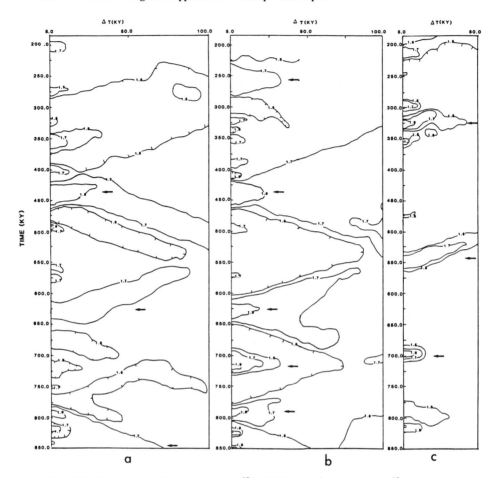

Fig. 146. Contoured semblance plots of $\delta^{18}O$ TYPE record versus other $\delta^{18}O$ records in the time range 200–850 KY. Contour interval is 0.2 and contours below 1.5 are not depicted. Arrows indicate high semblance (coherence) events; (a) TYPE and 502B, (b) TYPE and 504, and (c) TYPE and 552A.

indicative of a possible common isotopic event recorded at the same time in the two locations. The event recorded at 435 KY is the most coherent, persisting to a window width of >100 KY. The two records are also relatively coherent (semblance >1.5) in the time intervals 285–390, 460–475, and 660–690 KY.

Semblance for TYPE and site 504 [Fig. 146(b)] shows relatively high coherence (semblance >1.7) centered on 255, 435, 625, 720, and 790 KY, again indicative of common events in the two records. The events recorded at 435 and 625 KY are in accordance with those events identified in TYPE and 502B [Fig. 146(a)]. Again, the event at 435 KY is the most coherent. TYPE and 504 are relatively incoherent in the intervals 185–225, 285–310, and 465–600 KY. In the range of 650–775 KY, there is broad incoherence but a short-term coherent event is recorded, centered on 720 KY. The semblance plot for TYPE and site 552A [Fig. 146(c)] shows only minor coherence centered on 325, 550, and 700 KY. In a broad sense these records are incoherent.

Comparison of the time ranges of high coherence (Fig. 147) with the respective time ranges of the $\delta^{18}O$ records (Fig. 143) shows that these correlations are detectable by visual correlation. Why then the use of the semblance technique? Not only are we objectively quantifying the correlations, but also with further testing we may find that some types of correlations (i.e., long-term trends, shifts in records, magnitude of shifts, direction of shifts) may be more easily detected by this technique than by visual correlation.

C. Power Spectra, Cross-Spectra, and Spectral Coherency

In addition to locating common signals in time, we may also search for common isotopic signals in the frequency domain using frequency methods. For each $\delta^{18}O$ record, we have constructed power spectra, cross-spectra for each record versus TYPE, and spectral coherency at discrete frequencies for each record versus TYPE (Figs. 147–150). In each figure (Figs. 147–150), there are 136 harmonics (130 depicted) of spacing $1/T$, where T is the total length of the record(s) (1355 KY). Thus, the power levels at a given harmonic in each figure are comparable on a one-to-one basis. The zeroth harmonic has been suppressed from each plot for scaling clarity. Each spectrum has been smoothed by the triangular, Hamming–Tukey filter. Figure 147 shows the spectral analysis results for TYPE and 502B records for the time range 5–1355 KY. Some of the dominant frequencies for TYPE are labeled with their corresponding periods and uncertainties. Frequencies above harmonic 62 in TYPE were defined as noise previously. To be noted in the TYPE power spectrum is the general decay of power as the frequency increases [Fig. 147(a)]. Superimposed on this decay are significant spikes with periodicities of 850 ± 510, 153 ± 17, 97 ± 7, 72 ± 4, 50 ± 2, 40 ± 1, 32 ± 1, 27 ± 1, and 22–23 KY. Dominant peaking occurs roughly every 5–7 harmonic widths on either side of harmonic 14 (the most dominant harmonic at a period of 97 KY), with a width of 2–3 harmonics. Thus, a sequence of events in the TYPE record is occurring in a roughly cyclic manner, but neither the cycle time (different peaks) nor the duration of the events (width of peaks) is

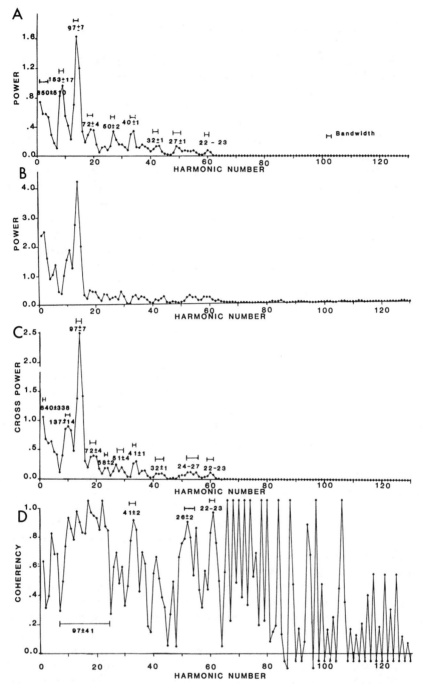

Fig. 147. (A) Power spectrum for TYPE, (B) power spectrum for $\delta^{18}O$ record from DSDP 502B, (C) cross-power spectrum for TYPE versus DSDP 502B, and (D) spectral coherency for TYPE versus DSDP 502B. Records are from the time range 5–1360 KY, and harmonic spacing is $7.38 \times 10^{-4}\ (KY)^{-1}$.

holding precisely constant in frequency. Also, there is a substantial dropoff in power above the dominant harmonic 14, indicating that the low-frequency component is stronger or better preserved in the record.

The power spectrum of DSDP 502B [Fig. 147(b)] shows roughly the same pattern of peaks and troughs as for TYPE [Fig. 147(a)]. However, the strength of the dominant harmonic 14 is increased in DSDP 502B, the higher harmonics have a lower and more constant power level, and the peak at harmonics 8–10 (153 ± 17 KY) in TYPE is shifted toward harmonics 10–12 (125 ± 12 KY) in DSDP 502B [Fig. 147(a) and (b)].

The similarities in spectral character of TYPE and 502B are quantified with the cross-spectrum and coherency analyses [Fig. 147(c) and (d)]. In the cross-spectrum, the dominant harmonic is 14 with other peaks occurring roughly every 4–5 harmonics with widths of 2–3 harmonics. Again, an abrupt decrease in power occurs after the dominant harmonic. Figure 147(d) illustrates the spectral coherency of the two records. Coherency of one indicates that the two records are perfectly correlated at that frequency, while coherency of zero means that the records are incoherent (coherency values just greater than one and just less than zero are flags for a slight round-off error in computation). Ranges of relatively good coherence are visible between harmonics 10–24, 32–34, 50–55, and 60–62, with valleys of incoherence between these peaks [Fig. 147(d)]. These harmonic ranges correspond to periods 97 ± 41, 41 ± 2, 26 ± 2, and 22–23 KY, respectively. Harmonics 63 and above show considerable computational roundoff error because of the way we defined the noise in the TYPE record. Overprinted on these broad trends are spikes of coherence every 3–5 harmonics with a width of 2–3 harmonics. Thus, common periodic events have been recorded by the isotopic data in both TYPE and 502B.

In a similar manner, the power spectra, cross-spectrum, and spectral coherency were constructed for TYPE and DSDP 504 for the time range 185–1540 KY (Fig. 148). The TYPE power spectrum [Fig. 148(a)] shows roughly the same power spectrum as for the time range 5–1360 KY used with DSDP 502B (Fig. 147). Dominant peaking occurs at harmonic 15 with smaller peaks every 5–7 harmonics on either side, of width 2–3 harmonics. The power spectrum for DSDP 504 [Fig. 148(b)] shows the dominant harmonic 15 superimposed on a slowly decaying background with increasing frequency. This power spectrum also shows the sharp drop-off in power after the dominant harmonic and many of the same peaks. The sharper peaks and lower troughs in Fig. 148(b) than in Fig. 148(a) indicate there is better resolution at preferred discrete frequencies in DSDP 504 than in TYPE. The cross-spectrum for TYPE versus DSDP 504 [Fig. 148(c)] is smoother than the power spectrum for 504, as a result of its crossing with TYPE. Peaks can be traced from the cross spectrum directly to the

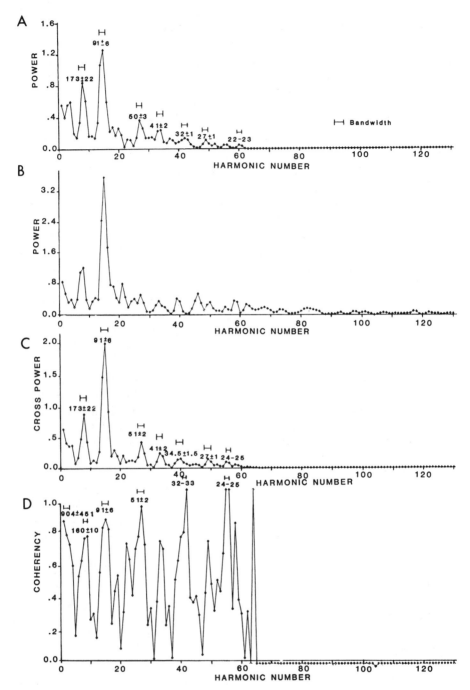

Fig. 148. (A) Power spectrum for TYPE, (B) power spectrum for δ^{18}O record from DSDP 504, (C) cross-power spectrum for TYPE versus DSDP 504, and (D) spectral coherency for TYPE versus DSDP 504. Records are from the time range 185–1540 KY, and harmonic spacing is 738×10^{-4} (KY)$^{-1}$.

corresponding harmonics in the individual power spectra, and these peaks are reflected in the coherency plot [Fig. 148(d)] at harmonics 1–3, 8–9, 14–16, 26–28, 41–42, and 55–56 (904 ± 451, 160 ± 10, 91 ± 6, 51 ± 2, 32–33, and 24–25 KY, respectively). (Recall that above harmonic 63 there is considerable roundoff error.) The raggedness of the TYPE versus 504 coherency plot, when compared to that of TYPE versus 502B [(Fig. 147(d)], is related to the sharper peaks in the power spectra for DSDP 504.

Figure 149 shows the power spectra, cross-spectrum, and spectral coherency for TYPE and DSDP 552A, evaluated in the time range 5–1360 KY. Figure 149(a) is identical to Fig. 147(a) but is shown for ease of comparison. The power spectrum for DSDP 552A [Fig. 149(b)] shows the dominant harmonics 12 and 13 (109 ± 8 KY) with a sharp drop-off in power at higher harmonics. Few peaks in 552A correlate with those shown in TYPE [Fig. 149(a)], giving rise to a cross-spectrum with a much different character than the individual power spectra [Fig. 149(c)]. The incoherence of much of the spectra for TYPE and 552 is shown by a significant power level decrease in the cross-spectrum as seen also in the coherency plot [Fig. 149(d)]. Relatively good coherence between TYPE and 552A occurs only at harmonics 1, 11–12, and 17–18 (1355, 117 ± 5, and 77.5 ± 2.5 KY, respectively).

From our comparison of each individual record with the TYPE record (Figs. 147–149), DSDP 502B seems to be the most coherent with the TYPE record when viewed across all harmonics, i.e., DSDP 502B is recording more of the template isotopic signal than 504 and 552. The record of site 552A, based on the $\delta^{18}O$ of a benthic foraminiferal species, seems the least correlated to our TYPE record. Of course, we must also bear in mind that these results are dependent on the accuracy of the original assumptions of constant sedimentation rates between time horizons (for example, as shown in Table VI, 502B has better time horizon control than the other two sites) and that the benthic foraminifera (552A) are recording the same events as the planktonic foraminifera (TYPE, 502B, 504).

Alternatively, we may assume that the isotopic signal *is* faithfully recorded in all records but in some distorted sense. We may then wish to define the frequencies at which all the isotope records are correlated, in a global sense. To address the global aspect of isotopic records in a preliminary fashion, the spectral coherencies of Figs. 147(d), 148(d), and 149(d) have been arithmetically averaged at each harmonic from the three DSDP records (502B, 504, 552A) versus TYPE [Fig. 150(a)]. Figure 150(b) shows the same average after a smoothing operation (triangular filter with a three-harmonic width with coefficients, 0.15, 0.70, 0.15) was applied to the coherency values. To be noted in Fig. 150(a) and (b) is the relatively good coherence at harmonics 1 (period = 1355 KY), 8–27 (period = $110 + 60$ KY), 32–34 ($41 \pm$

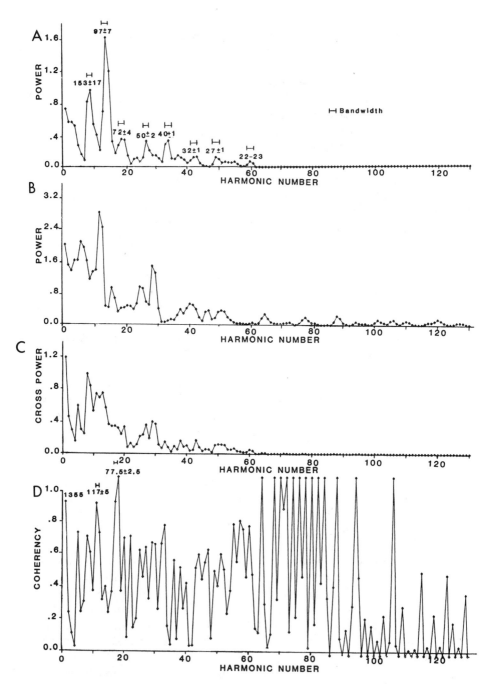

Fig. 149. (A) Power spectrum for TYPE, (B) power spectrum for $\delta^{18}O$ record from DSDP 552A, (C) cross-power spectrum for TYPE versus DSDP 552A, and (D) spectral coherency for TYPE versus DSDP 552A. Records are from the time range 5–1360 KY, and harmonic spacing is 7.38×10^{-4} $(KY)^{-1}$.

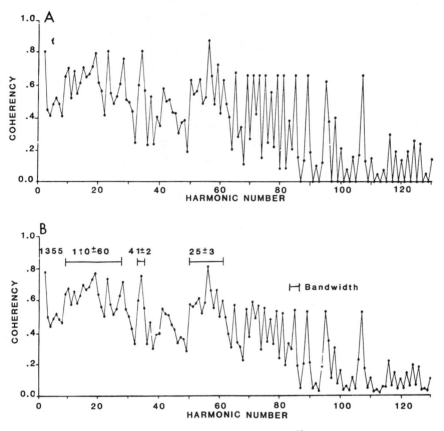

Fig. 150. (A) Average spectral coherency for TYPE versus $\delta^{18}O$ records from DSDP cores 502B, 552A, and 504. (B) The same as (A) but after a triangular smoothing filter, has been applied. Harmonic spacing is 7.38×10^{-4} $(KY)^{-1}$. The numbered bars indicate the corresponding signal peaks in thousands of years.

2 KY and 49–60 $(25 \pm 3$ KY) with broad incoherent troughs occurring between peaks. Superimposed on this trend are peaks of 2–3 harmonic width of higher coherence. This coherency can be used to define a global frequency signal for $\delta^{18}O$ records of the Pleistocene. The coherency of this global data base can be improved with other $\delta^{18}O$ records to determine the best global signal. This defined global signal can be subtracted from individual records, thereby producing a residual record representing a regional or local signature (after definition of noise level) that may be used for regional analyses (e.g., temperature fluctuations and stratigraphic correlation).

Results of this initial investigation point favorably to using time series analysis for interpretation of isotopic records. Construction of our preliminary TYPE record gives us a template with which to compare records. Our TYPE record originated from a single record and evolved via application of some data-determined filters into a noiseless record. While the final level of noise was determined by us, the decision should be guided as much as possible by quantitative methods and reasoning.

Using this TYPE record we were able to extract common $\delta^{18}O$ signals from several DSDP sites employing several techniques. Semblance allowed us to determine isotopic events recorded in TYPE, 502B, and 504 centered on the times 435 and 625 KY. Autocorrelation and cross-correlation techniques identified a common periodic signal of 95 ± 30 KY in TYPE, 502B, 552A, and 504, very comparable to that obtained from the earth's orbital eccentricity parameter (Milankovitch, 1946; Berger, 1976; Kominz et al., 1979). This periodicity is also reflected in the power spectra, cross-spectra (versus TYPE), and spectral coherency (versus TYPE), within resolution limits, for TYPE and 502B (97 ± 7 KY), 504 (91 ± 6 KY), and 552A (1177 ± 5 KY) (Figs. 147–149). This periodicity is also seen in a global sense (with less resolution) in the average spectral coherency (110 ± 60 KY) (Fig. 150). Other global periodicities of approximately 1355, 41 ± 2, and 25 ± 3 KY are also seen in the average (global) spectral coherency. The latter two and that at 110 ± 60 KY are in accord with the orbital periodicities determined by Hays et al. (1976), Kominz et al. (1979), and Imbrie and Imbrie (1980). Table VIII provides a summary of events identified in the four records by these techniques.

Table VIII
Main Periodicities Identified by Coherency

δ¹⁸O Records	>900	200–100	100–90	90–50	50–40	40–30	30–20
			Periodicities[a]				
TYPE vs 502B			97 ± 41		41 ± 2		22–23 26 ± 2
TYPE vs 504	904 ± 451	160 ± 10	91 ± 6	51 ± 2		32–33	24–25
TYPE vs 552A	1355	117 ± 5		77.5 ± 2.5			
Global average (stacked)	1355	110 ± 60			41 ± 2		25 ± 3
Global auto-correlation			95 ± 30				
Global cross-correlation			95 ± 30				

[a] Dominant spectral coherency peak and those determined by auto- and cross-correlations (KY).

V. Summary

Obviously there are still many topics to address with isotopic and other geochemical records. Alternative assumptions to provide time-depth conversions are available to that of the constant sedimentation rates between the time horizons used here. Power spectra and semblance techniques can be used to determine the *relative* distortion of a record against a template. With these methods comes the assumption of a common signal in the records. In addition, what of the interpolation techniques to produce an equally spaced time record from unequally spaced data? Assumptions other than linear connectivity of adjacent data points need to be addressed, for instance, by the use of rectangular or triangular interpolation (Davis, 1986) to determine if a signal is independent of processing techniques. Different noise identification methods and noise reduction techniques need to be investigated, including (1) frequency filtering by power level (white noise filtering) rather than the on–off (low-pass) filters used here and (2) maximum entropy algorithm and CABFAC/Q-MODE analyses. We also need to improve our TYPE section using identified global signal information to get the best template for comparison with other records.

This chapter illustrates how the use of these quantitative analysis methods in interpreting isotopic records provides a powerful alternative to visual and graphic correlation techniques currently used in correlating isotopic records. When used in conjunction with these empirical techniques, interpretations and correlations carry more objective significance. At the same time, we stress that care must be taken to interpret the results within the constraints of the resolution each method provides. The assumptions must be clearly acknowledged when applying the methods and interpreting the results accordingly.

Chapter 21 | Summary

It is difficult to provide a succinct summary of all the topics that have been touched upon in this synthesis of the character and global nature of oxygen and carbon isotope records of the Tertiary. As an introduction to the empirical and quantitative approaches to isotope chronostratigraphy, we have also shown how this strategy can lead to precise stratigraphic subdivisions of exploration wells and possible paleobathymetric models. While we realize that it will be impossible to anticipate all of the questions that any synthesis generates, we hope that this book serves as a focal point for the proper use of quantitative techniques of signal extraction and data manipulation in isotope chronostratigraphy. Obviously, isotope chrono-stratigraphy will not work in all depositional systems. In some basins or time periods, one may have to settle with either simple chemostratigraphic or diagenetic information, using stable isotope records in a stratigraphic context. However, the development of an isotope chronostratigraphic framework is particularly attractive for critical test wells. Isotope chrono-stratigraphy can provide the detailed framework in which results could be integrated with a large body of other information and then used to select other wells, or intervals in other wells, for the purpose of regional synthesis and depositional modeling of exploration basins.

We feel strongly that isotope chronostratigraphy is an innovative approach in studying the stratigraphy of exploration wells. We also believe that the empirical and quantitative approaches being developed will lead to an increasing number of breakthroughs in exploration efforts. These efforts are at this very moment continuing to be made in the general areas of (1) enhanced stratigraphic resolution of exploration wells; (2) precise and accurate determination of regional tie lines; (3) extension of shallow-water exploration strategies into deeper water environments; and (4) calibration of biostratigraphic and lithostratigraphic frameworks.

We also believe that isotope chronostratigraphy will become, perhaps within the next three to five years, an indispensible tool for the industrial stratigrapher and biostratigrapher. Isotope chronostratigraphy has advantages over many other stratigraphic frameworks in that the isotope record has the potential of providing not only datums closely spaced in time but also ones that are synchronous on a global basis. Isotope chronostratigraphy also offers the important possibility of bridging the gap between shallow- and deep-water environments, independent of facies changes in many cases.

The next step is to move away from visual (empirical) correlations of isotope records and apply well-established quantitative methods of signal correlation (i.e., spectral analysis, Fourier series analysis, cross correlation techniques, autocorrelation techniques, maximum entropy methods, and semblance methods). This book represents part of our attempt to move in this quantitative direction. The results, which the Isotope Stratigraphy Group at the University of South Carolina have obtained to date, have been truly encouraging, and we look forward to continuing our efforts to test and refine the concept of isotope chronostratigraphy in the future.

References

Ables, J. (1974). Digital dispersion spectroscopy: A technique for the study of very rapid variability of celestial radio sources. *Astron. Astrophy., Suppl. Ser.* **15**, 383–392.

Ahr, W. M., and Berg, R. R. (1982). *In* "Depositional and Diagenetic Spectra of Evaporites—A Core Workshop" (C. R. Handford, R. G. Loucks, and G. R. Davies, eds.), SEPM Core Workshop No. 3, pp. 305–395. Soc. Econ. Paleontol. Mineral., Los Angeles, California.

Akaike, H. (1973). Information theory and an extension of the maximum likelihood principle. *In* "Information Theory" (B. N. Petrov and F. Csaki, eds.), pp. 207–281. Akadémiai Kiadó, Budapest.

Akaike, H. (1974). A new look at the statistical model identification. *IEEE Trans. Autom. Control* **AC-19**, 176–723.

Akaike, H. (1977). On entropy maximization principle. *In* "Application of Statistics" (P. R. Krishnaiah, ed.), pp. 27–41. North-Holland Publ., Amsterdam.

Akaike, H. (1978). Time series analysis and control through parametric models. *In* "Applied Time Series Analysis" (D. F. Findley, ed.), pp. 1–23. Academic Press, New York.

Akers, W. H., and Holck, A. J. J. (1957). Pleistocene beds near the edge of the continental shelf, southeastern Louisiana. *Geol. Soc. Am. Bull.* **68**, 983–992.

Aki, K. (1960). Study of earthquake mechanism by a method of phase equilization applied to Rayleigh and Love waves. *J. Geophys. Res.* **65**, 729–740.

Aki, K., and Richards, P. G. (1980). "Quantitative Seismology: Theory and Methods." Freeman, San Francisco, California.

Al-Chalabi, M. (1979). Velocity determination from seismic reflection data. *In* "Developments in Geophysical Exploration Methods" (A. A. Fitch, ed.), pp. 1–68. Applied Science Publishers, Barking, England.

Alvarez, W., Ascero, F., Michelle, H. V., and Alvarez, L. W. (1982). Iridium anomaly approximately synchronous with terminal Eocene extinctions. *Science* **216**, 886–888.

Amaldi, E. (1979). Einstein and gravitational radiation. *In* "Gravitational Radiation, Collapsed Objects and Exact Solutions" (C. Edwards, ed.), pp. 246–298. Springer-Verlag, Berlin and New York.

Anderson, T. F., and Arthur, M. A. (1983). Stable isotopes in sedimentary geology. *SEPM Short Course* **10**.

320   References

Anderson, T. F., and Steinmetz, J. C. (1981). Isotopic and biostratigraphical records of calcareous nannofossils of a Pleistocene core. *Nature (London)* **294**, 741-744.

Anderson, T. F., and Steinmetz, J. C. (1983). Stable isotopes in calcareous nannofossils: Potential applications and deep sea paleoenvironmental reconstructions during the Quaternary. *Utrecht Micropaleontol. Bull.* **30**, 183-204.

Anstey, N. A. (1970). "Seismic Prospecting Instruments." Vol. 1. Brontraeger, Berlin.

Arnold, L. (1974). "Stochastic Differential Equations." Wiley, New York.

Arthur, M. A. (1983). The carbon cycle—controls on atmospheric CO_2 and climate in the geologic past. *In* "Climate in Earth History," pp. 55-67. Nat. Acad. Press, Washington, D.C.

Arthur, M. A., Scholle, P. A., and Hasson, P. (1979). Stable isotopes of oxygen and carbon in carbonates from Sites 398 and 116 of the Deep Sea Drilling Project. *Initial Rep. Deep Sea Drill. Proj.* **47**, 447-491.

Backman, J. (1979). Pliocene biostratigraphy of DSDP Sites 111 and 116 from the North Atlantic Ocean and the age of Northern Hemisphere Glaciation. *Stockholm Contrib. Geol.* **32**(3), 115-137.

Baker, P., Gieskes, J., and Elderfield, H. (1982). Diagenesis of carbohydrates in deep-sea sediments—Evidence from Sr/Ca ratios and interstitial dissolved Sr^{2+} data. *J. Sediment. Petrol.* **52**, 71-82.

Banik, N. C., Lerche, I., and Shuey, R. T. (1985a). Stratigraphic filtering. Part 1. Derivation of O'Doherty-Anstey formula. *Geophysics* **50**, 2768-2774.

Banik, N. C., Lerche, I., Resnick, J. R., and Shuey, R. T. (1985b). Stratigraphic filtering. Part II. Model spectra. *Geophysics* **50**, 2775-2783.

Barron, J. A. (1980). Lower Miocene to Quaternary diatom biostratigraphy of DSDP Leg 57, off northwestern Japan. *Initial Rep. Deep Sea Drill. Proj.* **57**, 641-686.

Barron, J. A., and Keller, G., (1982). Widespread Miocene deep-sea hiatuses: Coincidence with periods of global cooling. *Geology* **10**, 577-581.

Bartlett, M. S. (1966). "An Introduction to Stochastic Processes." 2nd ed. Cambridge Univ. Press, London and New York.

Beard, J. H. (1969). Pleistocene paleotemperature record based on planktonic foraminifers, Gulf of Mexico. *Gulf Coast Assoc. Geol. Soc. Trans.* **19**, 535-553.

Beard, J. H., Sangree, J. B., and Smith, L. A. (1982). Quaternary chronology, paleoclimate, depositional sequences and eustatic cycles. *Am. Assoc. Petrol. Geol. Bull.* **66**, 158-169.

Beauchamp, K. G. (1973). "Signal Processing Using Analog and Digital Techniques." Allen & Unwin, London.

Beauchamp, K. G. (1975). "Exploitation of Seismograph Networks." Nordhoff International Publishing, Leyden, Netherlands.

Bélanger, P. E., and Matthews, R. K., (1984). The foraminiferal isotope record across the Eocene-Oligocene boundary at DSDP site 540. *Initial Rep. Deep Sea Drill. Proj.* **77**, 589-592.

Bélanger, P. E., Curry, W. E., and Matthews, R. K. (1981). Core top evaluation of benthic foraminiferal isotopic ratios for paleo/oceanographic interpretations. *Palaeogeogr., Palaeoclimatol., Palaeoecol.* **33**, 205-220.

Bendat, J. S., and Piesol, A. G. (1971). "Random Data: Analysis and Measurement Procedures." Wiley (Interscience), New York.

Bender, M. L., and Graham, D. (1981). On late Miocene abyssal hydrography. *Mar. Micropaleontol.* **6**, 451-464.

Bender, M. L., and Keigwin, L. D., Jr. (1979). Speculations about the upper Miocene change in abyssal Pacific dissolved carbonates $\delta^{13}C$. *Earth Planet. Sci. Lett.* **45**, 383-393.

Benignus, V. A. (1969). Estimation of the coherence spectrum and its confidence interval using the fast Fourier transform. *IEEE Trans. Audio Electroacoust.* **AU-17**, 145-150.

Berger, A. L. (1976). Obliquity and precession for the last 5,000,000 years. *Astron. Astrophys.* **51**, 127.

Berger, W. H. (1979). Stable isotopes in foraminifera, foraminiferal ecology and paleoecology. *SEPM Short Course* **6**, 156-198.

Berger, W. H. (1985). On the time scale of deglaciation: Atlantic deep-sea sediments and Gulf of Mexico. *Palaeogeogr. Palaeoclimatol. Palaeoecol.* **50**, 167-184.

Berger, W. H., Killingley, J. S., and Vincent, E. (1978). Stable isotopes in deep sea carbonates: Box core ERDC-92, west equatorial Pacific. *Oceanol. Acta* **1**, 203-216.

Berger, W. H., Killingley, J. S., and Vincent, E. (1985a). Timing of deglaciation from an oxygen isotope curve for Atlantic deep-sea sediments. *Nature (London)* **314**, 156-158.

Berger, W. H., Killingley, J. S., Metzler, C. V., and Vincent, E. (1985b). Two-step deglaciation: ^{14}C-dated high resolution [18]O records from the tropical Atlantic ocean. *Quat. Res.* **23**, 258-271.

Berggren, W. A. (1972). Late Pliocene-Pleistocene glaciation. *Initial Rep. Deep Sea Drill. Proj.* **12**, 953.

Berggren, W. A., Kent, D. V., and Flinn, J. J. (1985a). Jurassic to Paleogene. Part 2. Paleogene geochronology and chronostratigraphy. *In* "The Chronology of the Geological Record" (N. J. Snelling, ed.), Geol. Soc. Mem. No. 10, pp. 141-198. Geol. Soc., Oxford.

Berggren, W. A., Kent, D. V., and VanCouvering, J. A. (1985b). The Neogene. Part 2. Neogene geochronology and chronostratigraphy. *In* "The Chronology of the Geological Record" (N. J. Snelling, ed.), Geol. Soc. Mem. No. 10, pp. 211-260. Geol. Soc., Oxford.

Berggren, W. A., and Van Couvering, J. A. (1974). The late Neogene: Biostratigraphy, geochronology and paleoclimatology of the last 15 million years in marine and continental sequences. *Paleogeogr. Paleoclimatol. Paleoecol.* **16**, 1-216.

Bernard, H. A., and LeBlanc, R. G. (1965). Resume of the Quaternary geology of the northwestern Gulf of Mexico provence. *In* "The Quaternary of the United States" (H. P. Wright, Jr. and D. G. Frey, eds.), pp. 137-185. Princeton Univ. Press, Princeton, New Jersey.

Berryman, L. H., Goupillaud, P. L., and Waters, K. H. (1958). Reflections from multiple transition layers. *Geophysics* **23**, 223-243.

Bezdek, J. C. (1981). "Pattern Recognition with Fuzzy Objective Function Algorithms." Plenum, New York.

Bezdek, J. C., Ehrlich, R., and Full, W. E. (1984). The fuzzy c-means clustering algorithm. *Comput. Geosci.* **10**, 2.

Biolzi, M. (1983-1984). Stable isotopic study of Oligocene-Miocene sediments from DSDP site 354, south Atlantic. *Mar. Micropaleont.* **8**, 121-131.

Blackman, R. B., and Tukey, J. W. (1959). "The Measurement of Power Spectra." Dover, New York.

Blair, D. G. (1979). Gravity wave antenna-transducer systems. *In* "Gravitational Radiation. Collapsed Objects and Exact Solutions" (C. Edwards, ed.), pp. 299-313. Springer-Verlag, Berlin and New York.

Blanc, P. L., Rabusierd, D., Vergnaud-Grazzini, C., and Duplessy, J. C. (1980). North Atlantic deep water formed by the later middle Miocene. *Nature (London)* **283**, 553-555.

Bloom, A. M., Broecker, W. S., Chappell, J. M. A., Matthews, R. K., and Mesolella, K. J. (1974). Quaternary sea level fluctuations on a tectonic coast: New ^{230}Th/^{234}U dates from the Huon Peninsula, New Guinea. *Quat. Res.* **4**, 185-205.

Boersma, A., and Shackleton, N. J. (1977a). Tertiary oxygen and carbon isotope stratigraphy, site 357 (mid latitude south Atlantic). *Initial Rep. Deep Sea Drill. Proj.* **39**, 911-924.

Boersma, A., and Shackleton, N. J. (1977b). Oxygen and carbon isotope through the Oligocene, DSDP site 366, equatorial Atlantic. *Initial Rep. Deep Sea Drill. Proj.* **39**, 957-962.

Boersma, A., Shackleton, N. J., Hall, M. A., and Given, Q. C. (1979). Carbon and oxygen isotope records at DSDP site 384 (North Atlantic) and some Paleocene paleotemperatures and carbon isotope variations in the North Atlantic. *Initial Rep. Deep Sea Drill. Proj.* **43**, 695-717.

Boon, J. D., III, Evans, D. A., and Hennigar, H. F. (1982). Interpretation of grain shape information from Fourier analysis of digitized two-dimensional images. *Math. Geol.* **14**, 412-417.

Box, G. E. P. (1979). Robustness in the strategy of scientific model building. *In* "Robustness in Statistics" (R. L. Launer and G. N. Wilkinson, eds.), pp. 201-236. Academic Press, New York.

Brigham, E. O. (1974). "The Fast Fourier Transform." Prentice-Hall, Englewood Cliffs, New Jersey.

Broecker, W. S., and Van Donk, J. (1970). Insolation changes, ice volumes, and the ^{18}O record in deep sea cores. *Rev. Geophys. Space Phys.* **8**, 169-198.

Broecker, W. S., Thurber, D. L., Goddard, J., Ku, T. L., Matthews, R. K., and Mesolella, K. J. (1968). Milankovitch hypothesis supported by precise dating of coral reefs and deep-sea sediments. *Science* **159**, 297.

Brown, P. J., Ehrlich, R., and Colguhoun, D. (1980). Origin of patterns of quartz sand types on the southeastern United States continental shelf and its implications on contemporary shelf sedimentation-Fourier grain shape analysis. *J. Sediment. Petrol.* **50**, 1095-1100.

Brunner, C. A., and Keigwin, L. D. (1981). Late Neogene biostratigraphy and stable isotope stratigraphy of a drilled core from the Gulf of Mexico. *Mar. Micropaleontol.* **6**, 397-418.

Buchart, B. (1978). Oxygen isotope paleotemperatures from the Tertiary in the North Sea area. *Nature (London)* **275**, 121-123.

Burckle, L. H., Keigwin, L. D., Jr., and Opdyke, N. D. (1982). Middle and late Miocene stable isotope stratigraphy: Correlation to the paleomagnetic reversal record. *Micropaleontology* **28**, 329-334.

Burg, J. P. (1967). Maximum entropy spectral analysis. Presented at the 37th Meeting of the Society of Exploration Geophysicists, Oklahoma City.

Burg, J. P. (1972). The relationship between Maximum Entropy Spectra and Maximum Likelihood Spectra. *Geophysics* **37**, 275-276.

Buttkus, B. (1975). Homomorphic filtering. *Geophys. Prospect.* **23**, 712-748.

Capon, J. (1969). High-resolution frequency-wavenumber spectrum analysis. *Proc. IEEE* **57**, 1408-1418.

Capon, J., and Goodman, N. R. (1970). Probability distributions for estimators of the frequency-wavenumber spectrum. *Proc. IEEE* **58**, 1785-1786.

Carter, G. C., Knapp, C. H., and Nuttall, A. H. (1973). Statistics of the estimate of the magnitude-coherence function. *IEEE Trans. Audio Electroacoust.* **AU-21**, 388-389.

Chayes, F. (1971). "Ratio Correlation." Univ. of Chicago Press, Chicago, Illinois.

Claerbout, J. F. (1979). "Fundamentals of Geophysical Data Processing With Applications to Petroleum Prospecting." McGraw-Hill, New York.

CLIMAP Project Members (1976). The surface of the Ice-Age Earth. *Science* **191**, 1131-1137.

Cole, T. W. (1973). Reconstruction of the image of a confined source. *Astron. Astrophys.* **24**, 41-45.

Collins, D. M. (1982). Electron density images from imperfect data by iterative entropy maximization. *Nature (London)* **298**, 49-51.

Cooper, H. W., and Cook, R. E. (1984). Seismic data gathering. *Proc. IEEE* **72**, 1266-1275.

Corliss, B. H., and Keigwin, L. D., Jr. (1986). "Eocene-Oligocene paleoceanography, Mesozoic and Cenozoic oceans," *Geodyn. Ser.* Vol. 15, pp. 101-1118. Am. Geophy. Union, Washington, D.C.

Corliss, B. H., Aubry, M. P., Berggren, W. A., Fenner, J. M., Keigwin, L. D., Jr., and Keller, G. (1984). The Eocene/Oligocene boundary event in the deep sea. *Science* **226**, 806–810.

Cox, H. (1973b). Resolving power and sensitivity to mismatch of optimum array processes. *J. Acoust. Soc. Am.* **54**, 771–785.

Craig, H. (1953). The geochemistry of the stable carbon isotopes. *Geochim. Cosmochim. Acta* **3**, 53–76.

Craig, H. (1957). Isotopic standards for carbon and oxygen and correction factors for mass-spectrometric analysis of carbon dioxide. *Geochim. Cosmochim. Acta* **12**, 133–149.

Craig, H. (1965). The measurement of oxygen isotope paleotemperatures. *Proc. Spoleto Conf. Stable Isotopes Oceanogr. Stud. Paleotemp.* **3**, 1–24.

Craig, H., and Gordon, L. I. (1965). Deuterium and oxygen 18 variations in the ocean and the marine atmosphere. *In* "Stable Isotopes in Oceanographic Studies and Paleotemperatures" (E. Tongiori, ed.), pp. 9–310. Consiglio Nazionale delle Ticherche, Laboratorio di Geologia Nucleare, Pisa.

Crouch, J. K. (1981). Noethwest margin of California continental borderland: Marine geology and tectonic evolution. *Am. Assoc. Petrol. Geol. Bull.* **65**, 191–218.

Curry, W. B., and Lohmann, G. P. (1982). Carbon isotopic changes in benthic foraminifera from the western south Atlantic: Reconstruction of glacial abyssal circulation patterns. *Quat. Res.* **18**, 218–235.

Curry, W. B., and Lohmann, G. P. (1983). Reduced advection into Atlantic Ocean deep eastern basins during last glacial maximum. *Nature (London)* **306**, 577–580.

Curry, W. B., and Matthews, R. K. (1981). Equilibrium ^{18}O fractionation in small size fraction planktic foraminifera: Evidence from recent Indian Ocean sediments. *Mar. Micropaleontol.* **6**, 327–337.

Dansgaard, W., and Tauber, H. (1969). Glacier oxygen-18 content and Pleistocene ocean temperatures. *Science* **166**, 499–502.

Davidon, W. C. (1968). Variance algorithm for minimization. *Comput. J.* **10**, 406–410.

Davis, J. C. (1986). "Statistics and Data Analysis in Geology." Wiley, New York.

Deines, P. (1970). Mass spectrometer correction factors for the determination of small isotopic composition variations of carbon oxygen. *Int. J. Mass. Spectrom. Ion Phys.* **4**, 283.

Deregowski, S. M. (1978). Self-matching deconvolution in the frequency domain. *Geophys. Prospect.* **25**, 252–290.

Deuser, W. G., Ross, E. H., and Waterman, L. S. (1976). Glacial and pluvial periods: Their relationship revealed by Pleistocene sediments of the Red Sea and Gulf of Aden. *Science* **191**, 1168–1170.

Devereaux, I. (1966). Oxygen isotope paleotemperature measurements on New Zealand Tertiary fossils. *N. Z. J. Sci.* **10**, 988–1011.

Dorman, F. H. (1966). Australian Tertiary paleotemperatures. *J. Geol.* **74**, 49–61.

Douglas, R. G., and Savin, S. M. (1971). Isotopic analyses of planktonic foraminifera from the Cenozoic of the northwest Pacific, Leg 6. *Initial Rep. Deep Sea Drill. Proj.* **6**, 1123–1127.

Douglas, R. G., and Savin, S. M. (1973). Oxygen and carbon isotope analyses of Cretaceous and Tertiary foraminifera from the Central North Pacific. *Initial Rep. Deep Sea Drill. Proj.* **17**, 591–605.

Douglas, R. G., and Savin, S. M. (1976). Oxygen and carbon isotope analyses of Tertiary and Cretaceous microfossils from Shatsky Rise and other sites in the north Pacific Ocean. *Initial Rep. Deep Sea Drill. Proj.* **32**, 509–520.

Drever, R. W., Hough, J., Bland, R., and Lessnoff, G. W. (1973). Signal processing methods for short bursts of gravitational radiation. *Nature (London)* **246**, 340–344.

Drexler, J. W., Rose, W. I., Jr., Sparks, R. S. J., and Ledbetter, M. T. (1980). The Los Chocoyos Ash, Guatemala: A major stratigraphic marker in Middle America and in three ocean basins. *Quat. Res.* **13**, 327–345.

Dudley, W. C. (1976). Paleoceanographic applications of calcareous nannoplankton grown in culture. Ph. D. Dissertation, University of Hawaii, Honolulu (unpublished).

Dudley, W. C., Duplessy, J. C., Blackwelder, P. L., Brand, L. E., and Guillard, R. R. L. (1980). Coccoliths in Pleistocene-Holocene nannofossil assemblages. *Nature (London)* 285, 222-223.

Duplessy, J. C. (1978). Isotope studies. *In* "Climatic Change" (J. Griffin, ed.), pp. 46-67. Cambridge Univ. Press, London and New York.

Duplessy, J. C., and Shackleton, N. J. (1984). Carbon-13 in the world ocean during the last interglaciation and the penultimate glacial maximum. Reevaluation of the possible biosphere response to the earth's climatic changes. *Prog. Biometerorol.* 3, 48-54.

Duplessy, J. C., Lalou, C., and Vinot, A. C. (1970). Differential isotopic fractionation in benthic foraminifera and paleotemperatures reassessed. *Science* 168, 250-521.

Duplessy, J. C., Moyes, J., and Pujol, C. (1980). Deep water formation in the North Atlantic during the last ice age. *Nature (London)* 286, 479-482.

Duplessy, J. C., Shackleton, N. J., Matthews, R. K., Prell, W., Ruddiman, W. F., Caralp, M., and Hendey, C. H. (1984). ^{13}C record of benthic foraminifera in the last interglacial ocean: Implications for the carbon cycle and global deep water circulation. *Quat. Res.* 21, 225-243.

Dziewonski, A. M., Hales, L. and Lapwood, E. R. (1977). Parametrically simple Earth models consistent with geological data. *Phys. Earth Planet. Inter.* 10, 12-48.

Ehrlich, R., Brown, P. J., Yarus, J. M., and Przygocki, R. S. (1980). The origin of shape frequency distributions and the relationship between size and shape. *J. Sediment. Petrol.* 50, 475-484.

Ehrlich, R., Crabtree, S. J., Kennedy, S. K., and Cannon, R. L. (1984). Petrographic image analysis. I. Analysis of Reservoir pore complexes. *J. Sediment. Petrol.* 54, 1365-1376.

Elderfield, H., and Gieskes, J. M. (1982). Sr isotopes in interstitial waters of marine sediments from Deep Sea Drilling Project cores. *Nature (London)* 300, 493-497.

Emiliani, C. (1954a). Depth habitats of some species of pelagic foraminifera as indicated by oxygen isotope ratios. *Am. J. Sci.* 252, 149-158.

Emiliani, C. (1954b). Temperatures of Pacific bottom waters and polar superficial waters during the Tertiary. *Science* 119, 853-855.

Emiliani, C. (1955a). Pleistocene temperatures. *J. Geol.* 63, 538-578.

Emiliani, C. (1955b). Pleistocene temperature variations in the Mediterranean. *Quaternaria* 2, 87-98.

Emiliani, C. (1958). Paleotemperature analysis of core 280 and Pleistocene correlations. *J. Geol.* 66, 264-275.

Emiliani, C. (1961a). Cenozoic climatic changes as indicated by the stratigraphy and chronology of deep-sea cores of *Globigerina* ooze facies. *Ann. N.Y. Acad. Sci.* 95, 521.

Emiliani, C. (1961b). Temperature decrease in surface sea water and high latitudes of abyssal-hadel water and open oceanic basins during the past 75 million years. *Deep-Sea Res.* 8, 144-147.

Emiliani, C. (1964). Paleotemperature analysis of the Caribbean cores A254-BR-C and CP-28. *Geol. Soc. Am. Bull.* 75, 129-144.

Emiliani, C. (1966). Paleotemperature analysis of the Caribbean cores P6304-8 and P6304-9 and a generalized temperature curve for the last 425,000 years. *J. Geol.* 74, 109-126.

Emiliani, C. (1970). Pleistocene paleotemperatures. *Science* 168, 822-825.

Emiliani, C. (1971a). Isotopic paleotemperature and shell morphology of *Globigerinoides ruber* in the type section for the Pliocene/Pleistocene boundary. *Micropaleontology* 17(2), 233-238.

Emiliani, C. (1971b). The amplitude of Pleistocene climatic cycles at low latitudes and the isotopic composition of glacial ice. *In* "The late Cenozoic Glacial Ages" (K. K. Turekian, ed.), pp. 1983-197. Yale Univ. Press, New Haven, Connecticut.

Emiliani, C. (1971c). Depth habitats of growth stages of pelagic foraminifera. *Science* **173**, 1122-1124.

Emiliani, C. (1972). Quaternary paleotemperatures and the duration of the high-temperature intervals. *Science* **178**, 398-401.

Emiliani, C., Gartner, S., Lidz, B., Eldridge, K., Elvey, D. K., Huang, T. C., Stipp, J. J., and Swanson, M. F. (1975). Paleoclimatological analysis of Late Quaternary cores from the northeastern Gulf of Mexico. *Science* **189**, 1083-1088.

Emrich, K., Ehhalt, D. H., and Vogel, J. C. (1970). Carbon isotope fractionation during precipitation of calcium carbonate. *Earth Planet. Sci. Lett.* **8**, 363-371.

Epstein, S., and Lowenstam, H. A. (1953). Temperature shell-growth relations of recent and interglacial Pleistocene shallow water biota from Bermuda. *J. Geol.* **61**, 424-438.

Epstein, S., and Mayeda, T. (1953). Variation of ^{18}O content of waters from natural sources. *Geochim. Cosmochim Acta* **4**, 213-224.

Epstein, S., Buchsbaum, R., Lowenstam, H., and Urey, H. C. (1951). Carbonate water isotope temperature scale. *Geol. Soc. Am. Bull.* **62**, 417-425.

Epstein, S., Buchsbaum, R., Lowenstam, H., and Urey, H. C. (1953). Revised carbonate-water isotopic temperature scale. *Geol. Soc. Am. Bull.* **64**, 1315-1325.

Ericson, D. B., and Wollin, G. (1968). Pleistocene climates and chronology in deep-sea sediments. *Science* **162**, 1227-1234.

Ericson, D. B., Ewing, M., Wollin, G., and Heezen, B. C. (1961). Atlantic deep sea sediment cores. *Geol. Soc. Am. Bull.* **72**, 193.

Fairbanks, R. G., and Matthews, R. K. (1978). The marine oxygen isotopic record in Pleistocene coral, Barbados, West Indies. *Quat. Res.* **10**, 181-196.

Falls, W. F. (1980). Glacial meltwater inflow into the Gulf of Mexico during the last 150,000 years: Implications for isotope stratigraphy and sea level studies. M.S. Thesis, University of South Carolina, Columbia.

Faulkner, E. A., and Buckingham, M. J. (1972). Comment on "Can a pulse excitation smaller than kT be detected?" *Electron. Lett.* **8**, 152-153.

Feller, W. (1966). "An Introduction to Probability Theory and Its Applications." Wiley, New York.

Ferm, J. B. (1984). Dolomitization of Lower to Middle Ordovician carbonates in the Appalachian Basin, West Virginia. Master's Thesis, University of South Carolina, Columbia (unpublished).

Fillon, R. H. (1984). Continental glacial stratigraphy, marine evidence of glaciation, and insights into continental-marine correlations. *In* "Principles of Pleistocene Stratigraphy Applied to the Gulf of Mexico" (N. Healy-Williams, ed.), pp. 149-206. IHRDC Press, Boston, MA.

Fillon, R. H., and Williams, D. F. (1984). The dynamics of meltwater discharge from the Northern Hemisphere ice sheets during the last deglaciation. *Nature (London)* **310**, 674-677.

Fischer, A. G., and Arthur, M. A. (1977). Secular variations in the pelagic realm. *Soc. Econ. Paleontol. Mineral.* **25**, 19-50.

Fisk, H. N. (1944). Geological investigations of the Alluvial Valley of the Lower Mississippi River, Corps of Engineers, Mississippi River Commission. *Tulsa Geol. Soc. Dig.* **15**, 50-55 (abstr.).

Folk, R. L. (1966). A review of grain-size parameters. *Sedimentology* **6**, 73-93.

Folk, R. L., and Ward, W. C. (1957). Brazos River Bar: A study in the significance of grain size parameters. *J. Sediment. Petrol.* **27**, 3-26.

Fomalont, E. B. (1968). The east-west structure of radio sources at 1425 MHz. *Astrophy. J.*, *Suppl. Ser.* **15**, 203-273.

Foster, M. R., and Guinzy, N. J. (1967). The coefficient of coherence: Its estimation and use in geophysical data processing. *Geophysics* **32**, 602-616.

Frieden, B. R. (1975). *In* "Picture Processing and Digital Filtering" (T. S. Huang, ed.), Springer-Verlag, Berlin and New York.

Friedman, G. M. (1967). Dynamic processes and statistical parameters compared for size frequency distributions of Beach and river sands. *J. Sediment. Petrol.* **37**, 327-354.

Full, W. E., and Ehrlich, R. (1982). Some approaches for location of centroids of quartz grain outlines to increase homology between Fourier amplitude spectra. *Math. Geol.* **14**, 43-54.

Full, W. E., Ehrlich, R., and Klovan, J. E. (1981). EXTENDED QMODEL—Objective definition of external end members in the analysis of mixtures. *Math. Geol.* **13**, 331-344.

Full, W. E., Ehrlich, R., and Bezdek, J. C. (1982). FUZZY QMODEL: A new approach for linear unmixing. *Math. Geol.* **14**, 259-270.

Full, W. E., Ehrlich, R., and Kennedy, S. K. (1984). Optimal configuration and information content of sets of frequency distributions. *J. Sediment. Petrol.* **54**, 117-126.

Garrison, R. E. (1981). Diagenesis of oceanic carbonate sediments: A review of the DSDP perspective. *Spec. Publ.—Soc. Econ. Paleontol. Mineral.* **32**, 181-207.

Garrison, R. E., and Douglas, R. G., eds. (1981). "The Monterey Formation and Related Siliceous Rocks of California." Soc. Econ. Paleontol. Mineral., Los Angeles, California.

Gartner, S., Chen, M. P., and Stanton, R. J. (1983-1984), Late Neogene nannofossil biostratigraphy and paleoceanography of the northeastern Gulf of Mexico and adjacent areas. *Mar. Micropaleontol.* **8**, 17-50.

Gary, A. C. (1985). A preliminary study of the relationship between test morphology and bathymetry in recent *Bolivina albatrossi* Cushman, Northwestern Gulf of Mexico. *Trans.— Gulf Coast Assoc. Geol. Soc.* **35**, 381-386.

Gibbons, G. W., and Hawking, S. W. (1971). Theory of the detection of short bursts of gravitational radiation. *Phys. Rev.* D **4**, 2191-2197.

Giffard, R. (1976). Ultimate sensitivity limit of a resonant gravitational wave antenna using a linear motion detection. *Phys. Rev.* D **14**, 2478-2486.

Gilbert, E. N., and Morgan, S. P. (1955). Optimum design of directive antenna arrays subject to random variations. *Bell Syst. Tech. J.* **34**, 637-663.

Gold, B., and Rader, C. M. (1969). "Digital Processing of Signals." McGraw-Hill, New York.

Goodman, N. R. (1963). Statistical analysis based on a certain multivariate complex Gaussian distribution (An introduction). *Ann. Math. Stat.* **34**, 152-177.

Goodney, D. E. (1977). Non-equilibrium fractionation of the stable isotopes of carbon and oxygen during precipitation of calcium carbonate by marine phytoplankton. Ph.D. Dissertation, University of Hawaii, Honolulu (unpublished).

Goodney, D. E., Margolis, S. V., Dudley, W. C., Kroopnick, P., and Williams, D. F. (1980). Oxygen and carbon isotopes of recent calcareous nannofossils as paleoceanographic indicators. *Mar. Micropaleo.* **5**, 31-42.

Govean, F. M., and Garrison, R. E. (1981). Significance of laminated and massive diatomites in the upper part of the Monterey Formation, California. *In* "The Monterey Formation and Related Siliceous Rocks of California" (R. G. Douglas, K. E. Pisciotto, C. M. Isaacs, and J. C. Ingle, eds.), pp. 181-198. Soc. Econ. Paleontol. Mineral., Los Angeles, California.

Graham, D. W., Corliss, B. H., Bender, M. L., and Keigwin, L. D. (1981). Carbon and oxygen isotopic disequilibria of recent deep sea benthic foraminifera. *Mar. Micropaleontol.* **6**, 483-497.

Groves, G. W., and Hannan, E. J. (1968). Time series regression of sea level on weather. *Rev. Geophys.* **6**, 129-174.

Gull, S. F., and Daniel, G. J. (1978). Image reconstruction from incomplete and noisy data. *Nature (London)* **272**, 686–690.

Hamming, R. W., and Tukey, J. W. (1949). "Measuring Noise Color." Bell Telephone Lab. Murray Hill, New Jersey.

Hannan, E. J. (1963). Regression for time series. *In* "Time Series Analysis" (M. Rosenblatt, ed.), pp. 17–37. Wiley, New York.

Hannan, E. J. (1970). "Multiple Time Series" Wiley, New York.

Haq, B. U., Worsley, J. R., Burckle, L. H., Douglas, R. G., Keigwin, L. D., Jr. *et al.* (1980). Late Miocene marine carbon isotopic shift and synchroneity of some phytoplanktonic biostratigraphic events. *Geology* **8**, 427–431.

Hassab, J. C., and Boucher, R. (1975). Analysis of signal extraction, echo detection and removal by complex cepstrum in the presence of distortion and noise. *J. Sound Vib.* **40**, 321–335.

Hays, J. D., and Shackleton, N. D. (1976). Globally synchronous extinction of the radiolarian *Stylotractus universus. Geology* **4**, 649–652.

Hays, J. D., Imbrie, J., and Shackleton, N. J. (1976). Variations in the Earth's orbit: Pacemaker of the ice ages. *Science* **194**, 1121–1132.

Healy-Williams, N. (1983). Fourier shape analysis of *Globorotalia truncatulinoides* from late Quaternary sediments in the Southern Indian Ocean. *Mar. Micropaleontol.* **8**, 1–15.

Healy-Williams, N. (1984). "Principles of Pleistocene Stratigraphy Applied to the Gulf of Mexico." IHRDC Press, Boston, Massachusetts.

Healy-Williams, N., and Williams, D. F. (1981). Fourier analysis of test shape of planktonic foraminifera. *Nature (London)* **289**, 485–487.

Healy-Williams, N., Ehrlich, R., and Williams, D. R. (1985). Morphometric and stable isotopic evidence for subpopulations of *Globorotalia truncatulinoides. J. Foraminiferal Res.* **15**, 242–253.

Hecht, A. (1977). The oxygen isotope record of foraminifera in deep-sea sediment. *In* "Foraminifera" (R. H. Hedley and C. G. Adams, eds.), vol. 2, pp. 1–43. Academic Press, London.

Hodell, D. A., Kennett, J. P., and Leonard, K. (1985). Climatically-induced changes in vertical water mass structure of the Vema Channel during the Pliocene: Evidence from DSDP Site 516A, 517 and 518. *Initial Rep. Deep Sea Drill. Proj.* **72**, 907–919.

Hodell, D. A., Williams, D. F., and Kennett, J. P. (1985). Reorganization of deep vertical water mass structure in the Vema Channel at 3.2 Ma: Faunal and isotopic evidence from DSDP Leg 72 (South Atlantic). *Geol. Soc. Am. Bull.* **96**, 495–503.

Hogbom, J. A. (1974). Aperture synthesis with a non-regular distribution of interferometer baselines. *Astron. Astrophy., Suppl. Ser.* **15**, 417–426.

Hsu, K. J., He, Q., McKenzie, J. A., Weissert, H., Perch-Nielsen, K., Oberhanlsi, H. *et al.* (1982). Mass mortality and its environmental and evolutionary consequences. *Science* **216**, 249–255.

Hudson, C. B., and Ehrlich, R. (1980). Determination of relative provenance contributions in samples of quartz sand using Q-mode factor analysis of Fourier grain shape data. *J. Sediment. Petrol.* **50**, 1101–1110.

Hudson, J. D. (1977). Stable isotopes and limestone lithifications. *J. Geol. Soc., London* **133**, 637–660.

Imbrie, J. (1963). "Factor and Vector Analysis Programs for Analyzing Geologic Data," Tech. Rep. (ONR Task No. 389-135). Office of Naval Research Geography Branch, Washington, D.C.

Imbrie, J., and Imbrie, J. Z. (1980). Modeling the climatic response to orbital variations. *Science* **207**, 943–953.

Imbrie, J., and Van Andel, T. H. (1964). Vector analysis of heavy mineral data. *Geol. Soc. Am. Bull.* **72**, 1131-1156.

Imbrie, J., Van Donk, J., and Kipp, N. G. (1973). Paleoclimatic investigation of a late Pleistocene Caribbean deep-sea core: Comparison of isotopic and faunal methods. *Quat. Res.* **3**, 10-38.

Imbrie, J., Hays, J. D., Martinson, D. G. *et al.* (1984). The orbital theory of Pleistocene climate: Support from a revised chronology of marine $\delta^{18}O$ record. *In* "Milankovitch and Climate" (A. Berger, J. Imbrie, J. Hays, G. Kukla, and B. Saltzman, eds.), Part 1, pp. 269-305. D. Riedel.

Inman, D. L. (1952). Measures for describing the size distribution of sediments. *J. Sediment. Petrol.* **22**, 25-145.

Isaacs, C. M. (1981a). Porosity reduction during diagenesis of the Monterey Formation, Santa Barbara coastal area, California. *In* "The Monterey Formation and Related Siliceous Rocks of California" (R. E. Garrison and R. G. Douglas, eds.), pp. 257-272. Soc. Econ. Paleontol. Mineral., Los Angeles, California.

Isaacs, C. M. (1981b). Guide to the Monterey Formation in the California Coastal area, Ventura to San Luis Obispo. *Am. Assoc. Pet. Geol.* **65**, 1-99.

Isaacs, C. M. (1984). Disseminated dolomite in the Monterey Formation, Santa Maria and Santa Barbara Areas, California. *In* "Dolomite in the Monterey Formation" (R. E. Garrison and M. Kastner, eds.), pp. 155-170. Soc. Econ. Paleontol. Mineral., Los Angeles, California.

Isaacs, C. M., Pisciotto, K. A., and Garrison, R. E. (1983). Facies and diagenesis of the Miocene Monterey Formation, California: A summary. *In* "Siliceous Deposits in the Pacific Region" (A. Iijima, J. R. Hein, and R. Siever, eds.), pp. 247-282. Elsevier, Amsterdam.

Jacka, A. D., Beck, R. H., St. Germain, L. C., and Harrison, S. C. (1968). Permian deep-sea fans of the Delaware Mountain Group (Guadalupean), Delaware Basin. *Soc. Econ. Paleontol. Mineral. (Permian Basin Sect.)*, Pub., **68-11**, 49-90.

Jaquet, J.-M., and Vernet, J.-P. (1976). Moment and graphic size parameters in sediments of Lake Geneva (Switzerland). *J. Sediment. Petrol.* **46**, 305-312.

Jaynes, E. T. (1959). "Probability Theory in Science and Engineering." Field Res. Lab., Socony Mobil Oil, Co., Inc.

Jeffreys, H. (1961). "Theory of Probability." 3rd ed. Oxford Univ. Press, London and New York.

Jenkins, G. M. (1963). Cross-spectral analysis and the estimation of linear open loop transfer functions. *In* "Time Series Analysis" (M. Rosenblatt, ed.), pp. 267-276. Wiley, New York.

Jenkins, G. M., and Watts, D. G. (1968). "Spectral Analysis and Its Applications." Holden-Day, San Francisco, California.

Joreskog, J. G., Klovan, J. E., and Reyment, R. A. (1976). "Geological Factor Analysis." Elsevier, Amsterdam.

Kablanow, R. I., II, and Surdam, R. C. (1983). Diagenesis and hydrocarbon generation in the Monterey Formation, Huasna Basin, California. *In* "Petroleum Generation and Occurrence in the Miocene Monterey Formation, California." (C. M. Issacs, R. E. Garrison, S. A. Graham, and W. A. Jensky, II, eds.), pp. 53-68. Soc. Econ. Paleontol. Mineral., Los Angeles, California.

Kailath, T. (1970). The innovation approach to detection and estimation theory. *Proc. IEEE* **58**, 680-695.

Kastner, M., and Gieskes, J. M. (1976). Interstitial water profiles and sites of diagenetic reactions, Leg 35, DSDP, Bellingshausen Abyssal Plain. *Earth Planet. Sci. Lett.* **33**, 11-20.

Keigwin, L. D., Jr. (1979). Late Cenozoic stable isotope stratigraphy and paleoceanography of DSDP sites from the east equatorial and central North Pacific Ocean. *Earth Planet. Sci. Lett.* **45**, 361-381.

Keigwin, L. D., Jr. (1980). Paleoceanographic change in the Pacific at the Eocene-Oligocene boundary. *Nature (London)* **287**, 722-725.

Keigwin, L. D., Jr. (1982a). Isotopic paleoceanography of the Caribbean and East Pacific: Role of Panama uplift in Late Neogene time. *Science* **217**, 350-353.

Keigwin, L. D., Jr. (1982b). Stable isotope stratigraphy and paleoceanography of sites 502 and 503. *Intitial Rep. Deep Sea Drill. Proj.* **68**, 445-453.

Keigwin, L. D., Jr. (1983). Stable isotopic results of upper Miocene and lower Pliocene foraminifera from Hole 552A. *Initial Rep. Deep Sea Drill. Proj.* **81**, 595-597.

Keigwin, L. D., Jr. (1987). Pliocene stable isotope record of DSDP 606: ^{18}O enrichment 2.4, 2.6 and 3.1 my ago. *Initial Rep. Deep Sea Drill. Proj.* **94**, 911-920.

Keigwin, L. D., Jr., and Corliss, B. H. (1986). Stable Isotopes in late middle Eocene to Oligocene foraminifera. *Geol. Soc. Am. Bull.* **97**, 335-345.

Keigwin, L. D., Jr., and Keller, G. (1984). Middle Oligocene cooling from equatorial Pacific DSDP site 77B. *Geology* **12**, 16-19.

Keigwin, L. D., Jr., and Shackleton, N. J. (1980). Uppermost Miocene carbon isotope stratigraphy of a piston core in the equatorial Pacific. *Nature (London)* **284**, 613-614.

Keigwin, L. D., Jr., Bender, M. L., and Kennett, J. P. (1979). Thermal structure of the deep Pacific ocean in the early Pliocene. *Science* **205**, 1386-1388.

Keller, G., and Barron, J. A. (1981). Integrated planktic foraminiferal and diatom biochronology for the Northeast Pacific and Monterey Formation. *In* "The Monterey Formation and Related Siliceous Rocks of California" (R. E. Garrison and R. G. Douglas, eds.), pp. 43-54. Soc. Econ. Paleontol. Mineral., Los Angeles, California.

Kemp, W. C., and Eger, D. T. (1967). The relationships among sequences with applications to geological data. *J. Geophys. Res.* **72**, 739.

Kennedy, S. K., Ehrlich, R., and Kana, T. W. (1981). The non-normal distribution of intermittent suspension sediments below breaking waves. *J. Sediment. Petrol.* **51**, 1103-1108.

Kennett, J. P. (1985). Miocene-Early Pliocene oxygen and carbon isotopic stratigraphy in the Southwest Pacific. *Initial Rep. Deep Sea Drill. Proj.* Leg 90.

Kennett, J. P., and Shackleton, N. J. (1975). Laurentide ice sheet meltwater recorded in Gulf of Mexico deep-sea cores. *Science* **188**, 147-150.

Kennett, J. P., and Shackleton, N. J. (1976). Oxygen isotopic evidence for the development of the psychrosphere 38 my ago. *Nature (London)* **260**, 513-515.

Kennett, J. P., Shackleton, N. J., Margolis, S. V., Goodney, D. E., Dudley, W. C., and Kroopnick, P. M. (1979). Late Cenozoic oxygen and carbon isotopic history and volcanic ash stratigraphy: DSDP site 284, south Pacific. *Am. J. Sci.* **279**, 52-69.

Kikuchi, R., and Soffer, B. H. (1977). Maximum entropy image restoration. I. The entropy expression. *J. Opt. Soc. Am.* **67**, 1656-1665.

Killingley, J. S. (1983). Effects of diagenetic recrystallization on $^{18}O/^{16}O$ values of deep-sea sediments. *Nature (London)* **310**, 504-507.

King, P. B. (1942). Permian of west Texas and southeastern New Mexico. *Am. Assoc. Pet. Geol. Bull.* **26**, 535-763.

Kleinpell, R. M. (1938). "Miocene Statigraphy of California." Am. Assoc. Pet. Geol., Tulsa, Oklahoma.

Klovan, J. E. (1966). The use of factor analysis in determining depositional environments from grain-size distributions. *J. Sediment. Petrol.* **36**, 115-125.

Klovan, J. E., and Imbrie, J. (1971). An algorithm and FORTRAN-IV program for large scale Q-mode factor analysis and calculation of factor scores. *Math. Geol.* **3**, 61-76.

Klovan, J. E., and Miesch, A. T. (1976). EXTENDED CABFAC and QMODEL computer programs for Q-mode factor analysis of compositional data. *Comput. Geosci.* **1**, 161-178.

Komesaroff, M. M., and Lerche, I. (1979). Extending the Fourier Transform—The positivity constraint. *In* "Image Formation from Coherence Functions in Astronomy" (C. van Schooneveld, ed.), pp. 241–247. D. Reidel Publ., Dordrecht, Netherlands.

Kominz, M. A., Heath, G. R., Ku, T. L., and Pisias, N. G. (1979). Brunhes time scales and the interpretation of climatic change. *Earth Planet. Sci. Lett.* **45**, 394–410.

Kroopnick, P. S. (1974). The dissolved O_2-CO_2-^{13}C system in the eastern equatorial Pacific. *Deep-Sea Res.* **21**, 211–227.

Kroopnick, P. S. (1980). The distribution of ^{13}C in the Atlantic Ocean. *Earth Planet. Sci. Lett.* **49**, 469–484.

Kroopnick, P. S., Margolis, P., and Wong, C. S. (1977), ^{13}C variations in marine carbonate sediments as indicators of the CO_2 balance between the atmosphere and the oceans. *In* "The Fate of Fossil Fuel CO_2 in the Oceans" (N. R. Andersen and A. Malahoff, eds.), pp. 295–322. Plenum, New York.

Kruge, M. A. (1983). Diagenesis of miocene biogenic sediments in Lost Hills Oil Field, San Joaquin Basin, California. *In*: "Diagenesis and Hydrocarbon generation in the Monterey Formation, Huasna Basin, California." (C. M. Isaacs, R. E. Garrison, S. A. Graham, and W. A. Jensky, II, eds.), pp. 39–50. Soc. Econ. Paleontol. Mineral., Los Angeles, California.

Ku, T. L. (1966). Uranium series disequilibrium in deep-sea sediments. Ph.D. Thesis, Columbia University, New York.

Ku, T. L., and Broecker, W. S. (1966). Atlantic deep-sea stratigraphy: Extension of absolute chronology to 320,000 years. *Science* **151**, 448.

Kullback, S. (1959). "Information Theory and Statistics." Wiley, New York.

Lacoss, R. T. (1971). Data adaptive spectral analysis methods. *Geophysics* **36**, 661–675.

Larner, K., Chambers, R., Yang, M., Lynn, W., and Wai, W. (1983). Coherent noise in marine seismic data. *Geophysics* **48**, 854–886.

Ledbetter, M. T. (1984a). Pleistocene magnetostratigraphy. *In* "Principles of Pleistocene Stratigraphy Applied to the Gulf of Mexico" (N. Healy-Williams, ed.), pp. 1–24. IHRDC Press, Boston, Massachusetts.

Ledbetter, M. T. (1984b). Late Pleistocene tephrochronology in the Gulf of Mexico region. *In* "Principles of Pleistocene Stratigraphy Applied to the Gulf of Mexico" (N. Healy-Williams, ed.), pp. 119–148. IHRDC Press, Boston, Massachusetts.

Lee, P. F. (1981). An algorithm for computing the cumulative distribution function for magnitude squared coherence estimates. *IEEE Trans. Acoust., Speech, Signal Process.* **ASSP-29**, 117–119.

Lee, R. (1974). "Entropy Models in Spatial Analysis," Discuss. Pap. Ser. University of Toronto, Department of Geography, Toronto, Canada.

Leonard, K. A., Williams, D. F., and Thunell, R. C. (1983). Pliocene paleoclimatic and paleoceanographic history of the South Atlantic Ocean: Stable isotopic records from Leg 72 DSDP Sites 516A and 517. *Initial Rep. Deep Sea Drill. Proj.* **72**, 895–906.

Letolle, R., and Renard, M. (1980), Sédimentologie-évolution des teneurs en ^{13}C des carbonates pélagiques aux limites Cretace-Tertiaire et Paleocene-Eocene. *C.R. Seances Acad. Sci., Ser. B.* **290**, 827–830.

Letolle, R., Renard, M., Bourbon, M., and Filly, A. (1978). ^{18}O and ^{13}C isotopes in Leg 44 carbonates: A comparison with the Alpine series. *Initial Rep. Deep Sea Drill. Proj.* **44**, 567–573.

Letolle, R., Vergnaud-Grazzini, C., and Pierre, C. (1979). Oxygen and carbon isotopes from bulk carbonates and foraminiferal shells at DSDP Sites 400, 401, 402, 403 and 406. *Initial Rep. Deep Sea Drill. Proj.* **48**, 741–755.

Leventer, A., Williams, D. F., and Kennett, J. P. (1982). Dynamics of the Laurentide ice sheet during the last glaciation: Evidence from the Gulf of Mexico. *Earth Planet. Sci. Lett.* **59**, 11–17.

Leventer, A., Williams, D. F., and Kennett, J. P. (1983). Relationships between anoxia, glacial meltwater and microfossil preservation in the Orca Basin, Gulf of Mexico. *Mar. Geol.* **53**, 23–40.

Levinson, N. (1949). The Wiener RMS error criterion in filter design and prediction. *In* "Extrapolation, Interpolation, and Smoothing of Stationary Time Series with Engineering Applications" by N. Wiener, Appendix B, pp. 129–148. M.I.T. Press, Cambridge, Massachusetts.

Lewis, J. B., and Schultheiss, P. M. (1970). Optimum and conventional detection using a linear array. *J. Acoust. Soc. Am.* **49**, 1083–1091.

Lines, L. R., and Clayton, R. W. (1977). A new approach to Vibroseis deconvolution. *Geophys. Prospect.* **25**, 417–433.

Lines, L. R., Clayton, R. W., and Ulrych, T. J. (1980). Impulse response models for noisy Vibroseis data. *Geophys. Prospect.* **28**, 49–59.

Loutit, T. S. (1981). Late Miocene paleoclimatology: Subantarctic water mass, southwest Pacific. *Mar. Micropaleontol.* **6**, 1–27.

Loutit, T. S., and Keigwin, L. D., Jr. (1982). Stable isotopic evidence for the latest Miocene sea level fall in the Mediterranean region. *Nature (London)* **300**, 163–166.

Loutit, T. S., and Kennett, J. P. (1979). Application of carbon isotope stratigraphy to late Miocene shallow marine sediments, New Zealand. *Science* **204**, 196–199.

Loutit, T. S., Pisias, N. G., and Kennett, J. P. (1983). Pacific Miocene carbon isotope stratigraphy using benthic foraminifera. *Earth Planet. Sci. Lett.* **66**, 48–62.

Loutit, T. S., Kennett, J. P., and Savin, S. M. (1983–1984). Miocene equatorial and southwest Pacific paleoceanography from stable isotope evidence. *Mar. Micropaleontol.* **8**, 215–233.

McCrea, J. M. (1950). On the isotopic chemistry of carbonates and a paleotemperature scale. *J. Chem. Phys.* **18**, 849–857.

McKenzie, J. A., Jenkyns, H. C., and Bennet, G. G. (1979–1980). Stable isotope study of the cyclic diatomite-claystones from the Tripoli Formation, Sicily: A prelude to the Messinian salinity crisis. *Palaeogr., Palaeoclimatol., Palaeocol.* **29**, 125–141.

McKenzie, J. A., Wiessert, H., Poore, R. Z. *et al.* (1984). Paleoceanographic implications of stable isotope data from upper Miocene-Lower Pliocene sediments from the southeast Atlantic (DSDP Site 519). *Initial Rep. Deep Sea Drill. Proj.* **73**, 719–724.

Maeder, D. (1972). Natural phase compensation in continuously wound delay lines using a split-core configuration. *Electron. Lett.* **8**, 128–129.

Makhoul, J. (1975). Linear prediction: A tutorial review. *Proc. IEEE* **63**, 561–580.

Malmgren, B., and Kennett, J. P. (1976). Principal component analysis of Quaternary planktonic foraminifera in the Gulf of Mexico: Paleoclimatic application. *Mar. Micropaleontol.* **1**, 299–306.

Mancini, E. A. (1979). Eocene/Oligocene boundary in southwest Alabama. *Trans.—Gulf Coast Assoc. Geol. Soc.* **29**, 282–286.

Mancini, E. A. (1981). Lithostratigraphy and biostratigraphy of Paleocene subsurface strata in southwest Alabama. *Trans.—Gulf Coast Assoc. Geol. Soc.* **31**, 359–368.

Margolis, S. V., Kroopnick, P. M., Goodney, D. E., Dudley, W. C., and Mahoney, M. E. (1975). Oxygen and cargbon isotopes form calcareous nannofossils as paleoceanographic indicators. *Science* **189**, 555–557.

Margolis, S. V., Kroopnick, P. M., and Goodney, D. (1977). Cenozoic and late Mesozoic paleoceanographic and paleoglacial history recorded in circum-Antarctic deep sea sediments. *Mar. Geol.* **25**, 131–147.

Martin, R. D., and Thomson, D. J., (1982). Robust-resistant spectrum estimation. *Proc. IEEE* **70**, 1097–1115.

Martinson, D. G., Menke, W., and Stoffa, P. (1982). An inverse approach to signal correlation. *J. Geophys. Res.* **87**, 4807–4818.

Mason, G. C., and Folk, R. L. (1958). Differentiation of beach, dune, and aeolian flat environments by size analysis, Mustang Island, Texas. *J. Sediment. Petrol.* **38**, 211-226.

Matalas, N. C., and Reiher, B. J. (1967). Some comments on the use of factor analysis. *Water Resour. Res.* **3**, 213-223.

Matthews, R. K. (1973). Relative elevation of late Pleistocene high sea-level stands: Barbados uplift rates and their implications. *Quat. Res.* **3**, 147-153.

Matthews, R. K. (1984a). "Dynamic Stratigraphy: An Introduction to Sedimentation and Stratigraphy." Prentice-Hall, Englewood Cliffs, New Jersey.

Matthews, R. K. (1984b). Oxygen isotope record of ice-volume history: 100 million years of glacio-eustatic sea-level fluctuation. *Mem.—Am. Assoc. Pet. Geol.* **36**, 97-107.

Matthews, R. K., and Poore, R. A. (1980). Tertiary ^{18}O record and glacioeustatic sea-level fluctuations. *Geology* **8**, 501-504.

Matthews, R. K., Curry, W. B., Lohmann, K. C., Sommer, M. A., and Poore, R. Z. (1980). Late Miocene Paleo/Oceanography of the Atlantic: Oxygen isotope data on planktonic and benthic foraminifera. *Nature (London)* **283**, 555-557.

Mazzullo, J., and Ehrlich, R. (1980). A vertical pattern of variation in the St. Peter Sandstone—Fourier grain shape analysis. *J. Sediment. Petrol.* **50**, 63-70.

Mazzullo, J., and Ehrlich, R. (1983). Grain shape variation in the St. Peter Sandstone: A record of eolian and fluvial sedimentation of an early Paleozoic cratonic sheet sand. *J. Sediment. Petrol.* **53**, 105-119.

Mazzullo, J., Ehrlich, R., and Hemming, M. A. (1984). Provenance and areal distribution of late Pleistocene and Holocene quartz sand on the southern New England continental shelf. *J. Sediment. Petrol.* **54**, 1335-1348.

Meissner, F. F. (1972). Cyclic sedimentation in Middle Permian strata of the Permian Basin, West Texas and New Mexico, *In:* "Cyclic Sedimentation in the Permian Basin," pp. 203-232. West Texas Geol. Soc., Midland, Texas.

Menke, W. (1984). "Geophysical Data Analysis: Discrete Inverse Theory." Academic Press, Orlando, Florida.

Mesolella, K. J., Matthews, R. K., Broecker, W. S., and Thurber, D. L. (1969). The astronomical theory of climatic change: Barbados data. *J. Geol.* **77**, 250.

Miesch, A. T. (1976a). Q-mode factor analysis of compositional data. *Comput. Geosci.* **1**, 147-159.

Miesch, A. T. (1976b). Q-mode factor analysis of geochemical and petrological data matrices with constant row sums. *Geol. Surv. Prof. Pap. (U.S.)* **574G**.

Milankovitch, M. (1941). Canon of Insolation and the Ice-Age Problem Beograd, Koninglich Serbische Akademie. 484 pp. (English Translation by Israel Program for Scientific Translation and published for the U.S. Department of Commerce and the National Science Foundation.)

Miller, K. G., and Curry, W. B. (1982). Eocene to Oligocene benthic foraminiferal isotopic record in the Bay of Biscay. *Nature (London)* **296**, 347-350.

Miller, K. G., and Fairbanks, R. G. (1983). Evidence for Oligocene-middle Miocene abyssal circulation changes in the western north Atlantic. *Nature (London)* **306**, 250-253.

Miller, K. G., and Fairbanks, R. G. (1985a). Oligocene and Miocene global carbon isotope cycles and abyssal circulation changes. *Geophys. Monogr., Am. Geophys. Union* **32**, 469-486.

Miller, K. G., and Fairbanks, R. G. (1985b). Cainozoic δ^{18}O record of climate and sea level. *S. Afr. J. Sci.* **81**, 248-249.

Miller, K. G., and Thomas, E. (1985). Eocene to Oligocene benthic foraminifera isotopic record, site 574 equatorial Pacific. *Initial Rep. Deep Sea Drill. Proj.* **85**.

Miller, K. G., Curry, W. B., and Ostermann, D. R. (1985). Late Paleogene (Eocene to Oligocene) benthic foraminiferal paleoceanography of the Goban Spur Region, DSDP Leg 80. *Initial Rep. Deep Sea Drill. Proj.* **80**, 505–538.

Miller, K. G., Aubre, M. P., Khan, M. J., Melillo, A. J., Kent, D. V., and Berggren, W. A. (1986). Oligocene to Miocene bio-, magneto-, isotopic stratigraphy fo the western North Atlantic. *Geology* **13**, 257–261.

Mix, A. C., and Fairbanks, R. G. (1985). Late Pleistocene history of north Atlantic surface and deep ocean circulation. *Earth Planet. Sci. Lett.* **73**, 231–243.

Mix, A. C., and Ruddiman, W. F. (1984). Oxygen-isotope analyses and Pleistocene ice volumes. *Quat. Res.* **17**, 279–313.

Mix, A. C., and Ruddiman, W. F. (1985). Structure and timing of the Last Deglaciation: Oxygen-isotope evidence. *Quat. Sci. Rev.* **4**, 59–108.

Moiola, R. J., Spencer, A. B., and Weiser, D. (1974). Differentiation of modern sand bodies by linear discriminant analysis. *Trans—Gulf Coast Assoc. Geol. Soc.* **24**, 324–332.

Moore, T. C., Jr., and Heath, G. R. (1977). Survival of deep sea sedimentary sections. *Earth Planet. Sci. Lett.* **37**, 71–80.

Moore, T. C., Jr., Van Andel, T. H., Sanchetta, C., and Pisias, N. (1978). Cenozoic hiatuses and pelagic sediments. *Micropaleontology* **24**, 113–138.

Moore, T. C., Jr., Pisias, N. G., and Keigwin, L. D., Jr. (1981). Ocean basin and depth variability of oxygen isotopes in Cenozoic benthic foraminifera. *Mar. Micropaleontol.* **6**, 465–481.

Morley, J. J., and Shackleton, N. J. (1978). Extension of the radiolarian *Stylatractus universus* as a biostratigraphic datum to the Atlantic Ocean. *Geology* **6**, 309–311.

Morse, P. M., and Feshbach, H. (1953). "Methods of Theoretical Physics." McGraw-Hill, New York.

Munk, W. H., and Cartwright, D. E. (1966). Tidal spectroscopy and prediction. *Philos. Trans. R. Soc. London Ser. A* **259**, 533–581.

Murata, K., Friedman, I., and Cremer, M. (1972). Geochemistry of diagenetic dolomites in Miocene marine formations of California and Oregon. *U.S. Geol. Survey Prof. Paper.* **724C**, 1–11.

Murphy, M. G., and Kennett, J. P. (1985). Development of latitudinal thermal gradients during the Oligocene: Oxygen isotopic evidence from the south west Pacific. *Initial Rep. Deep Drill. Proj.* **90**, 1347–1360.

Muza, J. P., Williams, D. F., and Wise, S. W., Jr. (1983). Paleogene oxygen isotope record for deep sea drilling sites 511 and 512, subantarctic south Atlantic Ocean: Paleotemperatures paleoceanographic changes and the Eocene/Oligocene boundary event. *Initial Rep. Deep Sea Drill. Proj.* **71**, 409–422.

Neff, E. (1983). Pre-late Pleistocene paleoclimatology and planktonic foraminiferal biostratigraphy of the Northwestern Gulf of Mexico. Ms. Thesis, University of South Carolina, Columbia (unpublished).

Neidell, N. S., and Taner, M. T. (1971). Semblance and other coherency measures for multichannel data. *Geophysics* **36**, 482–497.

Newman, W. I. (1977). A new method of multidimensional power spectral analysis. *Astron. Astrophys.* **54**, 369–380.

Newton, H. J., and Pagano, M. (1984). Simultaneous confidence bands for autoregressive spectra. *Biometrika* **71**, 197–202.

Ninkovich, D., and Shackleton, N. J. (1975). Distribution, stratigraphic position and age of ash layer 'L', in the Panama Basin Region. *Earth Planet. Sci. Lett.* **27**, 20–34.

Nuttall, A. H., and Carter, G. C. (1976). Bias of the estimate of magnitude-squared coherence. *IEEE Trans. Acoust., Speech, Signal Process* **ASSP-24**, 582–583.

Oberhansli, H., and Toumaekine, M. (1985). Paleogene oxygen isotopic history of site 522, 523 and 524 from the central south Atlantic. In "South Atlantic Paleoceanography" (K. J. Hsu and H. Weissert, eds.). pp. 125-147. Cambridge Univ. Press, London and New York.

Oberhansli, H., Grunig, A., and Herb, R. (1984). Oxygen and carbon isotope study in the Late Eocene sediments of Possagno (Northern Italy). Riv. Ital. Paleontol. Strat. **89**, 377-394.

Odin, G. S., Renard, M., and Vergnaud-Grazzini, C. (1982). Geochemical events as a means of correlation. "Numerical Dating in Stratigraphy." pp. 37-70. Wiley, New York.

Olausson, E. (1965). Evidence of climatic changes in North Atlantic deep-sea cores, with remarks on isotopic paleotemperature analysis. Program Oceanogr. **3**, 221-252.

Pallottino, G. V. (1974). Two sensor correlation technique for the detection of gravitational waves, Alta Freq. **43**, 1043-1044.

Pallottino, G. V. (1979). Data analysis algorithms for gravitational antenna signals. In "Gravitational Radiation. Collapsed Objects and Exact Solutions" (C. Edwards, ed.), pp. 341-369. Springer-Verlag, Berlin and New York.

Pallottino, G. V., and Pizzella, G. (1978). On the electrical equivalent circuits of gravitational-wave antennas. Nuovo Cimento Soc. Ital. Fis. B **45B**, 275-296.

Peng, T. H., Broecker, W. S., Kipphut, G., and Shackleton, N. (1977). Benthic mixing in deep sea cores as determined by ^{14}C dating and its implications regarding climate stratigraphy and the fate of fossil fuel Co_2. In "The Fate of Fossil Fuel CO_2 in the Oceans" (N. R. Andersen and A. Malahoff, eds.), pp. 355-373. Plenum, New York.

Papadopoulos, G. D. (1975). A statistical technique for processing radio interferometer data. IEEE Trans. Antennas Propag. **AP-23**, 45-53.

Papoulis, A. (1965). "Probability, Random Variables and Stochastic Processes." McGraw-Hill, New York.

Papoulis, A. (1973). Minimum bias windows for high resolution spectral estimates. IEEE Trans. Inf. Theory **IT-19**, 9-12.

Parzen, E. (1974). Some recent advances in time series modeling. IEEE Trans. Autom. Control **AC-19**, 723-730.

Pastouret, L., Chamley, H., Delibrias, G., Duplessy, J. C., and Thiede, J. (1978). Late Quaternary climatic changes in Western Tropical Africa deduced from deep-sea sedimentation off the Niger delta. Oceanol. Acta **1**, 217-232.

Peacock, K. L., and Treitel, S. (1969). Predictive deconvolution, theory and practice. Geophysics **34**, 155-169.

Pisarenko, V. F. (1970). Statistical estimates of amplitude and phase corrections. Geophys. J. **20**, 89-98.

Pisciotto, K. A. (1981). Review of secondary carbonates in the Monterey Formation, California. In "The Monterey Formation and Related Siliceous Rocks of California" (R. E. Garrison and R. G. Douglas, eds.), pp. 273-284. Soc. Econ. Paleontol. Mineral., Los Angeles, California.

Pisciotto, K. A., and Garrison, R. E. (1981). Lithofacies and depositional environments of the Monterey Formation, California. In "The Monterey Formation and Related Siliceous Rocks of California" (R. E. Garrison and R. G. Douglas, eds.), pp. 71-86. Soc. Econ. Paleontol. Mineral., Los Angeles, California.

Pisias, N. G., and Moore, J. C., Jr. (1981). The evolution of Pleistocene climate: A time series approach. Earth Planet. Sci. Lett. **52**, 450-458.

Pisias, N. G., and Shackleton, N. J. (1984). Modelling the global climate response to orbital forcing and atmospheric carbon dioxide changes. Nature (London) **310**, 757-759.

Pisias, N. G., Martinson, D. G., Moore, T. C., Jr., Shackleton, N. J., Pressl, W., Hays, J., and Boden, G. (1984). High resolution stratigraphic correlation of benthic oxygen isotope records spanning the last 30,000 years. Mar. Geol. **56**, 119-136.

Pisias, N. G., Shackleton, N. J., and Hall, M. A. (1985). Stable isotope and calcium carbonate records from hydraulic piston cored hole 574A: High-resolution records from the Middle Miocene. *Initial Rep. Deep Sea Drill. Proj.* **85**, 735–748.

Pitman, W. C., III (1978). Relationship between eustasy and stratigraphic sequences of passive margins. *Geol. Soc. Am. Bull.* **89**, 1389–1403.

Poore, R. Z., and Matthews, R. K. (1984a). Late Eocene-Oligocene oxygen and carbon isotope record from South Atlantic Ocean, Deep Sea Drilling project site 522. *Initial Rep. Deep Sea Drill. Proj.* **78**, 725–735.

Poore, R. Z., and Matthews, R. K. (1984b). Oxygen isotope ranking of Late Eocene and Oligocene Planktonic Foraminifers: Implication for Oligocene sea-surface temperatures and global ice-volume. *Mar. Micropaleontol.* **9**, 111–134.

Poore, R. Z., McCougall, K., Barron, J. A., Brabb, E. E., and Kling, S. A. (1981). Microfossil biostratigraphy and biochronology of the Type Relizian and Luisian stages of California. *In* "The Monterey Formation and Related Siliceous Rocks of California" (R. E. Garrison and R. G. Douglas, eds.), pp. 15–42. Soc. Econ. Paleontol. Mineral., Los Angeles, California.

Prell, W. L. (1980). A continuous high-resolution record of Quaternary evidence for two climatic regions: DSDP hydraulic piston core site 502. *Geol. Soc. Am. Abstr. Programs* **12**, 503.

Prell, W. L. (1982). Oxygen and carbon isotope stratigraphy for the Quaternary of Hole 502B: Evidence for two modes of isotopic variability. *Initial Rep. Deep Sea Drill. Proj.* **68**, 455–464.

Prell, W. L. (1984). Covariance patterns of foraminiferal $\delta^{18}O$: An evaluation of Pliocene ice volume changes near 3.2 million years ago. *Science* **226**, 692–694.

Prell, W. L., Hutson, W. H., Williams, D. F., Bé, A. W. H., Geitzenhauer, K., and Molfino, B. (1980). Surface circulation of the Indian Ocean during the last glacial maximum, approx. 18,000 years B.P. *Quat. Res.* **14**, 309–336.

Raikes, S. A., and White, R. E. (1984). Measurements of earth attenuation from down-hole and surface seismic recordings. *Geophys. Prospect.* **32**, 892–919.

Ramachandran, G. N., and Srinivasan, R. (1970). "Fourier Methods in Crystallography." Wiley (Interscience), New York.

Reed, W. E., Le Fever, R., and Moir, G. J. (1975). Depositional environment interpretation from settling velocity (psi) distributions. *Geol. Soc. Am. Bull.* **86**, 1305–1315.

Renard, M. (1985). Géochimie des carbonates pélagiques. *Ser. Doc. B.R.G.M.* **85**, 650–653.

Riester, D. D., Shipp, R. C., and Ehrlich, R. (1982). Patterns of quartz sand shape variation, Long Island littoral and shelf. *J. Sediment. Petrol.* **52**, 1307–1314.

Rietsch, E. (1980). Estimation of the signal-to-noise ratio of seismic data with application to stacking. *Geophys. Prospect.* **28**, 531–550.

Robinson, E. R. (1967). "Statistical Communication and Detection." Griffin, London.

Robinson, E. R. (1983). "Multichannel Time Series Analysis with Digital Computer Programs." Goose Pond Press, Houston, Texas.

Roehl, P. O. (1981). Dilation brecciation—a proposed mechanism of fracturing, petroleum expulsion and dolomitization in the Monterey Formation, California. *In* "The Monterey Formation and Related Siliceous Rocks of California" (R. E. Garrison and R. G. Douglas, eds.), pp. 285–316. Soc. Econ. Paleontol. Mineral., Los Angeles, California.

Rona, E., and Emiliani, C. (1969). Absolute dating of Caribbean cores P6304-8 and P6304-9. *Science* **163**, 66–68.

Roshalt, J. N., Emiliani, C., Geiss, J., Koczy, F. F., and Wangersky, P. J. (1961). Absolute dating of deep-sea cores by the Pa_{231}/Th_{230} method. *J. Geophys. Res.* **67**, 2907–2911.

Saito, T., and Van Donk, J. (1974). Oxygen and carbon isotope measurement of late Cretaceous and early Tertiary foraminifera. *Micropaleontology* **20**, 152–177.

Samson, J. C. (1983a). Pure states, polarised waves, and principal components in the spectra of multiple geophysical time series. *Geophys. J. R. Astron. Soc.* **72**, 647–664.

Samson, J. C. (1983b). The reduction of sample-bias in polarization estimators for multichannel geophysical data with anisotropic noise. *Geophys. J. R. Astron. Soc.* **75**, 289–308.

Sarnthein, M., Erlenkeuser, H., von Grafenstein, R., and Schroder, C. (1984). Stable-isotope stratigraphy for the last 750,000 years: 'Meteor' core 13519 from the eastern equatorial Atlantic. "Meteor" *Forschungs Ergeb. Reihe C* **38**, 9–24.

Savin, S. M. (1977). The history of the Earth's surface temperature during the past 100 million years. *Ann. Rev. Earth Planet. Sci.* **5**, 319–355.

Savin, S. M., and Douglas, R. G. (1973). Stable isotope and magnesium geochemistry of Recent planktonic foraminifera from the South Pacific. *Geol. Soc. Am. Bull.* **84**, 2327–2342.

Savin, S. M., and Stehli, F. G. (1974). Interpretation of oxygen isotope paleotemperature measurements: Effect of the $^{18}O/^{16}O$ ratio of sea water, depth stratification of foraminifera and selective dissolution. *Colloq. Int. C.N.R.S.* **219**, 183–191.

Savin, S. M., and Yeh, H.-W. (1981). Stable isotopes in ocean sediments. *In* "The Sea" (C. Emiliani, ed.), vol. 7, pp. 1521–1554. Wiley, New York.

Savin, S. M., Douglas, R. G., and Stehli, F. G. (1975). Tertiary marine paleotemperatures. *Geol. Soc. Am. Bull.* **36**, 1499–1510.

Savin, S. M., Douglas, R. G., Keller, G., Killingley, J. S., Shaughnessy, L., Sommer, M. A., Vincent, E., and Woodruff, F. (1981). Miocene benthic foraminifera isotope records: A synthesis. *Mar. Micropaleontol.* **6**, 423–450.

Schafer, R. W. (1969). "Echo Removal by Discrete Generalized Linear Filtering," Tech. Rep. No. 446, Res. Lab. Electron., Massachusetts Institute of Technology, Cambridge.

Schlanger, S. O., and Douglas, R. G. (1974). The pelagic ooze chalk limestone transition for marine stratigraphy. *Spec. Publ. Int. Assoc. Sedimentol.* **1**, 117–148.

Schneider, W. A. (1984). The common depth point stack. *Proc. IEEE* **72**, 1238–1254.

Schnitker, D. (1974). West Atlantic abyssal circulation during the past 120,000 years. *Nature (London)* **248**, 385–387.

Schnitker, D. (1979). Cenozoic deep water benthic foraminiferas Bay of Biscay. *Initial Rep. Deep Sea Drill. Proj.* **48**, 377–413.

Schnitker, D. (1980). Global Paleoceanography and its Deep Water Linkage to the Antarctic Glaciation," Vol. 16, pp. 1–20. Elsevier, Amsterdam.

Schoenberger, M., and Levin, F. K. (1978). Apparent attenuation due to intrabed multiples. II. *Geophysics* **43**, 730–737.

Schoenberger, M., and Levin, F. K. (1979). The effect of subsurface sampling on one-dimensional synthetic seismograms. *Geophysics* **44**, 1813–1829.

Scholle, P. A. (1977). Chalk diagenesis and its relation to petroleum exploration: Oil from chalks a modern miracle? *Am. Assoc. Pet. Geol. Bull.* **61**, 982–1009.

Scholle, P. A., and Arthur, M. A. (1980). Carbon isotope fluctuations in Cretaceous pelagic limestones: Potential stratigraphic and petroleum exploration tool. *Am. Assoc. Pet. Geol. Bull.* **64**, 67–87.

Shackleton, N. J. (1965). The high-precision isotopic analysis of oxygen and carbon in carbon dioxide. *J. Sci. Instrum.* **42**, 689–692.

Shackleton, N. J. (1967). Oxygen isotope analyses and Pleistocene temperature reassessed. *Nature (London)* **215**, 15–17.

Shackleton, N. J. (1969). The last interglacial in the marine and terrestrial record. *Proc. R. Soc. London, Ser. B* **174**, 135–154.

Shackleton, N. J. (1974). Attainment of isotopic equilibrium between ocean water and the benthonic foraminifera genus *Uvigerina*: Changes in the ocean during the last glacial. *Colloq. Int. C.N.R.S.* **291**, 203–209.

Shackleton, N. J. (1977a). The oxygen isotope stratigraphic record of the late Pleistocene. *Philos. Trans. R. Soc. London* **280**, 169-182.

Shackleton, N. J. (1977b). Carbon-13 in *Uvigerina*: Tropical rain forest history and the Equatorial Pacific carbonate dissolution cycles. *In* "The Fate of Fossil Fuel CO_2 in the Ocean" (N. R. Andersen and A. Malahoff, eds.), pp. 401-427. Plenum, New York.

Shackleton, N. J. (1982). The Deep-Sea record of climate variability. *Prog. Oceanogr.* **11**, 199-218.

Shackleton, N. J. (1985). Oceanic carbon isotope constraints on oxygen and carbon dioxide in the Cenozoic atmosphere. *Geophys. Monog., Am. Geophys. Union* **32**, 412-417.

Shackleton, N. J. (1987). The carbon isotope record of the Cenozoic history of organic carbon burial and of oxygen in the ocean and atmosphere. *In* "Marine Petroleum Source Rocks" (J. R. V. Brooks and A. J. Fleet, eds.). Oxford Univ. Press, London and New York.

Shackleton, N. J., and Boersma, A. (1981). The climate of the Eocene Ocean. *J. Geol. Soc., London* **138**, 153-157.

Shackleton, N. J., and Cita, M. (1979). Oxygen and carbon isotope stratigraphy of benthic foraminifers at Site 397: Detailed history of climatic change during the Neogene. *Initial Rep. Deep Sea Drill. Proj.* **47**, 433-459.

Shackleton, N. J., and Hall, M. A. (1984a). Stable isotope record of hole 504 sediments: High resolution record of the Pleistocene. *Init. Rep. Deep Sea Drill. Proj.* **69**, 431-441.

Shackleton, N. J., and Hall, M. A. (1984b). Carbon isotope data from Leg 74 sediments. *Initial Rep. Deep Sea Drill. Proj.* **74**, 613-619.

Shackleton, N. J., and Hall, M. A. (1984c). Oxygen and carbon isotope stratigraphy of Deep Sea Drilling Project Hole 552A: Pliocene/Pleistocene glacial history. *Initial Rep. Deep Sea Drill. Proj.* **81**, 599-609.

Shackleton, N. J., and Kennett, J. P. (1975a). Paleotemperature history of the Cenozoic and the initiation of Antarctic glaciation: Oxygen and carbon isotope analyses in DSDP Sites 277, 279 and 281. *Initial Rep. Deep Sea Drill. Proj.* **29**, 743-755.

Shackleton, N. J., and Kennett, J. P. (1975b). Late Cenozoic oxygen and carbon isotopic changes at DSDP Site 284: Implications for glacial history of the Northern Hemisphere. *Initial Rep. Deep Sea Drill. Proj.* **29**, 801-807.

Shackleton, N. J., and Matthews, R. K. (1977). Oxygen isotope stratigraphy of late Pleistocene coral terraces in Barbados. *Nature (London)* **268**, 618-620.

Shackleton, N. J., and Opdyke, N. D. (1973). Oxygen isotope and paleomagnetic stratigraphy of equatorial Pacific core V28-238: Oxygen isotope temperatures and ice volumes on a 10T year and 10Y year scale. *Quat. Res.* **3**, 39-55.

Shackleton, N. J., and Opdyke, N. D. (1976). Oxygen-isotope and paleomagnetic stratigraphy of Pacific core V28-239 late Pliocene latest Pleistocene. *Mem.—Geol. Soc. Am.* **145**, 449-464.

Shackleton, N. J., and Opdyke, N. D. (1977). Oxygen isotope and paleomagnetic evidence for early Northern Hemisphere glaciation. *Nature (London)* **270**, 216-219.

Shackleton, N. J., and Pisias, N. G. (1985). Atmospheric carbon dioxide, orbital forcing and climate. *Geophys. Monogr., Am. Geophys. Union* **32**, 303-317.

Shackleton, N. J., Hall, M. A., Line, J., and Cang, S. (1983a). Carbon isotope data in core V19-30 confirm reduced carbon dioxide concentrations in the ice age atmosphere. *Nature (London)* **306**, 319-322.

Shackleton, N. J., Imbrie, J., and Hall, M. A. (1983b). Oxygen and carbon isotope record of east Pacific core V19-30: Implications for the formation of deep water in the late Pleistocene north Atlantic. *Earth Planet. Sci. Lett.* **65**, 233-244.

Shackleton, N. J., Hall, M. A., and Boersma, A. (1984a). Oxygen and carbon isotope data from Leg 74 foraminiferas. *Initial Rep. Deep Sea Drill. Proj.* **74**, 599-612.

Shackleton, N. J., Moore, T. C., Jr., Rabinowitz, P. D. et al. (1984b). Accumulation rates in Leg 74 sediments. *Initial Rep. Deep Sea Drill. Proj.* **74**, 621–644.

Shackleton, N. J., Backman, J., Zimmerman, H. et al. (1984c) Oxygen isotope calibration of the onset of ice-rafting in the history of glaciation in the North Atlantic region. *Nature (London)* **307**, 620–623.

Shackleton, N. J., Hall, M. A., and Bleil, U. (1986). Carbon isotope stratigraphy, site 577. *Initial Rep. Deep Sea Drill. Proj.* **86**, 503–511.

Shannon, C. E. (1948). A mathematical theory of communication. *Bell Syst. Tech. J.* **28**, 379–423. reprinted in Shannon and Weaver, 1963.

Shannon, C. E., and Weaver, W. (1963). "The Mathematical Theory of Communication." Univ. of Illinois Press, Urbana.

Sharp, W. E. (1973). Entropy as a parity check. *Earth Res.* **1**, 27–30.

Sharp, W. E., and Fan, P. (1963). A sorting Index, *J. Geol.* **71**, 76–84.

Sherwood, J. W. C., and Trorey, A. W. (1965). Minimum-phase and related properties of the response of a horizontally stratified absorptive earth to plane acoustic waves. *Geophysics* **30**, 191–197.

Shokes, R. F., Trabant, P. K., Presley, B. J., and Reid, D. F. (1977). Anoxic hypersaline basin in the northern Gulf of Mexico. *Science* **196**, 1443–1446.

Shumway, R. H. (1971). On detecting a signal in N stationarily correlated noise series. *Technometrics* **13**, 1813–1829.

Smart, E., and Flinn, E. A. (1984). Fast frequency-wavenumber analysis and Fisher signal detection in real-time infrasonic array data processing. *Geophys. J. R. Astron. Soc.* **76**, 279–284.

Snee, R. D. (1983). Discussion of 'Developments in linear regression methodology: 1959–1982' by R. R. Hocking. *Technometrics* **25**, 230–237.

Soluhob, J. T., and Klovan, J. E. (1970). Evaluation of grain-size parameters in lacustrine environments. *J. Sediment. Petrol.* **40**, 81–101.

Spaeth, C., Hoefs, J., and Vetter, U. (1971). Some aspects of isotopic composition of belemnites and related paleotemperatures. *Geol. Soc. Am. Bull.* **82**, 3139–3150.

Stainforth, R. M., Lamb, J. L., Luterbacher, H., Beard, J. H., and Jeffords, R. M. (1975). Cenozoic Planktonic Foraminiferal Zonation and Characteristics of Index Forms. *The University of Kansas Paleontological Contributions,* Article 62, pp. 1–425.

Stein, R., and Sarnthein, M. (1984). Late Neogene events of atmospheric and oceanic circulation offshore Northwest Africa: High-resolution record from deep-sea sediments. *Palaecolog. Afr. Surrounding Isl.* **16**, 9–36.

Stelting, C. E., Barnes, N. E., and Rhodes, E. G. (1984). Cyclic sedimentation features in cores for Brushy Canyon Formation, Delaware Basin, Texas. *Permian Basin Symp.* 1984, p. 3.

Stoffa, P. L., Buhl, P., and Bryan, G. M. (1974). The application of homomorphic deconvolution of shallow-water marine seismology. Part I. Models. Part II. Real data. *Geophysics* **39**, 401–426.

Stone, D. G. (1984). Wavelet estimation. *Proc. IEEE* **72**, 1394–1402.

Stude, G. R. (1984). Neogene and Pleistocene biostratigraphic zonation of the Gulf of Mexico Basin. *Trans.—Gulf Coast Assoc. Geol. Soc.* **14**.

Swan, D., Clague, J., and Leternauer, J. L. (1979). Grain-size statistics. II. Evaluation of grouped moment measures. *J. Sediment. Petrol.* **49**, 487–500.

Swingler, D. N. (1979). A comparison between Burg's maximum entropy method and a non-recursive technique for the spectral analysis of deterministic signals. *J. Geophys. Res.* **84**, 679–685.

Taira, A., and Scholle, P. A. (1979). Discrimination of depositional environments using settling tube data. *J. Sediment. Petrol.* **49**, 787–800.

Taylor, T. T. (1955). Design of line-source antennas for marrow beam width and low side lobes. *IRE Trans. Antennas Propag.* **AP-3**, 16–28.

Temple, J. T. (1978). The use of factor analysis in geology. *J. Math. Geol.* **10**, 379–387.

Thierstein, H. R., and Berger, W. H. (1978). Injection events and ocean history. *Nature (London)* **276**, 461–466.

Thierstein, H. R., Geitzenauer, K. R., Molfino, B., and Shackleton, N. J. (1977). Global synchroneity of late Quaternary coccolith datum levels: Validation by oxygen isotopes. *Geology* **5**, 400–404.

Thomson, D. J. (1982). Spectrum estimation and harmonic analysis. *Proc. IEEE* **70**, 1055–1096.

Thunell, R. C. (1984). Pleistocene planktonic foraminiferal biostratigraphy and paleoclimatology of the Gulf of Mexico. *In* "Principles of Pleistocene Stratigraphy Applied to the Gulf of Mexico" (N. Healy-Williams, ed.), pp. 25–64. IHRCDC Press, Boston, Massachusetts.

Thunell, R. C., and Williams, D. F. (1983). The stepwise development of Pliocene—Pleistocene paleoclimatic and paleoceanographic conditions in the Mediterranean. *Utrecht Micropaleontol. Bull.* **30**, 111–127.

Tick, L. J. (1963). Conditional spectra, linear systems, and coherency. *In* "Time Series Analysis" (M. Rosenblatt, ed.), pt. 197–203. Wiley, New York.

Tillman, R. W. (1973). Multiple group discriminant analysis of grain size data as an aid in recognizing environments of deposition. *Ores Sediments, Int. Sedimentol. Cong. 8th, 1971,* Abstr., p. 102.

Tompkins, R. E., and Shephard, L. E. (1979). Orca Basin: Depositional processes, geotechnical properties and clay mineralogy of Holocene sediments within an anoxic, hypersaline basin, northwest Gulf of Mexico. *Mar. Geol.* **33**, 221–238.

Trabant, P. K., and Presley, B. J. (1978). Orca Basin: An anoxic depression on the continental slope, northwest Gulf of Mexico. *Stud. Geol. (Tulsa, Okla.)* **7**, 289–303.

Tribolet, J. M. (1978). Applications of short-time homomorphic signal analysis to seismic wavelet estimation. *IEEE Trans. Acoust. Speech, Signal Process.* **ASSP-26**, 343–353.

Tribolet, J. M. (1979). "Seismic Applications of Homomorphic Signal Processing." Prentice-Hall, Englewood Cliffs, New Jersey.

Tucker, R. W., and Vacher, H. L. (1980). Effectiveness of discriminating beach, dune, and river sands by moments and the cumulative weight percentages. *J. Sediment. Petrol.* **50**, 165–172.

Tukey, J. W. (1959). Equilization and pulse shaping techniques applied to the determination of initial sense of Rayleigh waves. *In* "Report on a Panel of Seismic Improvement" (L. V. Berkner, chmn.), Appendix 9. National Science Foundation.

Ulrych, T. J. (1971). Application of homomorphic deconvolution to seismology. *Geophysics* **36**, 650–660.

Ulrych, T. J., and Bishop, T. N. (1975). Maximum entropy spectral analysis and autoregressive decomposition. *Rev. Geophy. Space Phys.* **13**, 183–200.

Urey, H. C. (1947). The thermodynamic properties of isotopic substances. *J. Chem. Soc., London,* pp. 562–581.

Urey, H. C., Lowenstam, H. A., Epstein, S., and McKinney, L. R. (1951). Measurement of paleotemperatures and temperatures of the Upper Cretaceous of England, Denmark, and the southeastern United States. *Geol. Soc. Am. Bull.* **62**, 399–416.

Vail, P. R., and Hardenbol, J. (1979). Sea level changes during the Tertiary. *Oceanus* **22**, 71–79.

Vail, P. R., Mitchum, R. M., Jr., and Thompson, S., III (1977). Seismic stratigraphy and global changes of sea level. Part 4. Global cycles of relative changes of sea level. *Mem.—Am. Assoc. Pet. Geol.* **26**, 83–97.

van den Bos, A. (1971). Alternative interpretation of maximum entropy spectral analysis. *IEEE Trans. Inf. Theory,* **IT. 17**, 493–494.

Van Donk, J. (1970). The oxygen isotope record in deep sea sediments. Ph.D. Thesis, Columbia University, New York.

Van Donk, J. (1976), ^{18}O record of the Atlantic Ocean for the entire Pleistocene Epoch. *Mem.—Geol. Soc. Am.* **145**, 147-163.

van Schoonfeld, C., ed. (1979). "Image Formation from Coherence Functions in Astronomy." Reidel Publ., Dordrecht, Netherlands.

van Schooneveld, C., and Frijling, D. J. (1981). Spectral analysis: On the usefulness of linear tapering for leakage suppression. *IEEE Trans. Acoust., Speech, Signal Process.* **ASSP-29**, 323-329.

Van Trees, H. L. (1968). "Detection, Estimation and Modulation Theory." Wiley, New York.

Veeh, H. H., and Chappell, J. M. A. (1970). Astronomic theory of climatic change: Support from New Guinea. *Science* **167**, 862.

Veizer, J., and Hoffs, J. (1976). The nature of $^{18}O/^{16}O$ and $^{12}C/^{12}C$ secular trends and sedimentary carbonate rocks. *Geochim. Cosmochim. Acta* **40**, 1387-1395.

Veizer, J., Holser, W. T., and Wilgus, C. K. (1980). Correlation of $^{13}C/^{12}C$ and $^{34}S/^{32}S$ secular variations. *Geochim. Cosmochim. Acta* **44**, 579-587.

Vergnaud-Grazzini, C., and Saliege, J.-F. (1987). Evolt mos isotopic um melu oceanique ala transition Eocene-Oligocene and Atlantique sood et al paleo circulation profons and Atlantique. In press.

Vergnaud-Grazzini, C., Ryan, W., and Cita, M. B. (1977). Stable isotopic fractionation of carbon and oxygen in benthic foraminifera. *Earth Planet. Sci. Lett.* **8**, 247-252.

Vergnaud-Grazzini, C., Pierre, C., and Latolle, R. (1978). Paleoenvironment of the northwest Atlantic during the Cenozoic: Oxygen and carbon isotope analyses of DSDP sites 398, 400A and 401. *Oceanol. Acta* **1**, 381-390.

Vergnaud-Grazzini, C., Grably, M., Pujol, C., and Duprat, J. (1983). Oxygen isotope stratigraphy and paleoclimatology of southwestern Atlantic Quaternary sediments (Rio Grande Rise) at Deep Sea Drilling Project Site 517. *Initial Rep. Deep Sea Drill. Proj.* **72**, 871-884.

Vincent, E., Killingly, K. S., and Berger, W. H. (1980). The magnetic epoch-6 carbon shift: A change in the ocean $^{13}C/^{12}C$ ratio 6.2 million years ago. *Mar. Micropaleontol.* **5**, 185-203.

Ward, R. F., Kendell, C. G., and Harris, P. M. (1986). Late Permian (Guadalupian) Facies and their Association with Hydrocarbons—The Permian Basin, West Texas and New Mexico. *Am. Assoc. Pet. Geol. Bull.* **70**, 239-262.

Watts, A. B. (1982). Tectonic subsidence, flexure and global changes of sea level. *Nature (London)* **297**, 469-474.

Watts, A. B., and Steckler, M. S. (1979). Subsidence and eustacy at the continental margin of eastern North America. *In*: "Deep Drilling Results in the Atlantic Ocean: Continental Margins and Paleoenvironment." (M. Talwani, W. F. Hay, and W. B. F. Ryan, eds.), pp. 213-234. American Geophysical Union, Maurice Ewing Series No. 3.

Weaver, F. M., Casey, R. E., and Perez, A. M. (1981). Stratigraphic and paleoceanographic significance of Early Pliocene to middle Miocene radiolarian assemblages for northern to Baja California. *In* "The Monetrey Formation and Related Siliceous Rocks of California" (R. E. Garrison and R. G. Douglas, eds.), pp. 71-86. Soc. Econ. Paleontol. Mineral., Los Angeles, California.

Weissert, H. J., and Oberhansli, H. (1985). Pliocene oceanography and climate: An isotope record from the southwestern Angola Basin. *In* "South Atlantic Paleoceanography" (K. Hsu and H. Weissert, eds.), pp. 79-97. Cambridge Univ. Press, London and New York.

Weissert, H. J., McKenzie, J. A., Wright, R. C., Clark, M., Oberhansli, H., and Casey, M. (1984). Paleoclimatic record of the Pliocene at Deep Sea Drilling Project Sites 519, 521, 522 and 523 (Central South Atlantic). *Initial Rep. Deep Sea Drill. Proj.* **73**, 701-715.

Wernecke, S. J. (1977). Maximum entropy image reconstruction *Radio Sci.* **12**, 831-844.

Wernecke, S. J., and D'Addario, L. R. (1977). Two-dimensional maximum entropy reconstruction of radio brightness. *IEEE Trans. Compt.* **C26**, 351-364.

White, R. E. (1973). The estimation of signal spectra and related quantities by means of the multiple coherence function. *Geophys. Prospect.* **21**, 660–703.

White, R. E. (1977). The performance of optimum stacking filters in suppressing uncorrelated noise. *Geophys. Prospect.* **25**, 165–178.

White, R. E. (1980). Partial coherence matching of synthetic seismograms with seismic traces. *Geophys. Prospect.* **28**, 333–358.

White, R. E. (1984). Signal and noise estimation from seismic reflection data using spectral coherence methods. *Proc. IEEE* **72**, 1340–1356.

White, R. E., and O'Brien, P. N. S. (1974). Estimation of the primary seismic pulse. *Geophys. Prospect.* **22**, 627–651.

Wiener, N. (1949). "Extrapolation, Interpolation and Smoothing of Stationary Time Series with Engineering Applications." M.I.T. Press, Cambridge, Massachusetts.

Williams, D. F. (1984). Correlation of Pleistocene marine sediments of the Gulf of Mexico and other basins using oxygen isotope stratigraphy. *In* "Principles of Pleistocene Stratigraphy Applied to the Gulf of Mexico" (N. Healy-Williams, ed.), pp. 67–188. IHRDC Press, Boston, Massachusetts.

Williams, D. F. (1987). Correlation of Pleistocene marine sediments using oxygen stratigraphy. *In* "Subsurface Geology" (A. LeRoy and D. O. Le Roy, eds.), Colorado School of Mines, Golden.

Williams, D. F., and Healy-Williams, N. (1980). Oxygen isotopic-hydrographic relationships among Recent planktonic foraminifera from the Indian Ocean. *Nature (London)* **283**, 848–852.

Williams, D. F., and Ledbetter, M. T. (1979). Chronology of Late Brunhes biostratigraphy and late Cenozoic disconformities in the Vema Channel (South Atlantic). *Mar. Micropaleontol.* **4**, 125–136.

Williams, D. F., and Trainor, D. (1986). Application of Isotope Chronostratigraphy in the Northern Gulf of Mexico, *Trans.—Gulf Coast Assoc. Geol. Soc.* **36**, 589–600.

Williams, D. F., Sommer, M. A., and Bender, M. L. (1977). Carbon isotopic compositions of Recent planktonic foraminifera of the Indian Ocean. *Earth Planet. Sci. Lett.* **36**, 391–403.

Williams, D. F., Moore, W. S., and Fillon, R. H. (1981a). Role of glacial Arctic Ocean ice sheets in Pleistocene oxygen isotope and sea level records. *Earth Planet. Sci. Lett.* **56**, 157–166.

Williams, D. F., Rottger, R., Schmaljohnn, R., and Keigwin, L. (1981b). Oxygen and carbon isotopic fractionation and an algal symbiosis in the benthic foraminiferan, *Heterostegina depressa. Palaeogr., Palaeoclimatol., Palaeocol.* **33**, 231–251.

Williams, D. F., Healy-Williams, N., Thunell, R. C., and Leventer, A. (1983). Detailed stable isotope and carbonate records from the upper Maestrictian-lower Paleocene section of Hole 516F (Leg 72) including the Cretaceous-Tertiary Boundary. *Initial Rep. Deep Sea Drill. Proj.* **72**, 921–929.

Williams, D. F., Thunell, R. C., Hodell, D. A., and Vergnaud-Grazzini, C. (1985). Synthesis of late Cretaceous, Tertiary and Quaternary stable isotope records of the South Atlantic based on Leg 72 DSDP core material. *In* "South Atlantic Paleoceanography" (K. Hsu and H. Weissert, eds.), pp. 205–244. Cambridge Univ. Press, London and New York.

Williams, D. F., Thunell, R. C., Tappa, E., Rio, D., and Raffi, I. (1988). Chronology of the Pleistocene oxygen isotope record. 0–1.88 million years before present. *Palaeogr. Palaeoclimatol., Palaeoecol.* (in press).

Williamson, C. R. (1979). Deep-sea sedimentation and stratigraphic traps. Bell Canyon formation (Permian), Delaware Basin. *Soc. Econ. Paleontol. Mineral. (Permian Basin Sect.), Symp. Field Conf. Guidebook, Publ.* **79-18**, 39–74.

Woodruff, F., and Savin, S. M. (1985). $\delta^{13}C$ values of Miocene Pacific benthic foraminifera: Correlations with sea level and biological productivity. *Geology* **13**, 119-122.

Woodruff, F., Savin, S. M., and Douglas, R. G. (1980). Biological fractionation of oxygen and carbon isotopes by Recent benthic foraminifera. *Mar. Micropaleontol.* **5**, 3-11.

Woodruff, F., Savin, S. M., and Douglas, R. G. (1981). Miocene stable isotopic record: A detailed deep Pacific Ocean study and its paleoclimatic implications. *Science* **212**, 665-668.

Woods, J. W. (1972). Two-dimensional discrete Markovian fields. *IEEE Trans. Inf. Theory* **IT-18**, 232-240.

Wuenschel, P. C. (1960). Seismogram synthesis including multiples and transmission coefficients. *Geophysics* **25**, 106-129.

Zachos, J. C., Arthur, M. A., Thunell, R. C., Williams, D. F., and Tappa, E. J. (1985). Stable isotope and trace element geochemistry of carbonate sediments across the Cretaceous/Tertiary boundary at DSDP Hole 577, Leg 86. *Initial Rep. Deep Sea Drill. Proj.* **86**, 513-532.

Zimmerman, H. B., Shackleton, N. J., Backman, J. *et al.* (1984). History of Plio-Pleistocene climate in the northeastern Atlantic. Deep Sea Drilling Project 552A. *Initial Rep. Deep Sea Drill. Proj.* **31**, 861-875.

Ziolkowski, A. M., Lerwill, W. E., Hatton, L., and Haugland, T. (1980). Wavelet deconvolution using a source scaling law. *Geophys. Prospect.* **28**, 872-901.

Index

Signals (noise), 159, 174
 Phase sensitive detection of, 195–202, 268
South China Sea, 9
Spectral ratio methods, 184–185
Stratigraphy
 biostratigraphy, 1, 2, 4–9, 16, 18–22, 47,
  56, 61, 62, 65, 68, 69, 74, 77, 78, 82,
  92, 107, 108, 116, 119, 121–129, 145, 146
 geochemical, 4
 isotopic, 4, 5, 6, 18, 49, 61, 125, 128, 133, 134
 litho, 5, 6, 16, 22, 62, 128, 131, 145, 146
 paleomagnetic, 4, 9, 56, 57, 61, 62, 63,
  69, 74, 77, 78, 82, 93, 103
 seismic, 1, 4, 6, 8, 9, 107–117, 123, 145, 146

T

Tertiary (*see* Cenozoic), 1, 4, 5, 6, 7, 8, 9,
 11, 12, 15, 17, 25, 26, 31, 36, 37, 41,
 42, 43, 45, 46, 47, 49, 51–55, 56, 93,
 102, 113, 129, 130, 145
Trinidad (offshore), 143
Tephrochronology, 62, 68, 78
TYPE, 292, 300

W

Whole rock analyses, 35, 37, 46, 47, 106
Wiener filter, 266, 270, 296
Wiener-Kolmogorov, 203